SDGs and Textiles

Series Editor

Hafeezullah Memon, International Institute of Silk, Zhejiang Sci-Tech University, Hangzhou, China

The book series "SDGs and Textiles" addresses the strategies to achieve sustainable development goals (SDGs) in the present, past, and future. It presents books about the present and future policies of textile ministries of different countries, and books related to sustainability education around different parts of the world in the textile sector. Moreover, it would welcome the conference proceeding related to SDGs and Textiles. The series would cover books comparing the sustainability and SDGs of different institutions and countries. The individual book volumes in the series are thematic. The goal of each book is to give readers a comprehensive overview of a different area of sustainability in the textile sector. As a collection, the series provides valuable resources to a broad audience in academia, the research community, industry, and anyone looking to expand their knowledge of SDGs and Textiles.

Textiles and life are together – life cannot be separated from textiles as it is the most important need for human beings after food. In 2015, the United Nations General Assembly proposed 17 interlinked global goals to be achieved by 2030. Since then, academia and industry have paid much attention to achieving these goals. Textile found its close relation with almost all of these 17 goals.

SDG 1 - No Poverty: Poverty would never be overcome by a charity only; it is essential to develop people's skills to have a better and wealthy life. Thus, the textile can be considered an excellent discipline to achieve this goal by creating jobs and small and medium businesses.

SDG 2 - Zero Hunger: Through the effective utilization of advanced application of Agrotech Textiles, it is possible to have higher crop yields and save crops from rough weather, unexpected rains, floods, insects, etc.; thus, geotextiles play an essential in achieving this goal of sustainable development.

SDG 3 - Good Health & well-being: There has been much health consciousness after Covid19, and medical textiles assist in getting good health and well-being.

SDG 4 - Learning & Education: Textile or fashion has remained a significant discipline for societies for ages, and there has always remained much to explore in this field. Textile-related universities may play a vital role by offering free access to their education resources, training and spreading information among the locals.

SDG 5 - Gender Equality: The textile sector is one of the industrial sectors that accepted gender equality long ago; in particular, the garment sector has more females than males. Thus, the textile sector has been doing gender equality. Moreover, there has been a recent trend for Gender Neutral Clothing, which need worth studying and may further assist gender equality.

SDG 6 - Clean Water & Sanitation: Textiles could be achieved through filtration, and of course, textile is one of the critical materials for filtration.

SDG 7 - Affordable & Clean Energy: With the recent advancement in material science and engineering, the textile sector has come on the front for, not only by using this clean energy during textile production but also by assisting the production of this clean energy, either in the form of wind turbines blades made of textile composites or by energy harvesting from T-Shirts, etc.

SDG 8 - Decent Work: Recently, there has been much attention that the textile workers are not paid well, labor rights are not cared about, etc.

SDG 9 - Industry and Innovation: Textile Industry always follows innovation; the textile companies that do not chase innovation cannot survive in the market.

SDG 10 - Reduced Inequalities: Getting better life and well-being would help reduce inequalities in the textile industry.

SDG 11 - Sustainable Cities: Sustainable Textile Cities through Buildtech and transport textiles.

SDG 12 - Consumption and Production: Textile and garment consumption and production all come under.

SDG 13 - Climate Action: Oekotech or Ecotech Textile, waste management of textiles are upfront to achieve this goal of sustainable development.

SDG 14 - Life Below Water: Mitigating microfiber waste in rivers and oceans may come under the context of it. There has been much attention on this subject after passing the bill at the parliament level of the UK.

SDG 15 - Life on Land: Geotech or Geotextiles studies life on land.

SDG 16 - Peace, Justice, and Strong Institutions: Protective textiles are doing their best to achieve peace, justice, and strong institutions.

SDG 17 - Partnerships for the Goals: The application of textiles to achieve sustainable development goals is only an example. In all textiles sectors, combined efforts of all the goals are essential to achieve true sustainability.

Abu Sadat Muhammad Sayem · Parik Goswami
Editors

SDG 12 and Global Fashion Textiles Production

Springer

Editors
Abu Sadat Muhammad Sayem
Fashion Institute
Manchester Metropolitan University
Manchester, UK

Parik Goswami
University of Huddersfield
Huddersfield, UK

ISSN 2948-1236 ISSN 2948-1244 (electronic)
SDGs and Textiles
ISBN 978-981-96-9050-3 ISBN 978-981-96-9051-0 (eBook)
https://doi.org/10.1007/978-981-96-9051-0

This Springer imprint is published by the registered company Springer Nature Singapore Pte Ltd.
The registered company address is: 152 Beach Road, #21-01/04 Gateway East, Singapore 189721, Singapore

Preface

Textile materials can be categorised into three broad groups based on their applications, such as technical textiles, home textiles and fashion textiles. The fashion textiles are used as fashion items and are worn on or wrapped around human body. The scope of this book is mostly limited to this category and refers to the fashion textile industry, that is long celebrated for its creativity, innovation, and cultural impact. This book *SDG 12 and Global Fashion Textiles Production*, a part of the Springer Nature book series *SDGs and Textiles*, aims to explore how this industry can transition toward more sustainable and responsible practices. It brings together a diverse collection of scholarly and industry-informed perspectives from researchers, academics, and professionals across the globe. Each chapter addresses specific challenges and solutions, providing actionable insights into how the fashion textile industry can contribute meaningfully to achieving SDG 12.

In Chap. 1, the authors re-examine the **interconnections** between the broader sustainable development goals (SDGs) and the fashion textile industry. They present a compelling overview of the sector's environmental and social impacts, arguing that a more systemic, coordinated approach is necessary to align the industry's practices with the 17 SDGs, particularly SDG 12. This foundational chapter sets the stage by contextualising fashion's global footprint and its role in both contributing to sustainability challenges and addressing those.

Chapter 2 introduces the **Identify-Act-Evaluate (I-A-E)** framework, a practical model developed to foster environmental sustainability in textile production. This framework emphasises proactive waste minimisation, sustainable material choices, and performance evaluation through eco-indices. It aims to transform sustainability from a post-production add-on into a core philosophy embedded within the early stages of product development.

Chapter 3 explores the transformative potential of **sustainable biomaterials** in fashion. The authors discuss naturally coloured fibres, microbe-derived alternatives, organic cotton, and bio-based leathers, showcasing both the technological advancements and shifting consumer mindsets that are driving a move toward biologically inspired materials. These innovations signal a shift in material culture within the fashion industry.

The discussion continues in Chap. 4 with a critical look at **regenerated and synthetic bio-based fibres**, such as viscose, lyocell, PLA, and PHA. The chapter provides a thorough evaluation of these alternatives in terms of their environmental footprints, lifecycle analyses, and regulatory frameworks, positioning them as key players in the future of sustainable textile manufacturing.

Chapter 5 focuses on the **weaving sector**, chalking its evolution toward sustainability through energy-efficient machinery, waste reduction techniques, and innovations like near-net-shape weaving. The authors argue for an interdisciplinary approach that integrates technical advancements with ethical labour practices and transparent supply chains to meet both ecological and social sustainability goals.

In Chap. 6, **advanced knitting technologies** are presented as another frontier in sustainable fashion. Techniques like 3D and 4D knitting, seamless knitting, and automated production offer opportunities to significantly reduce material waste and energy use. Moreover, these technologies pave the way for smart textiles and multifunctional garments that combine innovation with environmental consciousness.

Chapter 7 addresses one of the most environmentally harmful aspects of textile production—**colouration**. This chapter explores green colouration techniques such as digital printing, spray dyeing, and biological processes that drastically reduce water and chemical use. By prioritising these methods, the industry can reduce its toxic outputs and carbon footprint while maintaining product quality.

In Chap. 8, the focus shifts to **plasma processing**—a dry, low-energy, and chemically neutral alternative to traditional textile treatments. Industrial-scale plasma technologies are positioned as transformative tools that enhance fabric performance while reducing ecological harm, offering a promising pathway for sustainable innovation at scale.

Chapter 9 examines the rise of **circularity** in fashion and textile manufacturing. Highlighting practices like closed-loop systems, zero-waste pattern design, and sustainable packaging, the chapter highlights how circular principles can be embedded across the value chain. It also outlines the roles of various stakeholders, from manufacturers to consumers, in driving the transition toward a regenerative fashion economy.

Chapter 10 presents a critical perspective on **greenwashing**, a practice that undermines genuine sustainability efforts. The chapter investigates how misleading marketing can deceive consumers, delay regulatory progress, and hinder the goals of SDG 12. It calls for stronger global regulations, improved transparency, and consumer education to combat this growing threat.

In Chap. 11, the authors focus on Sri Lankan apparel industry as a case study in **climate resilience and sustainability**. As a country highly vulnerable to climate change, Sri Lanka faces challenges that require integrating environmental and occupational health strategies. This chapter explores how aligning SDGs 8, 12, and 13 can create a more resilient and equitable production environment.

Finally, Chap. 12 critiques the prevailing **linear economic model** in fashion and proposes a radical rethinking through circular economy strategies. By analysing

Australia's clothing product stewardship scheme "*Seamless*", the authors demonstrate how policies on extended producer responsibility (EPR) and shifts in consumer behaviour can support systemic change toward more responsible consumption and production.

Importantly, this volume does not only critique current practices but proposes practical pathways forward. It reflects a growing recognition that innovation and sustainability must go together, and that transformation requires cooperation among researchers, industry stakeholders, policymakers, and consumers alike.

We hope this book serves as a timely contribution to the global dialogue on sustainability in the fashion textile industry. We believe this book will inspire further research, innovation, and action toward building a more just, resilient, and sustainable fashion ecosystem for future generations.

Manchester, UK

Abu Sadat Muhammad Sayem
CText FTI

Contents

1 **UN SDGs and the Fashion Textile Industry: Exploring the Interconnections** 1
Abu Sadat Muhammad Sayem, Md. Mahbubor Rahman, Jannatul Fardus, Sumaiya Mohsina, Kazi Wasima Tasnim, Minhajul Islam Jawad, and Parikshit Goswami

2 **Framework for Environmentally Sustainable Fashion and Textile Production to Achieve SDG 12** 29
Abu Sadat Muhammad Sayem, Julfikar Haider, Mohammad Abbas Uddin, Ayub Nabi Khan, M. M. Alamgir Sayeed, and Mahbub Hasan

3 **Sustainable Biomaterials for Fashion** 45
Abdullah Sayam, Hasan Mazharul Haq, Md. Shahin Hossen, and Abu Sadat Muhammad Sayem

4 **Advances in Bio-Based Regenerated and Synthetic Fibres and Their Sustainability Issues** 65
Dewan Murshed Ahmed, Mir Abdul Gaffar, Muntaha Mahin, Borhan Uddin Khan, Rishad Rayyaan, and Afsana Mobin

5 **Progress in Weaving Technology in Context of Sustainability** 101
Kazi Sowrov, Md. Faisal Mahmud, Md. Abdullah Al Faruque, and Md. Abdullah Al. Mamun

6 **Advanced Knitting Technology for Sustainable Fashion** 123
Dewan Murshed Ahmed, Afsana Mobin, Nadvi Mamun Pritha, Arif Istiak Ador, Safi Ahmed, and Md. Raju Ahmed

7 **Green Colouration Processes Targeting Sustainability** 141
Eshita Datta, Md. Razib Sheikh, Mohammad Mahmudur Rahman Khan, and Sadia Mehrin

8 Industrial-Scale Plasma Processing for Sustainable Textiles 167
Mehbuba Manir Nova, Md. Najmul Islam, Syeda Maliha Fairuz, Emtiaz Ahmed Pappu, Abdullah Sayam, Gaffar Hossain, and Abu Sadat Muhammad Sayem

9 Sustainable Practices and Circularity in Fashion and Textiles Manufacturing 193
Mazed Islam, Md Shamsuzzaman, Raju Ahmed, Md. Abdullah Al. Mamun, and Abu Sadat Muhammad Sayem

10 How Greenwashing Negates Sustainable Production and Consumption of Fashion Textiles 215
Md Nazmul Haque, Chunmin Lang, and Chuanlan Liu

11 Climate Change, Decent Work and Sustainable Production: An Insight from Sri Lankan Apparel Industry 239
Devathanthrige Janaka Chamara Harshana Senadeera, Abu Sadat Muhammad Sayem, Walter Leal, Grace Farhat, and Haruna Musa Moda

12 SDG 12—Towards Responsible Consumption and Production in the Context of Fashion and Textiles 263
Saniyat Islam and Jenny Underwood

Chapter 1
UN SDGs and the Fashion Textile Industry: Exploring the Interconnections

Abu Sadat Muhammad Sayem, Md. Mahbubor Rahman, Jannatul Fardus, Sumaiya Mohsina, Kazi Wasima Tasnim, Minhajul Islam Jawad, and Parikshit Goswami

Abstract The fashion textile industry is a primary industry that has existed since the beginning of the human invasion of the planet Earth, and is causing significant environmental, social and economic impacts for this only liveable planet in the universe. Standing just five years away from the official deadline of achieving the United Nations (UN) Sustainable Development Goals (SDGs), this chapter re-examines the interconnections between the SDGs and the fashion textile industry. The 2030 Agenda for Sustainable Development is an outstanding aspiration from the UN that calls on every human being to play their role in achieving 17 goals with 169 targets. This chapter will provide readers with a general understanding of how the fashion textile industry is related to these targets and what it can do to further achieve them.

Keywords Sustainable development goals (SDGs) · Carbon footprint · Water footprint · Land use · Apparel industry

---

A. S. M. Sayem (✉)
Manchester Fashion Institute, Manchester Metropolitan University, Manchester, UK
e-mail: asm.sayem@mmu.ac.uk

Md. M. Rahman · K. W. Tasnim · M. I. Jawad
Bangladesh University of Textiles, Dhaka, Bangladesh

J. Fardus
American International University—Bangladesh (AIUB), Dhaka, Bangladesh

S. Mohsina
University of South Asia, Dhaka, Bangladesh

P. Goswami
University of Huddersfield, Huddersfield, UK

A. S. Muhammad Sayem and P. Goswami (eds.), *SDG 12 and Global Fashion Textiles Production*, SDGs and Textiles, https://doi.org/10.1007/978-981-96-9051-0_1

1 Introduction

The historic definition of sustainable development was presented in the report known as "Our Common Future" by the World Commission on Environment and Development in 1987. This definition talked about the development that would sustain generation after generation. It is not just to meet the needs of the present generation only; rather, it will ensure the uncompromised opportunities for the future generations to meet their needs while keeping our planet healthy and liveable for always.

At the dawn of the second millennium, world leaders adopted the UN Millennium Declaration and set out eight time-bound goals, known as Millennium Development Goals (MDGs), with a deadline of 2015. Out of these goals, goal 7 directly focused on environmental sustainability, while goals 1, 2, and 3 covered economic and social sustainability inherently. In 2015, all United Nations (UN) member states adopted the "2030 Agenda for Sustainable Development", which encompasses 17 Sustainable Development Goals (SDGs) at its heart. When the MDGs era came to an end in 2015, the era of the 2030 Agenda for Sustainable Development began in 2016 with a plan spanned over fifteen years to achieve 17 goals including 169 targets to be met by the end of 2030. Table 1 summarises the evolution of SDGs and Fig. 1 illustrates the icons of the MDGs and SDGs.

According to the website of the UN Alliance for Sustainable Fashion (UNASF) (accessed in February 2025), the value of the global fashion industry is worth $2.4 trillion-dollar and it employs approximately 300 million people across the value chain [3]. It is apparent from Figs. 2 and 3 that both the worldwide revenue generation from

Table 1 UN SDGs evolution timeline

Timeline	Evolution of 2030 agenda for sustainable development and 17 SDGs
1992	UN member states agreed on Agenda 21 at the Earth Summit in Rio de Janeiro, Brazil to promote global sustainable development
2000	The Millennium Declaration at the Millennium Summit UN Headquarters in New York introduced eight Millennium Development Goals (MDGs) to reduce extreme poverty by 2015
2002	At the World Summit on Sustainable Development in South Africa, the Johannesburg Declaration on Sustainable Development and the Plan of Implementation was adopted
2012	At the United Nations Conference on Sustainable Development (Rio + 20) in Rio de Janeiro, Brazil, the process to develop a set of SDGs to build upon the MDGs was launched, and the UN High-level Political Forum on Sustainable Development was formed
2013	The General Assembly set up a working group to develop a proposal for the SDGs
2015	The General Assembly began the negotiation process on the post-2015 development agenda in January, and ultimately, the 2030 Agenda for Sustainable Development, comprising 17 SDGs, was officially adopted at the UN Sustainable Development Summit in September

Fig. 1 Icons of 8 MDGs (left) and 17 SDGs [1, 2]

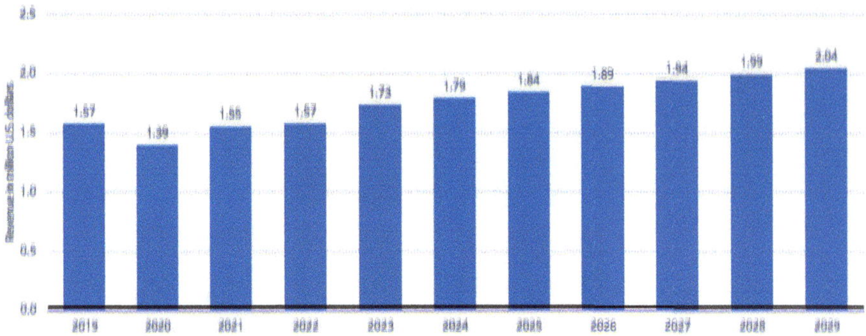

Fig. 2 Revenue of the apparel market worldwide from 2019 to 2029 (in trillion U.S. dollars) [4]

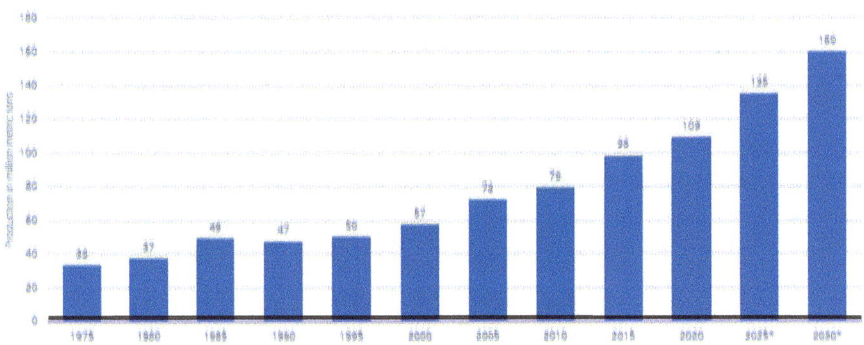

Fig. 3 Production of textile fibres worldwide from 1975 to 2020, with a forecast (marked with *) for 2025 and 2030 (in million metric tons) [5]

the apparel market and the global production of textile fibres are on an increasing trend.

The fashion textile supply chain is spread over multiple continents, and its value chain entails a series of complex activities, broadly fibre production, yarn and fabric production, garment production, consumption and end-of-life activities. Multiple stakeholders are connected to these supply chain and value chain activities (see Fig. 4), which have multifarious impacts on the animal kingdom, plant kingdom and the earth's ecosystem. This chapter aims to explore the implications of these activities for the UN Sustainable Development Goals (SDGs), which are discussed separately in the following sections.

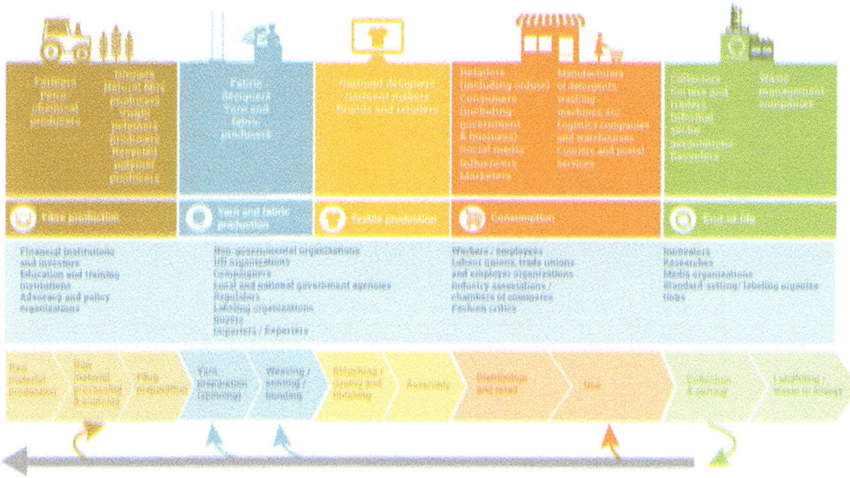

Fig. 4 Textile value chain activities and associated stakeholders [6]

2 SDGs and Fashion Textile Industry—The Interconnections

2.1 SDG 1: No Poverty

SDG 1 aims to end poverty in all forms from everywhere. However, it is forecasted that 575 million people worldwide will still be living in extreme poverty by 2030 if the current trend continues [7]. Historically, the textile and garment industry has contributed to poverty alleviation and economic development in various parts of the world. The Gherzi Textil Organisation from Switzerland and Gapminder Foundation from Sweden analysed economic data compiled from 1800 to 2018 and revealed that the textile industry is a serious contributor to reducing poverty in the world [8]. The World Bank acknowledges that the global textile and apparel sector has instigated the start of industrialisation in many developing countries and provided employment opportunities to thousands of low-income workers, including many women [9]. Among the least developed countries (LDCs), Bangladesh has experienced significant success in the apparel industry and export, which contributes to its poverty alleviation [10]. The country is now eyeing to LDC graduation in 2026. Vidanage et al. [11] report that the textile industry in Sri Lanka plays an essential role in reducing both income and multidimensional poverty in the country. Table 2 maps the relevance of the targets of SDG 1 to the fashion and textile industries.

Table 2 Targets of SDG 1 and their relevance to fashion textiles

Target No.	Relevance to fashion textiles		
	Directly relevant	Indirectly relevant	Not relevant
1.1	✓		
1.2	✓		
1.3			✓
1.4		✓	
1.5			✓
1.a		✓	
1.b	✓		

SDG 1: End poverty in all its forms everywhere (see: https://sdgs.un.org/goals/goal1)

2.2 SDG 2: Zero Hunger

SDG 2 aims to end hunger, achieve food security and improved nutrition, and promote sustainable agriculture. While the global fashion textile industry has a positive impact on poverty eradication, as discussed previously, it is causing a loss of biodiversity worldwide by having a high land footprint, meaning a high land use impact. Natural fibre production is entirely dependent on land use. 2.5% of the total arable land of the world is used for cotton cultivation [6]. It is estimated that the land use by fashion textile industry will increase by 35% by 2030 [12]. This will have significant conflict with the land use by food agriculture to feed the growing world population.

At the same, the use of technical textiles, known as Agrotech textiles, is helping to increase crop yield by preventing soil from drying out and by protecting crops from birds and insects. The use of geotextiles helps save land by preventing soil erosion. Thus, SDG 2 is very relevant to the fashion and textile industries which is mapped in the Table 3.

2.3 SDG 3: Good Health and Well-Being

SDG 3 focuses on achieving good physical and mental health. As the fashion and textile industry has been making positive contributions to poverty alleviation, it is also contributing to some extent to achieving SDG 3. When it comes to occupational health it is directly relevant to the fashion industry, especially when combined with SDG 8 (decent work). Historically, the textile industry is not known for providing a healthy working environment for its employees due to the complexity of the production process, lack of awareness among both employers and employees and weakness of regulatory systems, etc. However, the situation has started to change in the last decade, especially after the rollout of the UN agenda for sustainable development, the formation of the UNASF, activism from international NGOs like Greenpeace and

Table 3 SDG 2 targets and their relevance to fashion textiles

Target No.	Relevance to fashion textiles		
	Directly relevant	Indirectly relevant	Not relevant
2.1		✓	
2.2	✓		
2.3		✓	
2.4		✓	
2.5	✓		
2.a	✓		
2.b		✓	
2.c		✓	

SDG 2: End hunger, achieve food security and improved nutrition and promote sustainable agriculture (see: https://sdgs.un.org/goals/goal2)

ever-increasing consumer awareness. We can observe several good practices around the world. For example, Bangladesh Garments Manufacturers and Exporters Association (BGMEA) reports on its website that it runs eighth health care centres in different parts of the country, which provide free health-care services and medicines to more than 60,000 garment workers every year [13]. These centres also work on creating awareness of HIV/AIDS, tuberculosis, reproductive health and the use of contraceptives, which complement target 3.3 of SDG 3 (see Table 4). To provide critical health care for its workers either free of charge or at subsidised prices, it has set up hospitals in Chattogram and Dhaka, which host the major garment manufacturing hubs in Bangladesh. Through five schools, it is also providing free education to garment workers' children and provides stipends to meritorious students. With support from the Bangladesh government, the project has been launched to provide lactating mothers working in garment factories with extra monthly bonuses for a period of 36 months. It has set up a dormitory for 3000 beds in Chattogram, and another in Ashulia near Dhaka is under construction to accommodate migrant garment workers, who otherwise will live in unhealthy conditions in slums. It has also launched an online health toolkit (https://rmghealthtoolkit.com/) for workers and factory management [14]. The Forbes reports about a workforce nutrition project, "Strengthening Workers' Access to Pertinent Nutrition Opportunities (SWAPNO)", implemented in an apparel manufacturing facility during COVID pandemic, which provided healthy meals to workers, including Iron Folate Acid (IFA) tablets for women and educated them about balanced diet [15]. By replicating and increasing such initiatives in every apparel-producing country, the fashion industry can play a key role in achieving the SDG 3.

Table 4 Targets of SDG 3 and their relevance to fashion textiles

Target No.	Relevance to fashion textiles		
	Directly relevant	Indirectly relevant	Not relevant
3.1		✓	
3.2		✓	
3.3			✓
3.4		✓	
3.5		✓	
3.6			✓
3.7			✓
3.8			✓
3.9	✓		
3.a			✓
3.b			✓
3.c	✓		
3.d		✓	

SDG 3: Ensure healthy lives and promote well-being for all at all ages (see: https://sdgs.un.org/goals/goal3)

2.4 *SDG 4: Quality Education*

Like all industrial sectors, fashion and textile also rely on well-trained and educated manpower. SDG 4 comprises ten targets. Target 4.7, which calls for equipping all learners with relevant knowledge and skills to drive sustainable development, is directly relevant to the fashion and textile sector. Even the existing manpower, starting from farmers to managers, requires professional development training to ensure sustainable development of the sector. On the other hand, successful fashion businesses and their associations can contribute to achieving SDG 4 through initiatives from their social corporate responsibility (CSR) programs. For example, BGMEA operates two universities in Bangladesh which provide higher education in different disciplines, including fashion and textiles [13]. It also runs five free schools for the children of garment workers, which is a good example to follow elsewhere (Table 5).

2.5 *SDG 5: Gender Equality*

SDG 5 calls for equality which is an element of social sustainability. The global fashion and textile industries are under serious scrutiny when it comes to gender equality. Both in the upstream and downstream value chains of fashion and textiles, there are visible images of disparities between genders. Globally, the textile sector

Table 5 SDG 4 targets and their relevance to fashion textiles

Target No.	Relevance to fashion textiles		
	Directly relevant	Indirectly relevant	Not relevant
4.1		✓	
4.2		✓	
4.3		✓	
4.4	✓		
4.5		✓	
4.6		✓	
4.7	✓		
4.a			✓
4.b		✓	
4.c		✓	

SDG 4: Ensure inclusive and equitable quality education and promote lifelong learning opportunities for all (See: https://sdgs.un.org/goals/goal4)

employs millions of people, the majority of whom are women. However, these individuals often face a lack of labour rights, gender discrimination, and income disparities [16]. Lai [17] reported in Vogue Australia, citing a study of 2019, that only 12.5 and 4.8% of the Fortune 1000 and Fortune 500 listed clothing companies respectively had female CEOs, while 73% of entry-level employees and 78% of the students at world's leading fashion schools were women. Similarly, in manufacturing countries, the population of workers and entry-level are dominated by women, whereas both mid-level and top-level management positions are occupied by men. Globally, the fashion and textile industries need to do more to achieve the targets 5.1, 5.5 and 5c, which are directly relevant to SDG 5 (see Table 6). To reduce gender inequality and enhance economic inclusion, social safeguards should be strengthened, and equitable salaries should be guaranteed [18].

2.6 SDG 6: Clean Water and Sanitation

The world is seriously running behind the pace to achieve SDG 6, which calls for availability of safe water and appropriate sanitation systems for all. Billions of people worldwide are lacking these fundamental needs. According to the Sustainable Development Goals Report 2024 by the UN, 2 billion people will not have access to safe drinking water and 3 billion will not see any safely managed sanitation by 2030 if the world progresses at the current speed [19]. In this grim situation, the global fashion and textile industries are playing a negative role. On the one hand, it consumes a huge amount of fresh water, but on the other hand, it produces a similar amount of contaminated water. According to a report published by the European Parliament, as

Table 6 Targets of SDG 5 and their relevance to fashion textile industry

Target No.	Relevance to fashion textiles		
	Directly relevant	Indirectly relevant	Not relevant
5.1	✓		
5.2		✓	
5.3			✓
5.4			✓
5.5	✓		
5.6			✓
5.a		✓	
5.b		✓	
5.c	✓		

SDG 5: Achieve gender equality and empower all women and girls (See https://sdgs.un.org/goals/goal5)

much as 2,700 L of fresh water are required to make a single cotton t-shirt [20] (EU Report n.d.). This amount of water is enough to meet the drinking water needs of one person for 2.5 years. According to UNASF, the fashion textile industry as a whole consumes about 215 trillion litres of water per year which is approximately 20% of the wastewater generation around the world. It also releases approximately 9% of annual microplastic release to the oceans [3]. Figure 5 shows the freshwater uses and water scarcity footprint across the global apparel value chain, where water scarcity footprint means the contribution to overall water scarcity of a region through pollution of fresh waterways. By improving its water management, chemical management and wastewater management systems, the fashion industry can significantly impact on the targets 6.3, 6.4, 6.5 and 6.6 of SDG 6 (see Table 7).

2.7 *SDG 7: Affordable and Clean Energy*

Traditionally, governments play a key role in producing and ensuring energy supply in their countries. Therefore, the fashion and textile industries does not have direct relevance with the SDG targets 7.1–7b (see Table 8). However, being a primary industrial sector globally, it can have a prominent voice in influencing governments about electricity generation and transition towards renewable energy. According to UN SDG report 2024, the electricity generation capacity had nearly doubled in 2022, increasing to 424 watts per capita from 250 watts per capita in 2015 [19]. This development has been partly driven by the fashion and textile industries, which are inching towards installing more and more solar panels and energy-efficient technologies.

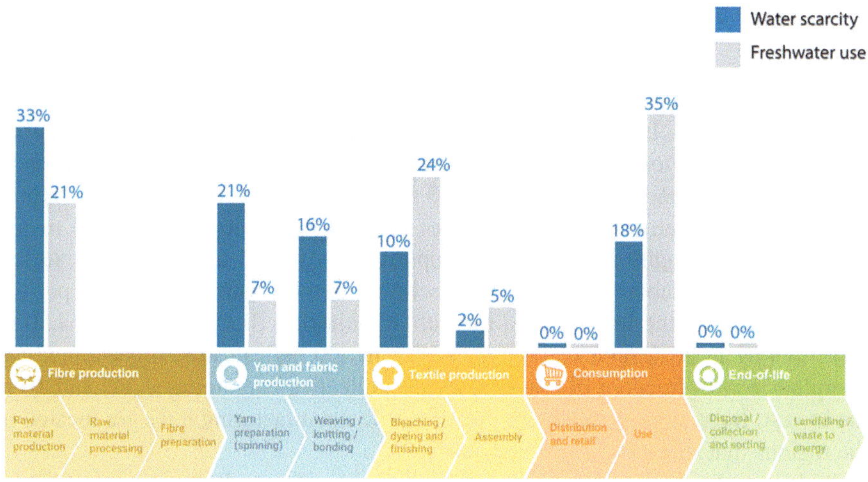

Fig. 5 Fresh water uses and water scarcity footprint across the global apparel value chain [6]

Table 7 SDG 6 targets and their relevance to fashion textiles

Target No.	Relevance to fashion textiles		
	Directly relevant	Indirectly relevant	Not relevant
6.1		✓	
6.2		✓	
6.3	✓		
6.4	✓		
6.5	✓		
6.6	✓		
6.a		✓	
6.b			✓

SDG 6: Ensure availability and sustainable management of water and sanitation for all (From https://sdgs.un.org/goals/goal6)

Table 8 Targets of SDG 7 and their relevance to fashion textile industry

Target No.	Relevance to fashion textiles		
	Directly relevant	Indirectly relevant	Not relevant
7.1		✓	
7.2		✓	
7.3			✓
7.a		✓	
7.b		✓	

SDG 7: Ensure access to affordable, reliable, sustainable and modern energy for all (See https://sdgs.un.org/goals/goal7)

2.8 SDG 8: Decent Work and Economic Growth

SDG 8 promotes inclusive and sustainable economic growth, full employment, and decent work for all. It is built on twelve targets. Ten (8.1–8.10) are about promoting and sustaining economic growth and improving employment by providing a protective working environment. The other two (8.a–8.b) are about developing trade support nationally and internationally. It includes productive employment related to job creation, labour rights and economic opportunities for all aspects of the sector. Compliance with labour rights is on a declining trend worldwide [19]. As per the scoring system of the International Labour Organisation (ILO), it fell from 4.50 in 2015 to 4.81 in 2022 on a scale of 0 to 10, where 0 indicates the highest national compliance with freedom of association and collective bargaining rights and 10 is the lowest. Interestingly, the major global textile manufacturing hubs are located in the regions that show the least compliance.

Although the textile and apparel industries contribute substantially to employment generation and poverty alleviation in many countries, reports on poor working conditions, including excessive working hours and low wages, workplace abuse and harassment, and unsafe work practices are not uncommon. Figure 6 illustrates the social risk distribution along the apparel value chain based on the social life cycle assessment of a low-cost garment. It highlights the prevalence of major social risks, encompassing the risks of excessive work and injury in the fibre, yarn, and fabric production sectors. Table 9 shows how the targets of SDG 8 are relevant to the fashion and textile industries.

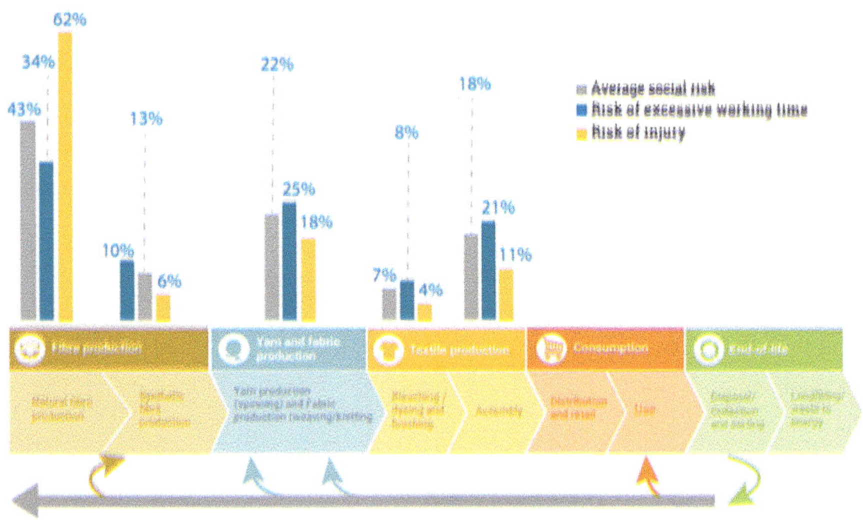

Fig. 6 Social risk distribution along the apparel value chain [6]

Table 9 SDG 8 targets and their relevance to fashion textiles

Target No.	Relevance to fashion textiles		
	Directly relevant	Indirectly relevant	Not relevant
8.1		✓	
8.2	✓		
8.3	✓		
8.4	✓		
8.5	✓		
8.6			✓
8.7	✓		
8.8	✓		
8.9			✓
8.10		✓	
8.a		✓	
8.b		✓	

SDG 8: Promote sustained, inclusive and sustainable economic growth, full and productive employment and decent work for all (See: https://sdgs.un.org/goals/goal8)

2.9 *SDG 9: Industry, Innovation and Infrastructure*

The focus of SDG 9 is on resilient infrastructure, sustainable industrialisation, and innovation, which are also the key elements for transforming the fashion and textile industry into a sustainable and circular sector. Achieving SDG 9 necessitates the integration of advanced technologies and innovative practices to enhance sustainability and efficiency. The fashion textile sector is resource-intensive and poses significant environmental and social impacts, as highlighted in Sects. 2.5, 2.6, 2.7, 2.8, 2.10, 2.11, 2.12, 2.13, 2.14, 2.15 and 2.16. Technological advancements and innovative business models are crucial for eradicating these negative impacts.

The rise in global CO_2 emissions was a record high of 37.4 gigatons in 2023 [19]. Industrial activities, including those of the fashion and textile industries, are a major contributor to this. The Fashion Industry Charter for Climate Action (FICCA) has set a target to achieve net-zero emissions by 2050 and reduce greenhouse gas (GHG) emissions by 30% by 2030, compared to a baseline of 2015 or later, in order to limit the rise in global temperature to 1.5 °C above pre-industrial levels [21]. Most of the targets of SDG 9 are directly relevant to the fashion and textile industries (see Table 10). Innovations such as waterless dyeing technologies (see in Chap. 7), bio-based materials (see in Chap. 3), and closed-loop recycling systems can contribute to reducing carbon emissions, water pollution, and waste generation by the fashion and textile industries. Additionally, Industry 4.0 solutions, including automation, digital traceability, and AI-driven supply chain optimisation, can enhance efficiency and transparency. The framework presented by Sayem et al. [22] offers a systematic

Table 10 Targets of SDG 9 and their relevance to fashion textile industry

Target No.	Relevance to fashion textiles		
	Directly relevant	Indirectly relevant	Not relevant
9.1	✓		
9.2	✓		
9.3	✓		
9.4	✓		
9.5	✓		
9.a	✓		
9.b	✓		
9.c		✓	

SDG 9: Build resilient infrastructure, promote inclusive and sustainable industrialization and foster innovation (See https://sdgs.un.org/goals/goal9)

approach to technology integration in fashion textile production, which supports the SDG target 9.4 (see Chap. 2 for further details).

2.10 SDG 10: Reduced Inequalities

SDG 10 aims to minimise differences in income, social inclusion, and equitable access to opportunities and is closely related to the fashion textile sector. The sector has a significant impact on global supply chain equality, worker rights, salaries, and economic and social frameworks in different countries. The trillion-dollar global fashion industry is often criticised for depending on cheap labour from impoverished nations, where workers are frequently exploited and paid poorly [23]. One example of global inequality is the gap between high-paid executives and low-paid garment workers in fashion capitals like Paris, Milan, and New York. Fairtrade policies and ethical sourcing contribute to reducing these economic inequalities [24].

By taking advantage of economically disadvantaged labour markets and consolidating wealth in high-income countries, fast fashion makes global inequality worse. To establish a more equitable supply chain, sustainable fashion entities emphasise ethical labour standards, fair salaries, and responsible acquisition of materials [25]. Corporate accountability-enforcing policies might help close these disparities.

Due to the growth of e-commerce and digital fashion, marginalised designers and craftspeople now have the chance to access international markets. However, in countries with lower incomes, inadequate technological infrastructures and internet access lead to a gap in accessibility that limits participation [26]. Equal opportunities could be maintained by bridging this gap with free internet access and digital literacy initiatives. To align fashion textiles with SDG 10, policymakers and business executives should uphold fair labour laws that guarantee equal pay and

Table 11 SDG 10 targets and their relevance to fashion textiles

Target No.	Relevance to fashion textiles		
	Directly relevant	Indirectly relevant	Not relevant
10.1			✓
10.2		✓	
10.3	✓		
10.4	✓		
10.5		✓	
10.6	✓		
10.7			✓
10.a	✓		
10.b		✓	
10.c			✓

SDG 10: Reduce inequality within and among countries (see https://sdgs.un.org/goals/goal10)

working conditions, encourage economic policies that support local craftsmen and Indigenous designers, improve supply chain transparency to prevent labour abuse, and invest in digital education and accessibility to empower marginalised fashion entrepreneurs (Table 11).

2.11 SDG 11: Sustainable Cities and Communities

The significance of resilient and environmentally friendly urban development is emphasised in SDG 11. Urban environmental conditions are significantly influenced by the fashion textile sector, which generates adverse effects through inadequate waste management and unsustainable economic activity. The manufacturing of fast fashion products generates a substantial quantity of textile waste, which ends up in uncontrolled urban garbage sites and landfills. The practice of uncontrolled waste management generates higher levels of pollution in urban areas, degrading both water quality and air quality. Cities that adopt circular economy policies through textile collection and sustainable material practice strengthen their ability to reduce environmental burden [27].

The textile sector has a significant impact on urban sustainability because it requires energy-intensive manufacturing techniques and consumes high quantities of water. Numerous textile production facilities operate in densely, thus causing environmental pollution along with populated urban centres, thus causing environmental pollution and natural resource depletion [28]. The implementation of water-efficient production methods, alongside ecologically safe chemicals and dyestuffs would alleviate these sustainability concerns while enhancing urban resilience. Unregulated industries, along with manufacturing facilities located in urban districts,

Table 12 Targets of SDG 11 and their relevance to fashion textiles

Target No.	Relevance to fashion textiles		
	Directly relevant	Indirectly relevant	Not relevant
11.1			✓
11.2			✓
11.3			✓
11.4		✓	
11.5			✓
11.6	✓		
11.7		✓	
11.a		✓	
11.b		✓	
11.c			✓

SDG 11: Make cities and human settlements inclusive, safe, resilient and sustainable (see https://sdgs.un.org/goals/goal11)

employ millions of workers who support fashion textile manufacturing activities. The root cause of urban social disparities lies in inadequate work wages coupled with hazardous factory conditions while extended to exclusionary work rights [29]. Law enforcement of fair working laws and support for regional workers lead to improved social involvement and financial stability in cities [30].

The drive for global fashion stems from inhabitants of urban centres who lead the movement of fast fashion trends through frequent modifications. Promoting eco-friendly consumer practices, such as renting fashion and clothing, upcycling fashion products, and supporting ethical brands, can significantly reduce the negative environmental impacts of the fashion industry [31]. Long-term sustainability may be facilitated in urban areas by raising awareness among society about purchasing environmentally friendly fashion products and enacting eco-friendly retail regulations. The fashion textile sector can substantially play a role in the creation of more equitable, resilient, secure, and green cities by confronting issues of urban trash, worker rights, and environmentally friendly manufacturing (Table 12).

2.12 *SDG 12: Responsible Consumption and Production*

SDG 12 aims to promote environmentally friendly business practices, reduce contaminants, and maximise the efficient use of resources. Global environmental issues and societal challenges, such as overconsumption of resources, contamination, and waste generation, are largely caused by the fashion and textile industries. Fast fashion, which promotes mass production and inexpensive clothing, has led the global fashion business to grow promptly. Nevertheless, this expansion has taken place at a high social and environmental cost, resulting in one of the most hazardous industries in the global market [32].

The fashion textile manufacturing sector is one of the major consumers of naturally produced supplies. It takes around 2,700 L of water to produce one cotton shirt [33]. Furthermore, the World Bank declares that nearly 20% of industrial water pollution worldwide is caused by dyeing and finishing procedures in textile industries [34]. SDG 12 advocates for enterprises to use water-efficient dyeing methods and reduce excessive material use to achieve environmentally friendly resource management. Approximately 92 million tons of textile scraps are produced annually due to fast fashion, resulting in a significant increase in textile-related waste [32]. SDG 12 promotes responsible consumption by encouraging both customers and companies to adopt a circular economy model that recycles, reuses, and repurposes resources. In an effort to reduce landfill trash, manufacturers are making investments in garment take-back programs and biodegradable textiles. Energy-intensive production processes, along with synthetic textile fibres derived from fossil fuels, are the means through which the textile sector is causing environmental degradation. SDG 12 promotes the use of energy-efficient manufacturing, clean energy sources, and low-impact resources like hemp and organic cotton.

One of the key factors behind sustainable fashion textile practice is consumer consciousness. The emergence of sustainable manufacturing practices, as well as promotional activities that utilise second-hand apparel and slow fashion items, is indicative of a shift in consumer purchasing habits. Policymakers, organisations, and consumers need to collaborate to transition to sustainable fashion to maintain continual economic and sustainable resilience (Table 13).

Table 13 SDG 12 targets and their relevance to fashion textiles

Target No.	Relevance to fashion textiles		
	Directly relevant	Indirectly relevant	Not relevant
12.1	✓		
12.2	✓		
12.3			✓
12.4	✓		
12.5	✓		
12.6	✓		
12.7	✓		
12.8	✓		
12.a	✓		
12.b		✓	
12.c	✓		

SDG 12: Ensure sustainable consumption and production patterns (From https://sdgs.un.org/goals/goal12)

2.13 SDG 13: Climate Action

SDG 13 calls for policies and initiatives that reduce the emission of greenhouse gases and enhance resilience to mitigate climate change. The fashion and textile manufacturing industries are a significant contributor to climate change because of its high waste generation, water use, and carbon footprint.

The fashion industry is responsible for 2–8% of global greenhouse gas (GHG) emissions, which is estimated to increase by more than 60% by 2030. It also consumes approximately 215 trillion litres of water per year, contributing to 20% of the world's wastewater generation [3]. Energy-intensive procedures, including farming, fibre processing, fabric manufacturing, dyeing, finishing, garment production, shipping, and waste disposal, are all integral to the textile processing and distribution workflow. Cotton cultivation requires extensive water usage, resulting in excessive water consumption by the industry. Textile dyeing and finishing procedures release dangerous chemicals into aquatic environments, thus leading to significant water contamination. The fashion textile industry owes its dependence on fossil fuels primarily to the manufacturing process of synthetic fibres that include polyester. The worldwide carbon dioxide emission levels are mainly influenced by energy-hungry industrial and transportation activities [35].

The fashion textile sector needs sustainable and eco-friendly production procedures to realise SDG 13. The reduction of carbon emissions depends on sustainable approaches that incorporate organic cotton and renewable raw materials, as well as the adoption of recycled fibre, alongside eco-friendly concepts and the implementation of renewable energy in the textile industry [36]. Local production methods, combined with ethical labour practices and carbon emission-minimising supply networks, offer substantial advantages in reducing transport-generated emissions [37]. Sustainable textile technologies, including waterless dyeing and biodegradable textiles, enable substantial reductions in environmental pollution (Table 14).

Table 14 Targets of SDG 13 and their relevance to fashion textile industry

Target No.	Relevance to fashion textiles		
	Directly relevant	Indirectly relevant	Not relevant
13.1		✓	
13.2	✓		
13.3		✓	
13.a		✓	
13.b		✓	

SDG 13: Take urgent action to combat climate change and its impacts (From https://sdgs.un.org/goals/goal13)

2.14 SDG 14: Life Below Water

The primary purpose of SDG 14 targets sustainable conservation efforts, along with the sustainable use of seas and oceans and marine resources, are to develop a sustainable future. The fashion textile sector has a substantially negative impact on the marine environment through chemical spills, irresponsible acquisition of materials derived from marine biodiversity, and microplastic contaminants. Microplastic contamination is one of the most significant challenges that connects the fashion sector to Sustainable Development Goal 14. During fabric and garment washing, synthetic fibres like acrylic, nylon, and polyester release microplastics that eventually find their way into sewage systems and, ultimately, the ocean. When marine life consumes these microplastics, they bioaccumulate in the food chain and can ultimately pose a risk to human health [38].

Chemicals, including various types of processing chemicals, dyestuffs, and finishing agents, are heavily utilised in the textile manufacturing industry. Marine biodiversity is seriously endangered by many of these chemicals, including heavy metals and azo dyes, which accumulate in wastewater and are released into rivers and, ultimately, into oceans [39]. The use of environmentally friendly dyes, chemicals and wastewater treatment techniques are tangible instances of sustainable practices that can assist in lowering ocean pollution. Seaweed-based clothing, fish bone and fish leather are a few examples of resources that fashion textile manufacturers often use. The irresponsible acquisition of these resources could disrupt marine habitats, leading to overfishing. Promoting alternatives, such as lab-grown or plant-based materials, and facilitating responsible purchasing can help mitigate the impact on marine biodiversity [40].

A significant proportion of textile waste produced by fast fashion ends up in landfills or the seas. Ocean contamination is made worse by discarded fishing nets, sometimes referred to as "ghost nets," and abandoned synthetic-fibre textiles. Reducing ocean pollution is greatly aided by circular fashion projects which employ biodegradable materials, recycling, and upcycling [32]. The long-term sustainability of marine habitats and the preservation of ocean health depend on the shift to environmentally friendly manufacturing and consumption practices. Governments, companies and customers must work together to promote substantial progress and a future in which fashion does not compromise marine ecology (Table 15).

2.15 SDG 15: Life on Land

SDG 15 aims to prevent the loss of biodiversity, manage forests sustainably, combat desertification and preserve, restore, and promote the sustainable use of terrestrial ecosystems. The fashion textile sector requires considerable land for cultivating cotton, producing wool, and manufacturing synthetic fibres, making it a significant user of environmental resources. Figure 7 illustrates the impact of various activities

Table 15 SDG 14 targets and their relevance to fashion textiles

Target No.	Relevance to fashion textiles		
	Directly relevant	Indirectly relevant	Not relevant
14.1	✓		
14.2		✓	
14.3		✓	
14.4		✓	
14.5			✓
14.6			✓
14.7			✓
14.a			✓
14.b			✓
14.c		✓	

SDG 14: Conserve and sustainably use the oceans, seas and marine resources for sustainable development (See https://sdgs.un.org/goals/goal14)

and processes on land use along the global apparel value chain. Conventional cotton farming degrades soil and impacts biodiversity due to the extensive use of water and pesticides [41]. The degradation of biodiversity is also a result of deforestation for textile raw materials, such as rayon and viscose, and other essential resources [42].

Soil and water resources have been affected by the substantial pollutants produced by fashion textile manufacturing and dyeing and washing operations. Whole ecosystems are negatively impacted by the toxic chemicals employed in the fabric manufacturing and processing stages, which harm wildlife and plants and reduce soil

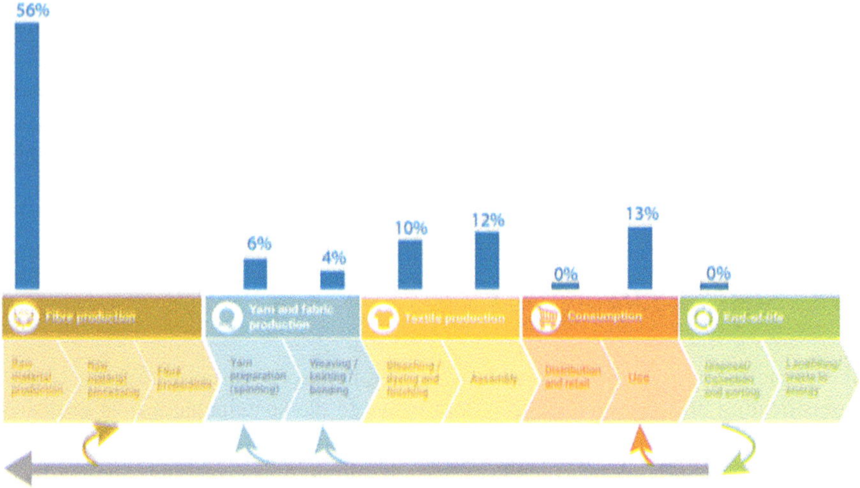

Fig. 7 The impact of land use at different stages of the apparel value chain globally [6]

Table 16 Targets of SDG 16 and their relevance to fashion textile industry

Target No.	Relevance to fashion textiles		
	Directly relevant	Indirectly relevant	Not relevant
15.1	✓		
15.2	✓		
15.3		✓	
15.4	✓		
15.5		✓	
15.6			✓
15.7			✓
15.8			✓
15.9	✓		
15.a		✓	
15.b		✓	
15.c			✓

SDG 15: Protect, restore and promote sustainable use of terrestrial ecosystems, sustainably manage forests, combat desertification, and halt and reverse land degradation and halt biodiversity loss (See https://sdgs.un.org/goals/goal15)

fertility [43]. The textile manufacturing sector produces extensive waste amounts that eventually reach landfills, where both greenhouse gas emissions and environmental deterioration occur [44].

Fashion brands, along with garment manufacturers, are now adopting sustainable techniques to fulfil the requirements of SDG 15. The textile industry minimises its harm through the implementation of organic cotton cultivation, along with sustainable agricultural practices, and responsible reforestation for textile materials [45]. The circular economy promotes textile recycling and upcycling, which helps decrease waste piles and functions as conservation practices for the Earth [46]. Natural ecosystems have served as a target for the impacts of the fashion and textile industries, thus demonstrating the critical need for eco-friendly methods to fulfil SDG 15. By utilising environmentally friendly products, adopting ethical purchasing practices, and implementing waste reduction measures, the industry fosters ecological sustainability and conserves biodiversity (Table 16).

2.16 *SDG 16: Peace, Justice and Strong Institutions*

SDG 16's main goals are to establish efficient institutions and to guarantee access to justice as well as to foster inclusive, peaceful communities at all levels. The fashion and textile industries maintain multiple connexions with this objective even though its core mission centres on human rights and governance together with an equitable society. Despite being one of the most notable employers in the world, the fashion sector is constantly under criticism for employing children, creating unsafe working conditions and avoiding compliance issues, especially in low-income countries. Sustaining fair pay, protecting workers' rights, and creating secure workplaces are all necessary to create an equitable and peaceful society [47]. The principles of SDG 16 contradict the unfair treatment experienced by numerous textile workers in countries like Bangladesh, Cambodia and India.

Corruption and lack of legitimacy in the fashion industry frequently lead to unfair employment conditions, environmental damage, and human rights violations. To execute fair trade regulations, ethical sourcing, and supply chain accountability, SDG 16 strongly emphasises legitimate entities [48]. Companies implementing third-party audits and blockchain technology contribute to increased industry accountability. Many women are employed in the textile industries, especially in underdeveloped nations. Nevertheless, many experience harassment, discrimination, and unfair compensation depending on their gender [49]. In the fashion industry, fostering inclusive societies entails guaranteeing diversity, social integration, and gender equality in leadership positions.

The fashion textile sector has a negative impact on underprivileged areas and is one of the most significant sources of pollution and environmental contamination. The exploitation of environmental resources, overconsumption of water, and the disposal of hazardous waste are frequently the outcomes of inadequate institutional governance. Establishing environmental regulations, accountability systems, and environmentally friendly approaches supports SDG 16 by upholding social justice and safeguarding human rights (Table 17).

2.17 SDG 17: Partnerships for the Goals

SDG 17 emphasizes the importance of strengthening implementation tools and revitalizing international partnerships to achieve long-term sustainability. This objective is closely related to the $2.5 trillion fashion and textile industries, which depends on global collaboration for distribution networks, equitable trading, ecological preservation, and social accountability [50].

The initial sourcing of supplies, product manufacturing, and shipment of finished items to various countries are all parts of the intricate global supply chain that comprises the global fashion industry. Governments, business enterprises, and non-governmental organisations (NGOs) should strengthen their links to ensure accountable sourcing, fair labour standards, and ethical manufacturing. Labour standards that promote sustainability and ethics exist because of initiatives such as the Fair Wear Foundation and Better Cotton Initiative (BCI) [51]. The integration of blockchain technology allows businesses to collaborate with brands to enhance supply chain transparency and responsibility, thus minimising the odds of questionable labour practices [52]. SDG 17 establishes that sustainable industrialisation requires both

Table 17 SDG 16 targets and their relevance to fashion textiles

Target No.	Relevance to fashion textiles		
	Directly relevant	Indirectly relevant	Not relevant
16.1			✓
16.2		✓	
16.3		✓	
16.4			✓
16.5			✓
16.6		✓	
16.7			✓
16.8		✓	
16.9		✓	
16.10			✓
16.a			✓
16.b	✓		

SDG 16: Promote peaceful and inclusive societies for sustainable development, provide access to justice for all and build effective, accountable and inclusive institutions at all levels (see https://sdgs.un.org/goals/goal16)

financial supports together with technological assistance to function effectively. Few clothing brands, together with textile mills in poverty-stricken nations cannot afford to purchase environmentally friendly production methods. The technological support and funding from international partnerships allow organisations to advance circular economy practises through textile recycling and bio-fibre utilisation. The Higg Index, developed by the Sustainable Apparel Coalition, provides fashion brands with a standardised measurement system to assess sustainability attributes [53]. Additionally, because of global financial partnerships, efforts in sustainable energy solutions for textile production like solar-powered industries and wastewater reuse technologies are rising [54].

Public–private collaborations are crucial for fostering innovation and raising consumer awareness to achieve SDG 17. Sustainable resources and ethical fashion are being advanced by fashion brands working with NGOs and research universities. The Fashion Pact, for instance, is a group of more than 60 international brands that strive to meet science-based goals for biodiversity, climate and ocean conservation [55]. Furthermore, groups like Fashion Revolution strive for brands to disclose their supply chains and environmental effects to promote transparency in the fashion business. The demand for ethical and environmentally friendly items is also increased by customer awareness initiatives backed by governments and fashion professionals [56] (Table 18).

Table 18 Targets of SDG 17 and their relevance to fashion textile industry

Target No.	Relevance to fashion textiles		
	Directly relevant	Indirectly relevant	Not relevant
17.1			✓
17.2			✓
17.3		✓	
17.4	✓		
17.5		✓	
17.6			✓
17.7	✓		
17.8			✓
17.9	✓		
17.10			✓
17.11		✓	
17.12	✓		
17.13			✓
17.14		✓	
17.15		✓	
17.16	✓		
17.17			✓
17.18		✓	
17.19		✓	

SDG 17: Strengthen the means of implementation and revitalize the Global Partnership for Sustainable Development (see https://sdgs.un.org/goals/goal17)

3 Conclusion

Contextual analysis revealed that several SGD targets are directly and indirectly relevant to fashion textiles industry. The mapping presented in the Tables 2–18 are according to the authors' understanding based on the review of relevant literature. As fashion and textiles are indispensable part of human society, the industry is also an important part of global economy and a major stakeholder of sustainable development. Undoubtably fixing this industry will speed up the achievement of SDGs. The next chapters of this book will give deeper dive into SDG 12 to understand its connections with fashion textiles.

References

1. News on Millennium Development Goals, UN Webpage. https://www.un.org/millenniumgoals/. Accessed 14 March 2025
2. The 17 Goals, Webpage of UN Department of Economic and Social Affairs. https://sdgs.un.org/goals. Accessed 14 March 2025
3. UNASF (UN Alliance for Sustainable Fashion). https://unfashionalliance.org/. Accessed 14 March 2025
4. Statista, Revenue of the apparel market worldwide from 2019 to 2029 (in trillion U.S. dollars) [Graph], in *Statista* (2024, December 13). https://www-statista-com.mmu.idm.oclc.org/forecasts/821415/value-of-the-global-apparel-market. Accessed 19 Feb 2025
5. Textile Exchange, Production of textile fibres worldwide from 1975 to 2020, with a forecast for 2025 and 2030 (in million metric tons) [Graph], in *Statista* (2024, September 30). https://www.statista.com/statistics/1250985/global-textile-fiber-production/. Accessed 19 Feb 2025
6. United Nations Environment Programme, *Sustainability and Circularity in the Textile Value Chain: Global Stocktaking* (2020). https://wedocs.unep.org/20.500.11822/34184. Accessed 14 March 2025
7. The Sustainable Development Goals Report 2023: Special Edition. https://unstats.un.org/sdgs/report/2023/. Accessed 14 March 2025
8. S. David, Paper cites textiles potential to alleviate poverty, in *Ecotextile News* (2019). https://www.ecotextile.com/2019022836437/news/legislation/paper-cites-textiles-potential-to-alleviate-poverty/. Accessed 14 March 2025
9. G.C. Lopez-Acevedo, R. Robertson, Sewing success?—employment, wages, and poverty following the end of the multi-fibre arrangement (English), in *Directions in Development Poverty* (World Bank Group, Washington, D.C., 2012). http://documents.worldbank.org/curated/en/488191468330919867
10. Q. Ahsan, Charting a path ahead for the RMG sector, in *The Daily Star* (2024). https://www.thedailystar.net/opinion/views/news/charting-path-ahead-the-rmg-sector-3784806. Accessed 14 March 2025
11. T.N. Vidanage, L.C. Weragoda, M. Karunarathna, W. Athukorala, Achieving sustainable development goals: the impact of the apparel and textile industry on the eradication of poverty in Sri Lanka, in *Textiles of Sri Lanka. SDGs and Textiles*, ed. by G.M. Ranathunga (Springer, Singapore, 2024). https://doi.org/10.1007/978-981-97-0569-6_5
12. Global Fashion Agenda & Boston Consulting Group, *Pulse of the Fashion Industry 2017* (2017). https://globalfashionagenda.org/resource/pulse-of-the-fashion-industry-2017/. Accessed 14 March 2025
13. Bangladesh Garments Manufacturers and Exporters Association (BGMEA), https://www.bgmea.com.bd/page/Sustainability_Wellbeing_Safety. Accessed 14 March 2025
14. RMG Health Toolkit, https://rmghealthtoolkit.com/. Accessed 14 March 2025
15. D. Ewing-Chow, Healthy meals are a dream come true for female garment workers in Bangladesh, in *Forbes* (2022). https://www.forbes.com/sites/daphneewingchow/2022/10/18/healthy-meals-are-a-dream-come-true-for-female-garment-workers-in-bangladesh/. Accessed 14 March 2025
16. G.D. Brown, Women garment workers face huge inequities in global supply chain factories made worse by COVID-19. New Solut.: J. Environ. Occup. Health Policy **31**(2), 113–124 (2021)
17. G. Lai, Fashion has a gender problem, so what can we do about it?, *Vogue Australia* (2021). https://www.vogue.com.au/culture/features/fashion-has-a-gender-problem-so-what-can-we-do-about-it/image-gallery/fe823dbebfa7e5beb621a71d2193c5e8. Accessed 14 March 2025
18. N. Kabeer, Gender equality, inclusive growth, and labour markets, in *Women's Economic Empowerment* (Routledge, 2021), pp. 13–48
19. The Sustainable Development Goals Report 2024, https://unstats.un.org/sdgs/report/2024/. Accessed 14 March 2025

20. European Parliament, *The Impact of Textile Production and Waste on the Environment (Infographics)*, European Parliament. https://www.europarl.europa.eu/topics/en/article/20201208STO93327/the-impact-of-textile-production-and-waste-on-the-environment-infographics. Accessed 14 March 2025
21. FICCA, *The Fashion Industry Charter for Climate Action* (UNFCCC secretariat (UN Climate Change), n.d.). https://unfccc.int/sites/default/files/resource/Industry%20Charter%20%20Fashion%20and%20Climate%20Action%20-%2022102018.pdf. Accessed 14 March 2025
22. A.S.M. Sayem, J. Haider, Md. Abbas Uddin, A.N. Khan, M.M. Alamgir Sayeed, M. Hasan, Chapter 2: framework for environmentally sustainable fashion and textile production to achieve SDG 12, in *UN SDG 12 and Global Fashion Textiles Production*, ed. by A.S. Muhammad Sayem, P. Goswami (Springer Nature, 2025)
23. Clean Clothes Campaign (2021). https://www.schonekleren.nl/. Accessed 1 Feb 2025
24. Fashion Revolution Week 2022 Roundup: Fashion Revolution (2022). https://www.fashionrevolution.org/fashion-revolution-week-2022-roundup/. Accessed 1 Feb 2025
25. R. Gautam, The narrative of circular economy and sustainability—a critical analysis of fashion industry. Circ. Econ. Sustain. **4**(4), 3183–3213 (2024)
26. R. Kaplinsky, E. Kraemer-Mbula, Innovation and uneven development: the challenge for low-and middle-income economies. Res. Policy, **51**(2), 104394 (2022). https://doi.org/10.1016/j.respol.2021.104394
27. R. Činčikaitė, Assessment of sustainable waste management: a case study in Lithuania. Sustainability **17**(1), 120 (2024)
28. A.G. Capodaglio, G. Olsson, Energy issues in sustainable urban wastewater management: use, demand reduction and recovery in the urban water cycle. Sustainability **12**(1), 266 (2019)
29. F. Barten et al., Contextualising workers' health and safety in urban settings: the need for a global perspective and an integrated approach. Habitat Int. **32**(2), 223–236 (2008)
30. G.F. Kebede, Entrepreneurship and the promises of inclusive urban development in Ethiopia, in *Urban Forum* (Springer, 2023), pp. 1–30
31. J. Ma et al., Sustainability in design: sustainable fashion design practices and environmental impact using mixed-method analysis. Bus. Strateg. Environ. **33**(7), 6889–6910 (2024)
32. E. MacArthur, *Foundation a new Textiles Economy: Redesigning Fashion's Future* (London, UK [Preprint], 2017)
33. K. Shirvanimoghaddam et al., Death by waste: fashion and textile circular economy case. Sci. Total Environ. **718**, 137317 (2020)
34. K. Singha et al., Harmful environmental effects for textile chemical dyeing practice, in *Green Chemistry for Sustainable Textiles* (Elsevier, 2021), pp. 153–164
35. B. Lin, R. Tan, Sustainable development of China's energy intensive industries: from the aspect of carbon dioxide emissions reduction. Renew. Sustain. Energy Rev. **77**, 386–394 (2017)
36. L. Simpliciano et al., *Fashion Transparency Index* (Fashion Revolution CIC, London, 2022) [Preprint]
37. McKinsey & Company, *The State of Fashion 2020* (2020). https://www.mckinsey.com/~/media/mckinsey/industries/retail/our%20insights/the%20state%20of%20fashion%202020%20navigating%20uncertainty/the-state-of-fashion-2020-final.pdf. Accessed 14 March 2025
38. A. Al Mamun et al., Microplastics in human food chains: food becoming a threat to health safety. Sci. Total Environ. **858**, 159834 (2023)
39. S. Dutta et al., Contamination of textile dyes in aquatic environment: adverse impacts on aquatic ecosystem and human health, and its management using bioremediation. J. Environ. Manag. **353**, 120103 (2024)
40. K. Fletcher, M. Tham, *Earth Logic: Fashion Action Research Plan* (JJ Charitable Trust, 2019)
41. F. Baudron, M. Corbeels, F. Monicat et al., Cotton expansion and biodiversity loss in African savannahs, opportunities and challenges for conservation agriculture: a review paper based on two case studies. Biodivers. Conserv. **18**, 2625–2644 (2009). https://doi.org/10.1007/s10531-009-9663-x

42. S. Radhakrishnan, Sustainable development goal: sustainable management and use of natural resources in textile and apparel industry, in *Sustainable Approaches in Textiles and Fashion: Circular Economy and Microplastic Pollution* (Springer, 2022), pp. 165–205
43. W.U. Khan et al., A critical review of hazardous waste generation from textile industries and associated ecological impacts. J. Indian Chem. Soc. **100**(1), 100829 (2023)
44. R. Gupta et al., Waste management in fashion and textile industry: recent advances and trends, life-cycle assessment, and circular economy, in *Emerging Trends to Approaching Zero Waste* (2022), pp. 215–242
45. C.R.S. de Oliveira et al., Textile re-engineering: eco-responsible solutions for a more sustainable industry. Sustain. Prod. Consum. **28**, 1232–1248 (2021)
46. C. Arana et al., SDG 15 life on land: A review of sustainable fashion design processes: upcycling waste organic yarns, in *Actioning the Global Goals for Local Impact: Towards Sustainability Science, Policy, Education and Practice* (2020), pp. 247–264
47. ILO, *World Employment and Social Outlook: Trends* (2021). https://www.ilo.org/sites/default/files/wcmsp5/groups/public/@dgreports/@dcomm/documents/publication/wcms_794452.pdf. Accessed 14 March 2025
48. K. Khurana, M. Ricchetti, Two decades of sustainable supply chain management in the fashion business, an appraisal. J. Fash. Mark. Manag. **20**(1), 89–104 (2016)
49. A.M. Thakker, D. Sun, Sustainable development goals for textiles and fashion. Environ. Sci. Pollut. Res. **30**(46), 101989–102009 (2023)
50. A.M. James, S. Mather, K.J. Sheridan, Rethinking the fashion value chain: how reshoring can create a localised product lifecycle and support sustainable economic growth, in *Novel Sustainable Alternative Approaches for the Textiles and Fashion Industry* (Springer, 2023), pp. 1–39
51. R.C. Silva et al., Fashion market niches for organic agroforestry cotton: market potential for promoting sustainable supply chains. Sustainability **15**(1), 700 (2022)
52. C. Vijai, W. Wisetsri, The role of blockchain technology for transparency in the fashion supply, in *Illustrating Digital Innovations Towards Intelligent Fashion: Leveraging* (2024), p. 231
53. S. Radhakrishnan, The sustainable apparel coalition and the Higg index, in *Roadmap to Sustainable Textiles and Clothing: Regulatory Aspects and Sustainability Standards of Textiles and the Clothing Supply Chain* (Springer, 2014), pp. 23–57
54. M. Sharma et al., Advancement in solar energy-based technologies for sustainable treatment of textile wastewater: reuse, recovery and current perspectives. J. Water Process Eng. **56**, 104241 (2023)
55. A. Pastran, E. Colli, H.M. Nor, Public policy and legislation in sustainable fashion, in *Sustainable Fashion and Textiles in Latin America* (2021), pp. 171–189
56. A. Khandual, S. Pradhan, Fashion brands and consumers approach towards sustainable fashion, in *Fast Fashion, Fashion Brands and Sustainable Consumption* (2019), pp. 37–54

Chapter 2
Framework for Environmentally Sustainable Fashion and Textile Production to Achieve SDG 12

Abu Sadat Muhammad Sayem, Julfikar Haider, Mohammad Abbas Uddin, Ayub Nabi Khan, M. M. Alamgir Sayeed, and Mahbub Hasan

Abstract A framework with three key actions—identify (I), act (A), and evaluate (E) to achieve complete environmental sustainability in fashion and textile production in line with SDG 12—has been developed as a part of a Global Challenge Research Fund (GCRF) project. Called the 'I–A–E framework', this framework emphasises zero or near-zero waste generation at the source and incorporates sustainability thinking in material and process selection rather than postproduction waste management after the generation of large amounts of wastes and effluents. The first step involves identifying present scenarios and points of action in the context of resource consumption, air pollution and greenhouse gas emissions, water pollution, and solid waste generation in product development and production. The next step is to act for sustainable development, which includes incorporating (1) a zero-waste philosophy during production, (2) sustainability thinking into material selection and production processes, and (3) sustainability in resource and waste management. The third component of the framework is to evaluate for celebrating achievement and moving forward by checking the eco-indices to amend targets or set new ones. This new framework was validated through workshops with stakeholders and roundtable

A. S. M. Sayem (✉)
Manchester Fashion Institute, Manchester Metropolitan University, Manchester, UK
e-mail: asm.sayem@mmu.ac.uk

J. Haider
Department of Engineering, Manchester Metropolitan University, Manchester, UK

M. A. Uddin
Bangladesh University of Textiles, Dhaka, Bangladesh

A. N. Khan
BGMEA University of Fashion and Technology, Dhaka, Bangladesh

M. M. Alamgir Sayeed
Textile Physics Division, Bangladesh Jute Research Institute, Dhaka, Bangladesh

M. Hasan
Materials and Metallurgical Engineering, Bangladesh University of Engineering and Technology, Dhaka, Bangladesh

A. S. Muhammad Sayem and P. Goswami (eds.), *SDG 12 and Global Fashion Textiles Production*, SDGs and Textiles, https://doi.org/10.1007/978-981-96-9051-0_2

discussion. It is an easy-to-follow toolkit that the fashion and textile industry will be able to implement in product development and production activities.

Keywords Textiles · Fashion · Sustainable development goal · Pollution · Greenhouse gas · Eco-indices

1 Introduction

According to the United Nations Alliance for Sustainable Fashion (UNASF), the fashion industry is responsible for approximately 2–8% of global greenhouse gas (GHG) emission, which is estimated to increase by more than 60% by 2030, consuming approximately 215 trillion litres of water per year and causing 20% of the wastewater generation around the world, approximately 9% of total annual microplastic release to the oceans and an annual material loss of US $100 billion due to underutilisation [1]. The Fashion Industry Charter for Climate Action (FICCA), established under the auspices of UN Climate Change, aims to achieve net-zero emissions by 2050, and the signatories are committed to reducing GHG emission by 30% by 2030 against a baseline of no earlier than 2015 to keep the increase in global warming under 1.5 °C above preindustrial levels [2].

The United Nation (UN) Sustainable Development Goal (SDG) 12 has 11 targets and 13 indicators [3, 4]. Four of these targets are directly, and 2 are indirectly related to fashion and textile production (Table 1). A 10 Year Framework of Programmes on Sustainable Consumption and Production Patterns (10YFP) was adopted at the Rio + 20 UN Conference on Sustainable Development in 2012 [5–7]. The 10YFP consists of six programmes: Sustainable Public Procurement, Consumer Information for Sustainable Consumption and Production, Sustainable Tourism, Sustainable Lifestyles and Education, Sustainable Buildings and Construction, and Sustainable Food Systems, which provide directions for member countries. There are several frameworks to achieve sustainability and circularity in the fashion industry, which have been proposed and presented (Table 2), but they do not provide effectively applicable guidance and recommendations for manufacturing plants, especially those are in South Asia. This chapter presents a simple and easy-to-follow framework for environmentally sustainable fashion and textile production developed under the auspices of the Global Challenge Research Fund (GCRF) Project during 2020/2021.

Table 1 Relationships between SDG 12 targets and fashion and textile production

		SDG 12 targets
Fashion and textile production	Directly related	12.2: Sustainable management and use of natural resources 12.4: Responsible management of chemicals and waste 12.5: Substantially reduce waste generation 12.6: Encourage companies to adopt sustainable practices and sustainability reporting
	Indirectly related	12.8: Promote universal understanding of sustainable lifestyles 12.A: Support developing countries scientific and technological capacity for sustainable consumption and production

Table 2 Prevailing frameworks related to the fashion industry

#	Existing frameworks on textile and fashion	Refs.
1	Framework of circular business models for fashion and textiles	[8]
2	Apparel industry sustainable strategies framework	[9]
3	Fixing fashion framework	[10]
4	Integrated framework for the sustainable performance of fast fashion firms	[11]
5	Fashion design for sustainability: a framework for participatory practice	[12]
6	Framework for achieving sustainable fashion consumption through collaboration	[13]
7	Design framework for enabling sustainability in the clothing sector	[14]
8	Conceptual framework for sustainable fashion consumption within the circular fashion system	[15]
9	Common framework for responsible purchasing practices	[16]

2 What We Know Thus Far

2.1 Fibre Production

Textile fibres have significant environmental footprints, especially in terms of energy demand and water depletion [17, 18]. Among natural fibres, cotton dominates the fashion market. The major hotspots associated with cotton fibres are the energy used in farming and the use of fertilisers or pesticides [19]. Compared with other natural fibres, such as jute and kenaf, cotton cultivation requires more energy and water [20]. In the case of animal fibres, wool and silk contribute to the environmental footprint. On-firm activities account for two-thirds of wool's total CO_2 emission [21]. On the other hand, for silk, cocoon processing, irrigation and drying have environmental consequences. Synthetic fibres such as nylon, polyester, acrylic, and elastane also leave footprints at different processing stages [22]. For regenerated cellulose fibres, energy consumption and the application of hazardous substances increase the environmental footprint [23].

2.2 Yarn Production

Energy demand is the largest eco-indicator for yarn manufacturing, and most energy is depleted for driving machinery. Air quality (PM10 and PM2.5) is another indicator that reflects the impact of spinning. Air conditioning in a spinning factory can consume both energy and water [24]. The spinning of melange yarn consumes energy as well as water, dyes and auxiliaries [24]. The yarn count is an important factor that directly impacts the environmental footprint [22].

2.3 Fabric Production

The use of non-renewable energy, the generation of fibre flies, the use of heavy metals containing knitting oils, and the generation of solid waste are the key environmental issues related to knitting [25]. It accounts for 5 MJ/kg of energy consumption, causing emissions such as CO_2 and particulates [26]. The GHG emissions from the Knitting process are lower than the GHG emissions from the fibre production and yarn spinning processes. In one calculation, it was found to be 7% of the total life cycle GHG emissions of knitted garments [27]. The generation of flying short fibres during knitting not only creates indoor air pollution, which in turn causes acute and chronic respiratory problems for employees but also affects the quality of the fabric due to fibre contamination [28].

Weaving requires additional yarn preparation processes, such as winding, warping and sizing, which increase the extent of the environmental impact of weaving compared with knitting. Areas of concern in weaving are solid wastes from yarn, size ingredients and fabric scraps; effluent from the sizing process; particulates from sizing, warping and weaving; volatile organic compounds (VoCs) from sizing; and high decibel noise from looms. According to a study performed in Finland, the energy consumption of weaving, including singeing and sizing subprocesses, is 5.4 MJ/kg of fabric [29]. Higher energy requirements have been reported in other studies [30]. Among the total energy consumption in a weaving mill, 36.3% is used by machines, 29.4% is used by compressors, 27.1% is used by air conditioners, 3.4% is used for illumination, and 3.8% is used for cleaning [31]. Water pollution associated with the sizing of warp yarns is a significant environmental issue related to weaving, accounting for 30–70% of the increased chemical oxygen demand (COD) load in wastewater [30]. In addition to sizing chemicals, lubricants and oils are used to lubricate looms [25]. Small fibrous particulates as well as VoCs are generated from the sizing process, and particulates are generated from the weaving and warping processes. Weaving machines are the main cause of noise pollution, accounting for an average noise level of 100–120 dB [32]. Further sustainability issues in weaving industries involve the handling of solid wastes from yarn, fabric scraps and empty beam frames, maintaining proper humidity and temperature at the weaving floor.

2.4 Colourations

The textile industry consumes a large share of the global chemical stock. Hundreds of different chemicals, which fall broadly under the categories of dyes and pigments, scouring and bleaching agents, auxiliary chemicals, coatings and washing agents, are used for processing and producing textile materials. The fourth edition of the colour index published by the Society of Dyers and Colourists (SDC) lists over 34,500 dyes and pigments listed under 11,570 colour index generic names [33]. The use of organic dyes during the conventional dyeing process has an adverse effect on the environment [34]. Among the different groups of dyes used in textile colouration, direct and reactive synthetic dyes generally have less environmental impact than do vat or sulphur synthetic dyes. Textile colouration is recognised as one of the largest water-consuming and water-polluting industries in the world [35]. A large volume of effluent with a deep colour, various organic and inorganic chemicals, and chlorinated compounds is discharged during various textile wet processes.

2.5 Garment Production

The growing environmental impact can be traced back to the significant rise in the consumption and production of clothing items. A pair of jeans produces 11.5 kg of CO_2 equivalent, 57% of which is produced during the garment production phase [36]. A woman's sweatshirt has a carbon footprint of approximately 13.5 kg, of which 5 kg is produced during the manufacturing process [37].

2.6 Distribution Chain

Transportation activities between manufacturing units and retail stores are responsible for more than half of the total carbon emissions in the fashion supply chain [38]. Packaging and finishing items are necessary to protect products during distribution and maintain branding. However, they are in the forms of polybags, hangtags, printed levels, paper, plastic stands, boards and cartons; in many cases, they are landfilled or incinerated after one single use.

3 Framework

The framework presented here was initially developed by the project team utilising the latest information derived through a literature review and tacit knowledge, and professional experience from the industry. The initial version of the framework is

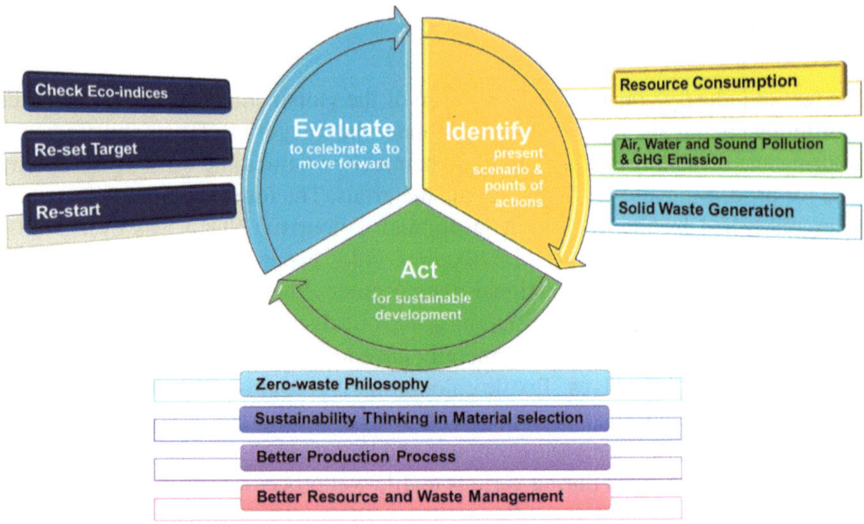

Fig. 1 Proposed I–A–E framework for textile and fashion sustainability

presented before participants (n = 30), including middle- to senior-level professionals from the textile and apparel industry in Bangladesh, representatives from national and international nongovernmental organisations (NGOs) and academics in two participatory workshops, and then it is fine-tuned to incorporate key feedback from the participants. Finally, the framework is presented in a round table discussion with representatives from international fashion retailers and national and international organisations to further validate it.

The proposed framework in Fig. 1 recommends three key actions—identify (I), act (A), and evaluate (E)—to achieve complete environmental sustainability in fashion and textile production in line with SDG 12. These key actions are described below.

(A) Identify present scenario and points of action
(B) Act for sustainable development
(C) Evaluate to celebrate and to move forward.

3.1 *Identify Present Scenario and Points of Action*

For any producer of fibre, yarn, fabric or garment, the starting point to achieve sustainability is to first know their present status in terms of resource consumption, pollution and waste generation (Table 3). Section 2 provides a general picture of the environmental impact of fashion and textile production. The UN environment programme identified sustainability hotspots along the textile and apparel value chain [39].

Table 3 Sources of resource consumption, pollution and solid waste generation along the textile and apparel value chain

Point of actions	Specific areas	Production phase
Resource consumption	Raw material	Fibre, yarn, fabric, garment
	Energy	Fibre, yarn, fabric, colouration, garment
	Water	Fibre, yarn, fabric, colouration, garment
	Chemicals	Colouration, industrial garment wash
Air pollution and GHG emissions	GHG	Fibre, yarn, fabric, colouration
	VoC	Fabric and colouration
	Fly (material dust)	Yarn and fabric
	Overall air quality (PM2.5/10)	Fibre, yarn, fabric, colouration
Water pollution	Effluent	Synthetic fibre, fabric, colouration, wash
Sound pollution	Noise	Synthetic fibre, yarn, fabric, colouration, wash
Solid waste generation	Short fibre	Fibre and yarn
	Waste yarn	Yarn and fabric
	Waste fabric	Fabric, colouration and garments
	Trims and packaging	Garments and others

We recommend the use of a mapping tool to identify the environmental performance of manufacturing facilities. For example, the Higg Facility Environmental Module (Higg FEM) from Sustainable Apparel Coalition helps assess the Environmental Management Systems, Energy Use and Greenhouse Gas Emissions, Water Use, Wastewater, Emissions to Air, Waste Management and Chemical Management of a facility.

3.2 *Act for Sustainable Development*

Once manufacturers identify the environmental hotspots within their production cycles, it is now time to act for sustainable development. We recommend four principles of action. They are:

(i) Adopt Zero-waste philosophy
(ii) Incorporate sustainability thinking in material selection
(iii) Apply better production process
(iv) Follow better resource and waste management.

3.2.1 Adopt Zero-Waste Philosophy

It is important to plan product development and production with a zero-waste target in mind. It is true that 100% zero-waste may never be possible in textile and apparel production, but having a zero-waste philosophy will enable manufacturers to gradually reduce waste and approach a near-zero-waste scenario. For example, recent research conducted at Manchester Metropolitan University, UK, demonstrated that the application of the zero-waste pattern cutting (ZWPC) approach in the mass production of selected apparels could achieve more than 98% fabric utilisation, whereas in traditional industrial practice, over 85% fabric utilisation is considered impossible [40]. Previously, the practice of ZWPC was thought to be applicable only in atelier environments for designing and making one or two garments at a time. The implementation of this approach in industry requires a new mindset and a new approach in apparel design and pattern cutting and the use of digital pattern cutting and marker-making techniques, including 2D/3D tools.

Like zero-waste approaches, zero-discharge approaches need to be identified for fibre, yarn and fabric production. Here, the "Road to Zero" initiative from ZDHC is worth mentioning (ZDHC n.d.). However, this initiative is still too far away from reaching small- and medium-scale manufacturers in the least developed and developing countries.

3.2.2 Sustainability Thinking in Material Selection

The use of sustainable and renewable materials in fibre production and other textile applications, such as dyeing, can play a major role in promoting sustainability in the textile and fashion industries [41]. The use of 100% recyclable and biodegradable materials must be ensured not only in fibre-making but also in other areas, such as colouration, accessories, and packaging. Product life cycle design and life cycle analysis are important at the early stage of product development and can help minimise the environmental impact and ensure material circularity. We recommend the use of any reliable lifecycle analysis (LCA) tool, such as the Higg Index—Material Sustainability Index (HIMSI) from the SAC, the Eco Index tool by the Outdoor Industry Association, OpenLCA, CCaLC, Granta EduPack software, etc., by product developers, designers and purchase teams to ensure that increasingly more sustainable materials can be fed into the production process.

3.2.3 Apply Better Production Process

Here, we present a non-comprehensive list of better alternatives that can be employed in the production of fibre to garments.

(a) ***Fibre and Yarn***

- Organic agricultural products have been reported to substantially minimise the impacts of growth stages on the environment [26].
- Compared with conventional cotton, naturally coloured cotton has better pest and disease resistance [21]. Furthermore, the use of coloured wool from sheep can help minimise the necessity of wool colouration [21].
- The crop rotation technique can minimise the damage caused by harmful living organisms and substances and thus can reduce the environmental footprint by countering the necessity of pesticides [42].
- Despite having some environmental impacts, the recycling of fibres from discarded wastages can spare the elevated impacts associated with the production of virgin fibres [18].
- To increase energy efficiency, in addition to process optimisation, the replacement of current machinery with energy-efficient machinery can be carried out [43–45].

(b) ***Fabric***

In case of knitting:

- The adoption of advanced technologies such as seamless knitting and 3D fabric production would reduce energy resource waste and the carbon footprint by removing post-knitting operations, reducing material waste, shortening downtime and lowering transport emissions [46, 47].
- Energy-efficient developments in needle technology for high-speed circular knitting are needed to save energy and resources [47].
- The use of semicombed yarn, as combed yarn consumes more energy than does the use of carded yarn of the same count in both knitting and weaving [48].
- The use of eco-friendly lubricants such as machine oil reduces the pollution load of the wastewater stream [47].
- Separating the knitting machine with polythene wrap would be an effective measure to prevent fibre contamination from the cotton fly on the knitting floor. An improvement in the atmospheric cleaning process inside a knitting room can lead to substantial changes but involves initial and running costs [28].

In case of weaving:

- In terms of sizing, chitosan is an excellent, economic and ecologically compatible alternative to synthetic sizing agents for cellulosic yarns, which are biodegradable and can also be recycled from wastewater [49].
- Keratin from chicken feathers, an accessible and environmentally friendly sizing agent, can be used to size polyester or polyester/cotton blended warp yarns and effectively contributes to negligible amounts of total ammonia and nitrogen in wastewater but is easily biodegradable in activated sludge [50]. Starch-based warp yarn sizing extracted from potatoes would be an economic and eco-friendly sizing agent for replacing synthetic sizing agents.
- Over the last 20 years, textile machinery manufacturers, especially weaving machine manufacturers, have made considerable progress in lowering noise

levels by adopting mechatronics, PLC operating systems, active noise control mechanisms, blocking and absorbing airborne sounds and vibration damping.
- The air suction system, blowing system and proper ventilation can maintain proper humidity and temperature and remove air contaminants from weaving sections.
- The rate of yarn breakage is related to the machine speed and yarn count. Therefore, the speed of machines needs to be controlled to minimise the yarn breakage rate and reduce solid waste.
- Empty cones and empty beam frames should be reused in the weaving process [51].

(c) ***Colouration***

- The use of energy efficient machinery and process optimisation can reduce environmental pollution during fabric colouration.
- Water-free dyeing techniques, such as supercritical fluid dyeing, should be made available by making the technology patent universal [52].

(d) ***Garments***

- Consider ZWPC technique in the mass production of garments to maximise fabric utilisation; see example in presented by Ramkalaon and Sayem [40].
- Integrate sustainability thinking in fashion product design and development.
- Build a sustainable fashion supply chain by embedding sustainability at every stage of fashion and textile production [53].

(e) ***Distribution Chain***

- Using biodegradable and less impactful materials for packaging and finishing items.
- Redesigning packaging to ensure recycling could be an approach to avoid landfilling packaging after a single use.

3.2.4 Follow Better Resource and Waste Management

- Wastewater treatment and recycling of treated water in production systems and in domestic use, such as for use in toilet flush.
- Rainwater harvesting for industrial use.
- Use of renewable energy to reduce energy consumption in the textile industry.
- Use direct solar energy for fibre drying and water heating in the textile industry.
- Shredding small fabric cut offs into fibres for recycling.

Notably, the list provided above is a noncomprehensive list, and novel alternative approaches, techniques and processes are expected to appear time to time on the market. The implementing party (the producer and manufacturer) will always need to keep an open eye in search of better alternatives for them.

3.3 Evaluate to Celebrate and to Move Forward

At the retailer and manufacturer level, effluent characterisation parameters, such as chemical oxygen demand (COD), dissolved oxygen (DO), biochemical oxygen demand (BOD), total dissolved solids (TDS), total suspended solids (TSS), potential hydrogen (pH) and color indices, are commonly tested, and the acceptable limits of these parameters are included in several national and international guidelines [54]. We recommend the evaluation of the following product eco-indices following the LCA methodology per BS EN ISO 14040: 2006 and PAS 2050:2011 [55] and include them in the guidelines of local authorities' control parameters:

- carbon footprint (or global warming potential);
- water footprint;
- acidification potential;
- eutrophication potential;
- ozone layer depletion potential;
- photochemical smog; and
- human toxicity potential.

In addition, we also recommend mapping the environmental performance of manufacturing facilities, as mentioned in Sect. 3.1. Once measured, textiles and apparel manufacturers will be able to judge the environmental performance of their facilities and products. This will provide them with a benchmark to set a target for the next year and restart with I–A–E for the next year.

4 Conclusion

The framework was developed primarily targeting the textile and apparel industry in Bangladesh, which is the partner country of our GCRF project. However, this approach is applicable to all textile and apparel manufacturers located in any part of the world. Effective implementation is the key to the success of any framework. The workshop participants revealed that a combined effort from all stakeholders will need to come together to implement the proposed framework. The textile and apparel manufacturers will not be able to implement this alone unless different parties such as fashion retailers, local industry bodies such as Bangladesh Garments Manufacturers and Exporters Associations (BGMEA) and the relevant government and nongovernment organisations become involved. We recommend setting up working groups involving all stakeholders similar to the FICCA working groups for implementation of the proposed framework.

Acknowledgements We thankfully acknowledge "Research England" for the GCRF funding and the Research and Knowledge Exchange (RKE) Directorate of Manchester Metropolitan University (MMU) for facilitating the project. The authors are thankful to all research assistants (RAs), namely, Anubhav Malik and Anna Romachney from MMU; Humayra Akhter Himu, Tanvir Mahady Dip,

Bristi Sarker and Nusrat Jahan from Bangladesh University of Textiles (BUTEX); and Rayed Barkat and Mohammad Iftekhar Rahman from BUFT, for their contributions to the project activities.

References

1. UNASF (n.d.). https://unfashionalliance.org/. Accessed 04 July 2021
2. FICCA, *The Fashion Industry Charter for Climate Action* (UNFCCC secretariat (UN Climate Change), n.d.). https://unfccc.int/sites/default/files/resource/Industry%20Charter%20%20Fashion%20and%20Climate%20Action%20-%2022102018.pdf. Accessed 04 July 2021
3. SDGs, Goal 12, in *Ensure Sustainable Consumption and Production Patterns* (n.d.). https://sdgs.un.org/goals/goal12
4. SDG Tracker, Sustainable development goal 12, in *Ensure Sustainable Consumption and Production Patterns* (n.d). https://sdg-tracker.org/sustainable-consumption-production
5. UNEP, *10YFP—10 Year Framework of Programmes on Sustainable Consumption and Production Patterns* (n.d). https://www.unep.org/explore-topics/resource-efficiency/what-we-do/one-planet-network/10yfp-10-year-framework-programmes
6. EPA, *10-Year Framework of Programmes on Sustainable Consumption and Production* (n.d.). https://www.epa.gov/international-cooperation/10-year-framework-programmes-sustainable-consumption-and-production
7. 10YFP Secretariat, *The 10 Year Framework of Programmes on Sustainable Consumption and Production* (n.d.). https://sustainabledevelopment.un.org/content/documents/944brochure10yfp.pdf. Accessed 22 May 2023
8. L. Coscieme, S. Manshoven, J. Gillabel, F. Grossi, L.F. Mortensen, A framework of circular business models for fashion and textiles: the role of business-model, technical, and social innovation. Sustain.: Sci. Pract. Policy **18**(1), 451–462 (2022). https://doi.org/10.1080/15487733.2022.2083792
9. CSB, *Apparel Industry Sustainable Strategies Framework* (Center for Sustainable Business, NYU Stern, 2022). https://www.stern.nyu.edu/experience-stern/about/departments-centers-initiatives/centers-of-research/center-sustainable-business/research/return-sustainability-investment-rosi/apparel-industry-sustainable-strategies-framework. Accessed 22 May 2023
10. J. Berentson-Shaw, Fixing fashion: a framework to make sustainable fashion the industry standard, in *Huffpost* (2015) https://www.huffingtonpost.co.uk/jess-berentsonshaw/sustainable-fashion_b_8393590.html. Accessed 22 May 2023
11. M.J. Seo, M. Kim, A new integrated framework for the sustainable performance of fast fashion firms, in *International Textile and Apparel Association (ITAA) Annual Conference Proceedings* (2015), p. 132. https://core.ac.uk/download/pdf/128974465.pdf
12. D. Williams, *Fashion Design for Sustainability: A Framework for Participatory Practice* (Centre for Sustainable Fashion, University of the Arts London, n.d). https://ualresearchonline.arts.ac.uk/id/eprint/14350/1/LeNSin_full_paper._dilys%20williams.pdf
13. NFA, *Framework for Achieving Sustainable Fashion Consumption through Collaboration, Nordic Initiative, Clean and Ethical* (Nordic fashion Association, 2012). https://www.bsr.org/reports/nice-consumer-framework.pdf
14. A.D. Santos, F. Ceschin, S.B. Martins, C. Vezzoli, A design framework for enabling sustainability in the clothing sector. Lat. Am. J. Manag. Sustain. Dev. **3**(1), 47–65 (2016). https://doi.org/10.1504/LAJMSD.2016.078615
15. K. Vladimirova, S. Iran, J. Barber, M. Blazquez, M. Burcikova, C.E. Henninger, E. Johnson, C. Joyner Martinez, K. Laitala, I. Maldini, L. McNeil, K. Niinimaki, K. Onthank, M. Plonka, M. Sauerwein, S. Wallaschkowski, *Conceptual Framework for Sustainable Fashion Consumption Within the Circular Fashion System* (International Research Network on International Fashion Consumption, 2021). https://sustainablefashionconsumption.org/projects/sfc-framework/

16. MSI, *Common Framework for Responsible Purchasing Practices* (The MSI Working Group on Purchasing Practices, 2022). https://www.fairwear.org/stories/the-common-framework-for-responsible-purchasing-practices-out-now/
17. A.S.M. Sayem, J. Haider, An overview on the development of natural renewable materials for textile applications, in *Encyclopedia of Renewable and Sustainable Materials*, ed. by S. Hashmi, I.A. Choudhury (Elsevier, 2020), pp. 822–838, ISBN: 9780128131961. https://doi.org/10.1016/B978-0-12-803581-8.10983-X. https://www.sciencedirect.com/science/article/pii/B978012803581810983X
18. F.A. Esteve-Turrillas, M. de la Guardia, Environmental impact of recover cotton in textile industry. Resour. Conserv. Recycl. **116**, 107–115 (2017). https://doi.org/10.1016/j.resconrec.2016.09.034
19. J. Günther, N. Thevs, H.J. Gusovius, I. Sigmund, T. Brückner, V. Beckmann, N. Abdusalik, Carbon and phosphorus footprint of the cotton production in Xinjiang, China, in comparison to an alternative fibre (Apocynum) from Central Asia. J. Clean. Prod. **148**, 490–497 (2017). https://doi.org/10.1016/j.jclepro.2017.01.153
20. A.D. La Rosa, S.A. Grammatikos, Comparative life cycle assessment of cotton and other natural fibres for textile applications. Fibres, **7**(12) (2019). https://doi.org/10.3390/FIB7120101
21. H.L. Chen, L.D. Burns, Environmental analysis of textile products. Cloth. Text. Res. J. **24**(3), 248–261 (2006). https://doi.org/10.1177/0887302X06293065
22. N.M. Van Der Velden, M.K. Patel, J.G. Vogtländer, LCA benchmarking study on textiles made of cotton, polyester, nylon, acryl, or elastane. Int. J. Life Cycle Assess. **19**(2), 331–356 (2014). https://doi.org/10.1007/s11367-013-0626-9
23. L. Rosson, N. Byrne, Comparative gate-to-gate life cycle assessment for the alkali and acid pretreatment step in the chemical recycling of waste cotton. Sustainability (Switzerland) **12**(20), 1–15 (2020). https://doi.org/10.3390/su12208613
24. Y. Liu, H. Huang, L. Zhu, C. Zhang, F. Ren, Z. Liu, Could the recycled yarns substitute for the virgin cotton yarns: a comparative LCA. Int. J. Life Cycle Assess. **25**(10), 2050–2062 (2020). https://doi.org/10.1007/s11367-020-01815-8
25. S. Roos, G. Sandin, B. Zamani, G. Peters, *Environmental Assessment of Swedish Fashion Consumption* (2015)
26. G. Baydar, N. Ciliz, A. Mammadov, Life cycle assessment of cotton textile products in Turkey. Resour. Conserv. Recycl. **104**, 213–223 (2015). https://doi.org/10.1016/j.resconrec.2015.08.007
27. BSR, Apparel industry life cycle carbon mapping, in *Business for Social Responsibility, June 2009* (2009). https://www.bsr.org/reports/BSR_Apparel_Supply_Chain_Carbon_Report.pdf. Accessed 22 May 2023
28. N. Bhowmick, S. Ghosh, Role of yarn hairiness in knitting process and its impact on knitting room's environment. WSEAS Trans. Environ. Dev. **4**(4), 361–372 (2008)
29. E. Kalliala, P. Talvenmaa, Environmental profile of textile wet processing in Finland. J. Clean. Prod. **8**(2), 143–154 (2000). https://doi.org/10.1016/S0959-6526(99)00313-3
30. S.S. Muthu, Assessing the environmental impact of textiles and the clothing supply chain, in *The Textile Institute Book Series*, 2nd edn. (Woodhead Publishing, 2020)
31. E. Koç, E. Çinçik, Analysis of energy consumption in woven fabric production. Fibres Text. East. Eur. **79**(2), 14–20 (2010)
32. CTNP, Textile value chain, in *Control Techniques for Noise Pollution in Textile Industry* (2019). https://textilevaluechain.in/in-depth-analysis/articles/textile-articles/control-techniques-for-noise-pollution-in-textile-industry/. Accessed 22 May 2023
33. SDC, Colour Index: Fourth Edition Online, Society of Dyers and Colorizts (SDC), UK (n.d). https://colour-index.com/assets/files/upl/CI_leaflet.pdf
34. M. Yu, W. Li, Z. Wang, B. Zhang, H. Ma, L. Li, J. Li, Covalent immobilisation of metal-organic frameworks onto the surface of nylon—a new approach to the functionalization and colouration of textiles. Sci. Rep. **6**(February), 1–9 (2016). https://doi.org/10.1038/srep22796
35. L.J. Rather, S. Jameel, O.A. Dar, S.A. Ganie, K.A. Bhat, F. Mohammad, Advances in the sustainable technologies for water conservation in textile industries, in *Consumption, Footprint,*

and Life Cycle Assessment (2019), pp. 175–194. https://doi.org/10.1016/b978-0-08-102633-5.00010-5
36. G. Sandin, S. Roos, B. Spak, B. Zamani, G. Peters, Environmental assessment of Swedish clothing consumption—six garments, in *Sustainable Futures* (Mistra Future Fashion). G.Sandin-Environmental-assessment-of-Swedish-clothing-consumption.MistraFutureFashionReport-2019.05.pdf
37. N. Jungmichel, The carbon footprint of textiles, in *Systain Consulting Conference Presentation, Berlin, July 5th, 2010* (2010)
38. H. Karaosman, G. Morales-Alonso, A. Brun, From a systematic literature review to a classification framework: sustainability integration in fashion operations. Sustainability (Switzerland) **9**(1) (2017). https://doi.org/10.3390/su9010030
39. UNEP, Sustainability and circularity in the textile value chain, in *Global Stocktaking* (UNEP, 2020). https://www.oneplanetnetwork.org/sites/default/files/unep_sustainability_and_circularity_textile_value_chain_1.pdf. Accessed 22 May 2023
40. S. Ramkalaon, A.S.M. Sayem, Zero-waste pattern cutting (ZWPC) to tackle over sixty billion square metres of fabric wastage during mass production of apparel. J. Text. Inst. 1–11 (2020)
41. A.S.M. Sayem, J. Haider, An overview on the development of natural renewable materials for textile applications, in *Reference Module in Materials Science and Materials Engineering*, ed. by S. Hashmi (Elsevier, 2019)
42. H. Kazan, D. Akgul, A. Kerc, Life cycle assessment of cotton woven shirts and alternative manufacturing techniques. Clean Technol. Environ. Policy **22**(4), 849–864 (2020). https://doi.org/10.1007/s10098-020-01826-x
43. A.S.M.M. Hasan, M. Rokonuzzaman, R.A. Tuhin, Drivers and barriers to industrial energy efficiency in textile industries of Bangladesh. Energies **12**, 1775 (2019). https://doi.org/10.3390/en12091775
44. J. Malinauskaite, H. Jouhara, L. Ahmad, M. Milani, L. Montorsi, M. Venturelli, Energy efficiency in industry: EU and national policies in Italy and the UK. Energy **172**, 255–269 (2019). https://doi.org/10.1016/j.energy.2019.01.130
45. A. Sa, P. Thollander, M. Rafiee, Industrial energy management systems and energy-related decision-making. Energies **11**(10) (2018). https://doi.org/10.3390/en11102784
46. S. Chakraborty, A detailed study on environmental sustainability in knit composite industries of Bangladesh. Am. J. Environ. Prot. **5**(5), 121 (2016). https://doi.org/10.11648/j.ajep.20160505.13
47. E.J. Power, Sustainable developments in knitting. Int. J. Bus. Glob. **9**(1), 1–11 (2012). https://doi.org/10.1504/IJBG.2012.047519
48. Md.R. Khan, T.M. Dip, Md.E. Rashid, R.U. Haque, F.K. Neloy, S.M.M. Salehin, A.S.M. Sayem et al., Comparative analysis of production processes and quality parameters of two different semi-combed Yarns. J. Nat. Fibers. **20**(1) (2023). https://doi.org/10.1080/15440478.2022.2163333
49. T. Stegmaier, W. Wunderlich, T. Hager, A.B. Siddique, J. Sarsour, H. Planck, Chitosan—a sizing agent in fabric production—development and ecological evaluation. Clean - Soil, Air, Water **36**(3), 279–286 (2008). https://doi.org/10.1002/clen.200700013
50. N. Reddy, L. Chen, Y. Zhang, Y. Yang, Reducing environmental pollution of the textile industry using keratin as alternative sizing agent to poly(vinyl alcohol). J. Clean. Prod. **65**, 561–567 (2014). https://doi.org/10.1016/j.jclepro.2013.09.046
51. E. Alkaya, G.N. Demirer, Sustainable textile production: a case study from a woven fabric manufacturing mill in Turkey. J. Clean. Prod. **65**, 595–603 (2014). https://doi.org/10.1016/j.jclepro.2013.07.008
52. A.S. Muhammad Sayem, V. Rossback, Progress in supercritical fluid dyeing (SFD). Taiwan Text. Res. J. **14**(4), 18 (2004)
53. B. Shen, Sustainable fashion supply chain: lessons from H&M. Sustainability (Switzerland) **6**(9), 6236–6249 (2014). https://doi.org/10.3390/su6096236
54. M. Shamsuzzaman, Md. Abul Kashem, A.S. Muhammad Sayem, A.M. Khan, S.Md. Shamsuddin, Md. Mazedul Islam, Quantifying environmental sustainability of denim garments

washing factories through effluent analysis: a case study in Bangladesh. J. Clean. Prod. **290**, 125740 (2021). ISSN: 0959-6526. https://doi.org/10.1016/j.jclepro.2020.125740
55. Ccalc, *Carbon Calculation* (2013). http://www.ccalc.org.uk/about.php. Accessed 22 May 2023

Chapter 3
Sustainable Biomaterials for Fashion

Abdullah Sayam, Hasan Mazharul Haq, Md. Shahin Hossen, and Abu Sadat Muhammad Sayem

Abstract This chapter overviews current developments and new ideas for sustainable biomaterials, including natural fibres and bio-based leathers. By analysing the latest scientific research, technological developments, and business trends, its aim is to shed light on existing and emerging bio-fibres that can re-define the future of fashion and textile manufacturing. Changing consumer attitudes towards sustainability offers excellent opportunities for the widespread use of sustainable bio-fibres in the fashion textile industry.

Keywords Naturally coloured fibre · Bio-based leather · Microbe-derived fibres · Oraganic cotton

1 Introduction

While the global fashion textile industry is struggling to address its significant impacts on the environment and society, the emergence of sustainable biomaterials offers a ray of hope. This chapter provides an in-depth look at the current situation and new approaches to producing sustainable biomaterials that have the potential to transform the industry and reduce its ecological footprint [1]. Bio-fibres are derived from renewable materials such as plants, bacteria, algae, and even animals, effectively reducing dependence on non-renewable resources [2].

There are a variety of pathways, including biotechnology, plant science, and new agricultural technique, to produce sustainable bio-fibre [3]. Additionally, consumers'

A. Sayam
Ahsanullah University of Science and Technology, Dhaka, Bangladesh

H. M. Haq · Md. S. Hossen
Bangladesh University of Textiles, Dhaka, Bangladesh

A. S. M. Sayem (✉)
Manchester Metropolitan University, Manchester, UK
e-mail: asm.sayem@mmu.ac.uk

A. S. Muhammad Sayem and P. Goswami (eds.), *SDG 12 and Global Fashion Textiles Production*, SDGs and Textiles, https://doi.org/10.1007/978-981-96-9051-0_3

demand for high-quality and eco-friendly products is encouraging the use of bio-fibres by fashion retailers and textile manufacturers [4]. In the following sections, key drivers and challenges associated with sustainable bio-fibre production are explored along with new trends and opportunities.

2 Naturally Coloured Fibres

Naturally coloured fibres offer a unique aesthetic appeal while reducing the environmental impact associated with traditionally dyed textiles. This section reviews current developments and potential applications of naturally tinted fibres, focusing on naturally coloured cotton, wool, silk, and other emerging options.

2.1 Naturally Coloured Cotton

Few species of *Gossypium barbadense L.* and *Gossypium hirsitum* produce naturally coloured cotton fibres with various shades of brown and green [5, 9]. They are cultivated on a small scale in different parts of the world, although they have long been known to indigenous communities. Increasing their production and use will eliminate the need for industrial bleaching and dyeing processes [6]. Additionally, the advantages of naturally coloured cotton include high resistance to insects and diseases, which ensure compatibility with organic agriculture without the use of pesticides and herbicides. The structure of naturally coloured cotton, mainly green cotton, exhibits distinct morphological characteristics. It features a unique secondary cell wall structure with concentric rings and a smoother fibre structure with a thinner cell wall than white cotton. Chemical analysis indicates that naturally coloured cotton contains higher fat, lignin, and pectin, contributing to its lower moisture absorbency and weaker tensile properties. Despite these characteristics, naturally coloured cotton fibres can be processed similarly to white cotton for yarn and fabric production [7, 8]. Figure 1 represents the different physical appearance of naturally coloured cotton.

2.2 Naturally Coloured Wool

The naturally coloured wool comes from sheep breeds that produce fibres containing pigments ranging from black, brown, grey, and cream. These unique colours result from genetic traits inherited from ancestral breeds and have been selectively bred over many generations to achieve the desired characteristics. The association of sheep farmers in the UK, known as British Wool, reports there are thirteen breeds of sheep in the UK that naturally produce coloured fleeces [10]. The Shetland Sheep Society has identified eleven main colours in Shetland sheep [11]. Unlike dyed wool, which

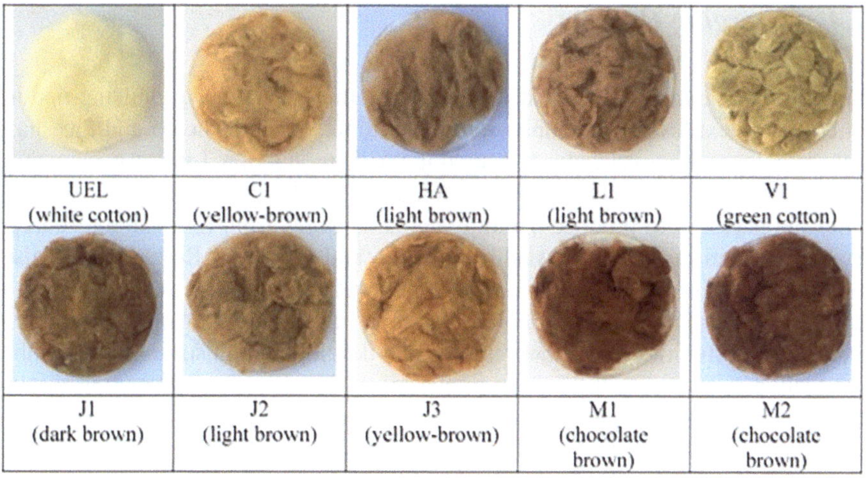

Fig. 1 Examples of physical appearance and code of naturally coloured cotton fibres [9] (reproduced with permission)

uses synthetic chemicals and water-intensive processes, naturally coloured wool retains its natural colour without requiring additional dyes. Innovations in sheep breeding techniques and genetic research offer the development of wool breeds with improved colour diversity, fibre quality, and sustainability characteristics. Breeders can use breeding and genetic mapping techniques to select desired traits such as colour intensity, fineness of fibre, and resistance to pests and diseases. Additionally, efforts to promote sustainable grazing practices and animal welfare standards contribute to the overall sustainability of coloured wool production systems.

2.3 Naturally Coloured Silk

Naturally coloured silk, harvested from silk-producing insects such as silkworms and spiders, offers a fascinating palette of colours. The wild silkworms of species *A. pernyi, A. yamamai and saturniidae* produce silk cocoons with brown, green and golden yellow colours [12]. These colours are inherent to the silk fibre itself and result from the unique biochemical composition of silk glands [12]. However, these colours are not permanent and get lost after degumming. Research in this field investigates the genetic basis for pigmentation in natural silk, focusing on identifying genes responsible for pigment production and colour change. By understanding the basic mechanisms of silk dyeing, researchers are trying to develop breeding strategies and biotechnological measures to improve naturally dyed silk's colour stability, filament quality, and production efficiency.

2.4 Emerging Options

In addition to cotton, wool, and coloured silk, there are other promising options in coloured textiles. For example, some plant fibres, such as hemp and flax, have natural colour variations that can be used in textile production. Likewise, innovative approaches such as microbial fermentation, biotechnology and genetic engineering offer the potential to produce naturally coloured fibres from renewable resources. Scientists from the Commonwealth Scientific and Industrial Research Organisation (CSIRO) discovered a technique for impregnating colour-producing genes in cotton plants to grow naturally coloured cotton [48]. These new options represent exciting opportunities for further research and development in sustainable bio-fibre production. Harnessing the unique properties of natural materials and embracing biological design principles opens up new opportunities to create sustainable fabrics that resonate with consumers and contribute to a more sustainable fashion industry. Overall, research on naturally coloured textiles highlights the interconnectedness of nature and textile production. Through interdisciplinary collaboration and innovation, we can harness the power of nature to drive positive changes in the fashion and textile industry and pave the way for a more sustainable and regenerative future.

3 Next Generation Bio-based Fibres

As the fashion industry continues to explore more sustainable alternatives, developing next-generation bio-based fibres stands out as a key innovation in reducing the environmental footprint of textile production. These bio-based fibres are derived from renewable biological sources, such as plants, microorganisms, and even cultivated animal cells, providing a significant departure from conventional petroleum-based fibres like polyester [13]. These alternatives reduce dependency on non-renewable resources and offer improved biodegradability and the potential for closed-loop production systems.

Key plant-derived fibres include innovative materials such as Infinna by Infinite Fibre [14], Agra Loop by Circular Systems [15], and Bananatex [16], all of which effectively harness natural resources. In parallel, developments in microbe-derived materials like Celium by Polybion [17] and Shorai by Bucha Bio [18] demonstrate how microorganisms can be utilised to create eco-friendly fibres. Additionally, new approaches to producing fibres from cultivated animal cells, such as those by VitroLabs [19] and Qorium [20], offer an alternative to traditional leather, addressing environmental and ethical concerns. Figure 2 represents different pathways in biobased fibre production.

The emergence of these next-generation bio-based fibres highlights the industry's commitment to innovation, aiming to balance performance, sustainability, and aesthetics. As consumer awareness around the environmental impact of textiles

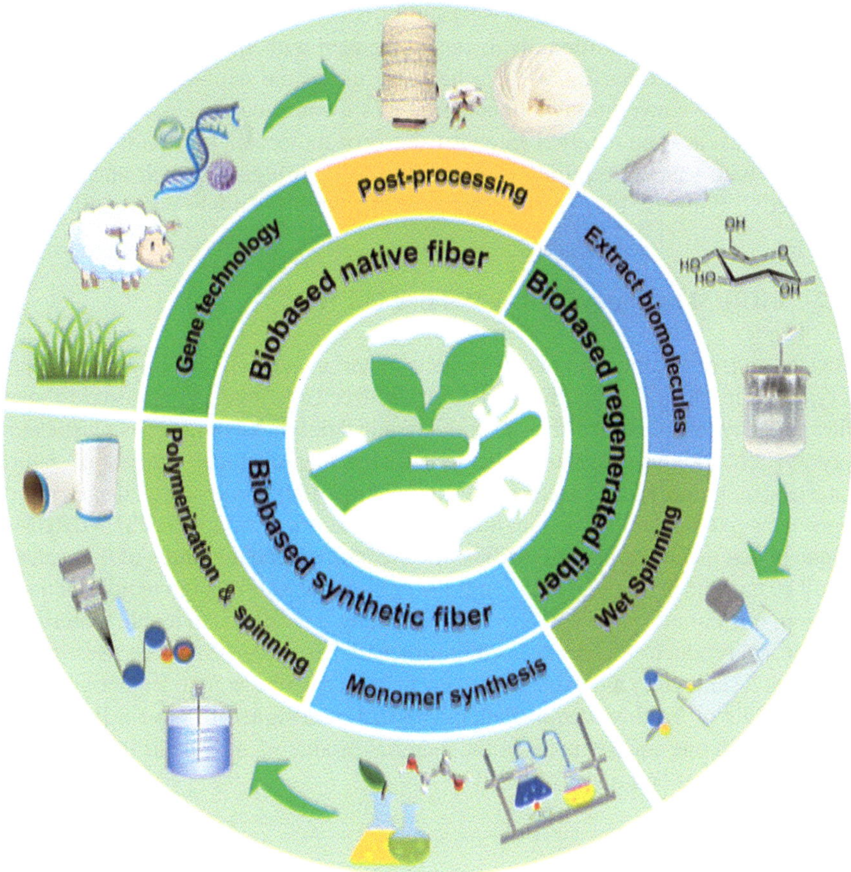

Fig. 2 Comprehensive pathways in biobased fibre production [21], reproduced with permission

grows, these fibres will likely play an increasingly prominent role in the global effort to create a more sustainable fashion ecosystem.

3.1 *Plant Derived Fibres*

3.1.1 Sorona by DuPont

Sorona, a derivative of polytrimethylene terephthalate (PTT), is a bio-derived fibre developed by DuPont and created from non-edible plant glucose starch extract [22, 23]. Sorona fibre offers high strength, stretch, comfort, and good thermal and dimensional stability while consuming 40% less energy and discharging 56% less greenhouse gases during production [24]. It has emerged as a new class of fibres with various uses, combining the high wearability benefits while lowering reliance on petroleum resources. Additionally, it can be blended with all kinds of fibres used in apparel and home textiles [22, 24]. Furthermore, the spatial semi-crystalline structure of PTT with the convolution of bonds produces a conspicuous zigzag shape structure that translates the external forces at the very molecular level and thus injects resiliency with shape retention properties resulting in higher stretch and recovery compared to synthetic PET (polyethene terephthalate) based fabric [25, 26].

3.1.2 MIRUM by Natural Fiber Welding (NFW)

MIRUM replaces leather and other synthetic materials typically utilised in the upholstery, car, and fashion industries with a plastic-free substitute. As opposed to many other vegan types of leather that use petrochemical polymers or noxious plastics like polyvinyl chloride and polyurethane, MIRUM is manufactured entirely of natural materials, including plant-based oils, waxes, natural rubber, and recycled agricultural byproducts like rice hulls, coconut husk fibres, and cork powder [27, 28]. MIRUM is unique in that it can be customised. The texture, colour, and thickness of the material can all be tailored by brands to fulfil unique performance and design requirements. In addition, it is compostable and recyclable, making it a fully sustainable material. In addition, its production process is environmentally friendly because it uses just the water that is inherently available in the raw materials. Compared to leather generated from animals or synthetic equivalents, MIRUM is designed to have a low carbon footprint, with one square meter of the material producing as little as 0.84 to 2.1 kg of CO_2 equivalent emissions [29, 30].

3.1.3 Infinna by Infinited Fiber

Infinna, developed by the Infinited Fiber Company, is a groundbreaking sustainable bio-fibre made entirely from textile waste, offering a circular solution to the textile industry. This regenerated fibre is created using patented technology that transforms cellulose-rich waste (such as old clothes, cardboard, wastepaper, and wood pulp) into a high-quality material that mimics the look and feel of cotton. Infinna is biodegradable, plastic-free, and can be recycled alongside other cotton-rich textiles at the end of its life cycle, reinforcing its role in promoting a circular economy. The process behind Infinna involves breaking down discarded textiles, capturing the cellulose, and regenerating it into a new fibre using a wet-spinning technique. This fibre can be used in various applications, from soft jersey fabrics to durable denim, and has been adopted by leading brands like H&M, Patagonia, and Adidas. Moreover, Infinna fibres can blend seamlessly with other materials, such as organic cotton and viscose [14, 31].

3.1.4 Agra Loop by Circular System

Agraloop BioFibre™, developed by Circular Systems, is a sustainable fibre from agricultural waste. This technology transforms crop residues, such as banana trunks, pineapple leaves, hemp stalks, and flaxseed oil by-products, into high-quality fibres suitable for textile production. The innovation behind Agraloop BioFibre not only helps reduce agricultural waste that would otherwise decompose or be burned, contributing to environmental issues but also introduces a circular, regenerative model for textile manufacturing. Agraloop's process relies on a closed-loop system that employs clean energy and recycled water while producing no waste. The fibres generated are fully biodegradable, with properties comparable to cotton, making them versatile for various applications, from fashion to durable home goods. Furthermore, Agraloop's low carbon footprint and eco-efficient process make it a promising solution for sustainable textiles [13, 15, 32].

3.1.5 Banantex

Bananas are cultivated in 129 countries and are the fourth-largest staple crop globally. Large amounts of waste (stems and leaves) are generated during these productions. This waste can produce eco-friendly fibres, offering a sustainable alternative to costly, non-biodegradable synthetic fibres [33]. Bananatex is a sustainable and innovative fabric made entirely from fibres derived from the Abacá banana plant, which is native to the Philippines. It was developed by the Swiss company QWSTION as a response to the growing demand for eco-friendly alternatives to synthetic materials like nylon and polyester. Its fibre extraction process requires minimal water and no harmful chemicals, resulting in a clean and low-impact production cycle. As a result, its production helps mitigate environmental problems, including plastic pollution and non-biodegradability [16].

3.1.6 Orange Fibre

Orange fibre represents a sustainable alternative to conventional textile raw materials by utilising the byproducts of citrus juice production. The food processing industry's significant amount of non-edible waste is repurposed to extract cellulose from citrus residues. This cellulose is spun into yarn suitable for producing high-quality, luxurious fabrics that are eco-friendly alternatives to wood pulp-based fibres. These fabrics offer a soft, silky texture and can be customised for opacity or shine depending on specific requirements [13, 34].

3.1.7 Pinatex

Pinatex is a sustainable bio-fibre derived from pineapple leaf fibres, a byproduct of the pineapple farming industry. This innovative material is an eco-friendly alternative to conventional leather, aligning with the rising demand for sustainable textiles. The production of Pinatex involves extracting fibres through scraping and retting, avoiding harmful chemicals. This sustainable approach utilises agricultural waste efficiently. Additionally, pineapple cultivation requires fewer pesticides than crops like cotton, enhancing its environmental credentials [35].

3.1.8 VEGATEX AppleSkin

VEGATEX is an alternative leather made from apple pomace, a by-product of apple processing in the juice and cider industries. Apple pomace is mixed with water-based polyurethane and layered with lyocell to make 100% animal-free leather [36].

3.2 *Mycelium-Based Fibres*

Mycelium-based fibres have emerged as innovative solutions in sustainable fashion and material production. Mycelium, the root-like structure of fungi, serves as a versatile and eco-friendly base for creating various bio-fibers. The vegetative portion of a fungus called mycelium can be formed into composite materials or pure mycelium. When grown under controlled conditions, mycelium can produce dense and durable material, making it a promising candidate for sustainable fibre production. These are also porous, elastic, inexpensive, and environmentally benign materials. Unlike traditional fibres derived from petroleum-based or agricultural sources, mycelium requires minimal energy, water, and resources to cultivate. Moreover, it is biodegradable and compostable, significantly reducing its ecological footprint at the end of its life cycle [37, 38].

3.2.1 Mylo by Bolt Threads

Mylo is a notable example of a mycelium-based bio-fibre designed for the fashion industry. Bolt Threads, a biotechnology company focused on sustainable material innovation, developed Mylo as an alternative to traditional leather. Mylo is created by growing mycelium in a lab, where the growth process can be optimised for scalability and efficiency. The material produced mimics the texture, flexibility, and appearance of leather while being entirely animal-free and sustainable.

The development of Mylo is a breakthrough in bio-fabrication because it addresses two significant issues in conventional leather production: animal hunting and environmental degradation. Traditional leather production involves resource-heavy processes, including raising livestock, tanning, and chemical treatments, all of which have significant environmental costs. On the other hand, Mylo uses a fraction of the resources and has a drastically lower environmental impact. It is grown within days, using renewable feedstocks, and is entirely biodegradable at the end of its lifecycle [39].

3.2.2 Reishi by MycoWorks

Reishi by MycoWorks is another groundbreaking mycelium-based material developed as a sustainable alternative to leather. MycoWorks, a California-based biotechnology company, has pioneered the creation of Reishi through its patented Fine Mycelium™ technology. Like Mylo, Reishi mimics the look and feel of leather while being produced in a manner that minimises environmental impact. However, Reishi's innovation lies in its customisation and material performance. The Fine Mycelium™ process enables MycoWorks to grow mycelium in a highly controlled environment, resulting in materials with fine-tuned properties, such as strength, flexibility, and texture. This level of control allows for a material that can rival traditional leather in durability and versatility. Additionally, MycoWorks' proprietary method allows for large-scale production, making Reishi a viable option for luxury and mass-market applications [40].

3.3 Microbe-Derived Fibres

Microorganisms have recently emerged as a promising source for developing sustainable, next-generation bio-based fibres. Unlike traditional materials, which often require extensive land use, water consumption, and chemicals, microbe-derived fibres can be cultivated in controlled environments using fermentation processes [41]. These fibres are not only biodegradable but also have the potential to mimic or surpass the performance of synthetic fibres, offering the fashion industry an eco-friendly and scalable alternative. By leveraging the capabilities of microbes, companies can produce materials that significantly reduce environmental impact while maintaining quality and versatility [42].

3.3.1 Celium by Polybion

Celium, developed by Polybion, represents a significant development in sustainable biomaterials as an alternative to traditional leather. This innovative material is produced through a microbial fermentation process, which utilises microorganisms to metabolise agro-industrial fruit waste into high-performance fibres (premium cultivated cellulose). Unlike conventional leather, which is often associated with significant environmental repercussions, Celium offers an environmentally responsible solution that does not compromise performance.

Furthermore, this unique production process reduces dependence on petrochemical-based fibres. It solves the growing waste management issue, transforming discarded materials into valuable resources for the fashion industry. Celium demonstrates comparable strength, flexibility, and aesthetic qualities to traditional leather, which makes it suitable for versatile applications in fashion, upholstery, and accessories. These materials can later be customised by colour, embossing, and water resistance without compromising exceptional strength. Polybion's approach aligns with the principles of a circular economy. As a fully plant-based, biodegradable material, Celium addresses the environmental concerns of the leather industry by offering a sustainable alternative that reduces carbon footprints and eliminates the need for animal-based production. Finally, its ability to translate identical properties of leather while being sustainable makes Celium a leading innovation in the bio-based textile and fashion industry [17].

3.3.2 Shorai by Bucha Bio

Shorai, which Bucha Bio develops, is an innovative microbe-derived bio-based material designed as a sustainable alternative to animal and plastic leather. Shorai is cultivated through fermentation, where microbes transform plant-based nutrients into bacterial nanocellulose, which is then processed into a flexible, durable, lightweight leather-like material. This process is both resource-efficient and scalable [18].

The key advantage of Shorai lies in its biodegradability. Unlike conventional synthetic leather materials, Shorai naturally decomposes at the end of its life cycle, making it a highly sustainable option for the fashion industry. Shorai can be customised for various applications, from textiles and accessories to packaging and interior design, aligning with circular design principles while reducing reliance on non-renewable materials. Moreover, Shorai's production process requires less water and energy than conventional textile manufacturing. Growing the material in a controlled lab environment further reduces the environmental impact of agriculture-based fibres. This makes Shorai an eco-friendly and resource-efficient high-performance substitute for leather and further applications for the fashion industry. As consumer demand for sustainable and ethical alternatives increases, Shorai can stand out as an innovative material that allows brands to reduce their environmental footprint without sacrificing quality or performance [18].

3.3.3 AirCarbon by Newlight

AirCarbon, developed by Newlight Technologies, is an innovative microbe-derived material that captures carbon emissions and converts them into high-performance fibres [43]. The core of this process lies in microorganisms that feed on air and greenhouse gases, such as methane and carbon dioxide, to produce a biopolymer known as polyhydroxybutyrate (PHB). This biopolymer can create fibres, plastics, and other materials, offering a sustainable alternative to petroleum-based products. What makes AirCarbon stand out is its ability to address two significant environmental challenges simultaneously: reducing carbon emissions and creating biodegradable materials. By using captured carbon from industrial and agricultural waste, the production of AirCarbon not only prevents additional carbon from entering the atmosphere but also transforms it into valuable, compostable and environmentally friendly materials.

AirCarbon can be moulded into various applications, from fashion accessories and packaging to furniture and electronics. Its versatility and sustainability credentials make it a compelling choice for industries looking to lower their carbon footprint. Additionally, AirCarbon's biodegradability ensures that, unlike conventional plastics, it will not persist in landfills or oceans for centuries. Newlight's approach of utilising greenhouse gases to produce materials highlights the potential of leveraging waste products to create sustainable solutions. AirCarbon exemplifies how innovation in biofabrication can contribute to a circular economy, transforming waste into valuable resources and ultimately helping mitigate climate change while providing practical, eco-friendly alternatives for diversified industries [43].

3.3.4 Microsilk by Bolt

Microsilk, developed by Bolt Threads, is a revolutionary microbe-derived fibre inspired by the properties of spider silk [44]. This bio-fabricated fibre is produced through fermentation, where genetically engineered yeast is used to create proteins similar to those found in spider silk. These proteins are then spun into fibres that mimic natural silk's incredible strength, elasticity, and softness, offering a sustainable alternative to traditional silk production.

The innovation behind Microsilk lies in its ability to replicate the desirable characteristics of spider silk, which has long been admired for its combination of strength and lightness. Unlike traditional silk, which requires the farming and harvesting silkworms, Microsilk is created entirely in a lab, eliminating the need for animal-derived materials. This phenomenon aligns with the growing demand for ethical, animal-free fibres in the fashion industry. Microsilk's production process is highly sustainable. It uses renewable inputs like sugar, water, and yeast, which are fermented to produce silk proteins. This bio-fabrication method reduces the environmental impact of conventional silk production and offers scalability, making it a viable option for large-scale textile manufacturing. In addition to its sustainability, Microsilk is highly customisable, allowing for different variations of strength, stretch, and texture depending on the application. This flexibility makes it suitable for various uses, from high-performance textiles in fashion to innovative materials in other industries, such as automotive or aerospace. By developing Microsilk, Bolt Threads has successfully combined nature's ingenuity with modern biotechnology, creating a fibre that not only matches but sometimes outperforms traditional silk's performance. Microsilk represents a significant step forward in sustainable material innovation, contributing to a future where high-quality textiles can be produced with minimal environmental impact and without animal exploitation [44].

3.3.5 QMilk by QMilk

QMilk is a unique, sustainable fibre created from casein; a protein found in milk that is no longer suitable for consumption [45]. Developed by QMilk, this fibre offers an innovative solution to food waste by transforming surplus or spoiled milk into a biodegradable and eco-friendly material. Creating QMilk fibres utilises upcycled milk that would otherwise go to waste, making it a highly resource-efficient alternative to traditional synthetic fibres. The production of QMilk fibres is designed to be environmentally friendly, using minimal water and energy. Production requires only about two litres of water per kilogram of fibre, significantly lower than that required for conventional textiles like cotton. Additionally, QMilk fibres are entirely free of chemical additives, ensuring they remain non-toxic and safe for the skin, making them ideal for applications such as clothing, home textiles, and even medical products. QMilk fibres are biodegradable and break down naturally at the end of their life cycle reducing the accumulation of textile waste in landfills. Despite being derived from milk, QMilk fibres exhibit impressive durability, softness, and antibacterial properties, making them a versatile option for various industries [45].

3.4 *Cultivated Animal Cells*

The cultivation of animal cells for material production is an emerging innovation in the fashion industry, offering a sustainable and ethical alternative to traditional animal-based textiles like leather. By cultivating cells in a controlled laboratory environment, companies can produce materials with the same characteristics as animal products without the environmental impact of farming or the ethical concerns surrounding animal welfare [19]. This approach significantly reduces the land, water, and greenhouse gas emissions associated with conventional leather production while still delivering high-quality materials that meet luxury fashion's performance and aesthetic standards.

3.4.1 Lab-Grown Leather by VitroLabs

VitroLabs is at the forefront of the cultivated animal cell movement, pioneering the development of lab-grown leather. Through a process known as cellular agriculture, VitroLabs cultivates animal cells to create genuine leather without the need for livestock. This innovative method starts by extracting a small sample of animal cells, which are then fed nutrients in a bioreactor to promote growth and tissue formation.

Over time, these cells form a structure that closely resembles the natural composition of leather.

The result is a material that maintains traditional leather durability, flexibility, and texture without ethical and environmental concerns. VitroLabs' cultivated leather is biodegradable, reducing the ecological impact of synthetic alternatives and offering a more sustainable option for industries that rely on leather. VitroLabs is addressing some of the most pressing challenges in the leather industry, such as the significant land and water usage and the harmful effects of tanning chemicals. Their process eliminates the prerequisite of slaughtering animals, making it an appealing option for brands and consumers looking to align with environmentally conscious practices. Moreover, the scalability of VitroLabs' technology presents an opportunity to revolutionise the leather industry by offering a large scale of sustainable and ethically sourced material. By combining cutting-edge biotechnology with a commitment to sustainability, VitroLabs contributes to the future of fashion by rethinking how leather can be produced to minimise its ecological footprint while maintaining luxury quality [19].

3.4.2 Lab-Grown Leather by Faircraft

Faircraft is an emerging leader in cellular agriculture, dedicated to developing lab-grown leather as a sustainable alternative to conventional leather [46]. Faircraft employs advanced biotechnology to address traditional leather production's ethical and environmental challenges. By cultivating animal cells in a controlled laboratory environment, the company replicates the structure and properties of animal-derived leather without the associated animal farming or slaughter, making it a transformative solution for the fashion industry.

The process employed by Faircraft begins with extracting a small, ethically sourced sample of animal cells. These cells are cultivated in a bioreactor, where they are supplied with nutrients to promote their growth and development. Over time, the cells organise into a matrix that mirrors the composition of natural leather. The resulting material offers the same durability, flexibility, and texture as conventional leather without compromising quality or performance. This innovative methodology ensures that Faircraft's leather is biologically equivalent to traditional animal leather and significantly more sustainable and harmless, in line with the growing consumer demand for ethical materials. One of the key environmental advantages of Faircraft's lab-grown leather lies in its reduced resource consumption. Conventional leather production requires vast amounts of water, leads to deforestation for grazing land, and contributes to significant greenhouse gas emissions.

In contrast, Faircraft's cultivated leather uses substantially fewer natural resources and generates minimal waste, making it a more environmentally responsible alternative. Furthermore, the scalability of the production process allows for industrial-level manufacturing, meeting the needs of brands and manufacturers seeking to reduce their environmental impact. Faircraft's dedication to sustainability, ethics, and innovation positions the company at the forefront of material science in the fashion industry. By offering an environmentally benign alternative to traditional leather, Faircraft contributes to a more eco-conscious and responsible future for fashion without sacrificing the luxury and quality consumers expect from leather products [46].

3.4.3 Bio-Fabricated Leather by Qorium

Qorium is a pioneering force within the cultivated leather industry, leveraging advanced tissue engineering techniques to develop bio-fabricated leather that provides a sustainable alternative to conventional leather [20]. Unlike traditional leather production, which is associated with animal slaughter and significant environmental degradation, Qorium's innovative approach involves cultivating animal cells within a controlled laboratory environment. This process addresses the growing demand for environmentally responsible materials in the fashion and textile sectors. Qorium's leather fabrication process begins with the extraction of a minimal sample of animal cells, which are then cultivated in a bioreactor under optimal conditions. The cells are nourished with the necessary nutrients to promote their growth and organisation into a leather-like structure. Through this cellular agriculture process, Qorium successfully replicates the complex collagen matrix and other proteins that give traditional leather its signature strength, flexibility, and texture. The result is a material that mirrors the physical properties of animal leather while eliminating the ethical concerns associated with its conventional production.

A defining feature of Qorium's innovation is its focus on producing premium-quality leather suitable for luxury fashion and high-performance applications. The company's proprietary technologies allow for precise control over the material's thickness, grain, and finish, enabling brands to achieve the same level of quality as traditional leather without environmental and ethical trade-offs. This high degree of customisation positions Qorium's biofabricated leather as a viable alternative for even the most demanding industries. In addition to being animal-free, Qorium's cultivated leather offers substantial environmental benefits. Traditional leather processing is resource-intensive, requiring large volumes of water, energy, and harmful chemicals such as those used in tanning.

In contrast, Qorium's bio-fabrication process significantly reduces the need for natural resources. It eliminates toxic chemicals, making it safer for the environment and workers in the production chain. By offering a scalable, high-quality, and sustainable alternative to traditional leather, Qorium is contributing to the transformation of the fashion industry, meeting the increasing demand for eco-friendly materials [20].

3.4.4 Zoa by Modern Meadow

Modern Meadow is a trailblazer in bio-fabrication, known for its pioneering work in developing sustainable alternatives to traditional animal-based materials [47]. One of the company's most significant innovations is Zoa®, a bio-fabricated leather alternative made from cultivated collagen. Collagen, the primary protein found in animal skin, is responsible for the strength and durability of leather. By cultivating collagen in a lab using advanced fermentation techniques, Modern Meadow can create a material that mimics the qualities of traditional leather without the need for animal farming. The production process begins by bioengineering yeast cells to produce collagen. These cells are grown in bioreactors where they multiply and secrete collagen, which is then harvested and assembled into sheets of material. This process allows for precise control over the characteristics of the final product, enabling Modern Meadow to produce leather-like materials that are customisable in terms of thickness, texture, and finish.

One of the key advantages of Zoa® is its environmental sustainability. Traditional leather production is resource-intensive, requiring significant amounts of water, land, energy, and toxic chemicals used in the tanning process. In contrast, Modern Meadow's bio-fabrication process requires far fewer resources and produces significantly less waste. Additionally, Zoa® is free from the toxic chemicals typically used in leather production, making it safer for the environment and workers. Modern Meadow's Zoa® material has garnered attention from the fashion and design industries due to its versatility, durability, and aesthetic appeal. The material can be used in various applications, including fashion, footwear, accessories, and furniture. Modern Meadow contributes to a future where high-quality materials can be produced with minimal environmental impact by offering a scalable alternative to traditional leather. Through its groundbreaking bio-fabrication technologies, Modern Meadow is helping to reshape the landscape of sustainable fashion and design, providing a glimpse into a future where innovative materials can meet the growing demand for ethical and eco-friendly products [47].

4 Better Alternatives by Improving Existing Fibres

While the development of next-generation bio-based fibres is exciting, improving the sustainability of existing fibres remains crucial for the fashion and textile industry. This involves adopting practices that minimise environmental impact and promote ethical production throughout the supply chain. One central area of focus is reducing water consumption, as the industry is known for its high water consumption in dyeing and finishing processes, which can be mitigated through water-efficient technologies [49]. Additionally, minimising chemical pollution by using eco-friendly dyes and finishing agents can significantly reduce harmful runoff, while innovations in biodegradable textiles and organic fibres further help to address the environmental challenges [50]. Exploring the use of recycled fibres is essential in transitioning to a circular economy, with biotechnological processes converting textile waste into valuable raw materials, reducing waste, and decreasing the demand for virgin materials [51]. Ethical production must also encompass fair labour practices and safe working conditions, acknowledging the social dimension of sustainability

[52]. Although there is growing interest in bio-based fibres, immediate efforts to improve the sustainability of existing fibres are crucial to reducing the industry's overall ecological footprint and ensuring social responsibility. Some of the better alternatives for commonly used fibres are explained in the following sub-sections.

4.1 Organic Cotton

Organic cotton is grown without synthetic pesticides and fertilisers, promoting healthier soil and ecosystems. It also prohibits genetically modified seeds, protecting biodiversity and farmers' rights. While organic cotton still requires significant water resources, it significantly reduces pollution and health risks associated with conventional cotton farming. This reduced impact on the environment and farmers' well-being makes it a more ethical choice for consumers [53].

Several organisations certify organic cotton, including the Global Organic Textile Standard (GOTS) and the Organic Content Standard (OCS). These standards ensure the integrity of organic cotton throughout the supply chain, from farm to finished product. By looking for these certifications, consumers can make informed decisions and support the growth of sustainable and responsible cotton production [54, 55].

4.2 Better Cotton Initiative (BCI) Cotton

The BCI promotes better practices in conventional cotton farming. It focuses on water stewardship, responsible pesticide and fertiliser use, soil health, and decent work conditions for farmers. BCI cotton is not organic but offers a more sustainable option than conventional cotton [56].

However, BCI has faced criticism for its mass-balance approach, where BCI cotton and conventional cotton can be mixed in the supply chain [57]. This raises concerns about traceability and the actual sustainability improvements achieved. Critics argue that this system lacks transparency and can potentially allow unsustainable cotton to be labelled as BCI cotton [58]. It is essential for consumers to be aware of these limitations when considering BCI-certified products and to seek out brands that prioritise transparency and traceability in their sourcing.

4.3 Cotton Made in Africa (CIMA)

The CMIA initiative supports African smallholder cotton farmers by promoting sustainable farming practices and fair-trade principles. CMIA cotton focuses on improving livelihoods, empowering women, and protecting the environment. This includes training farmers in efficient water use, reducing pesticide reliance, and ensuring fair prices [60].

Like BCI, CMIA uses a mass-balance system, meaning CMIA cotton may be mixed with conventional cotton during processing. This can raise concerns about traceability and the actual impact on individual farmers. While CMIA benefits cotton farmers in Africa, some argue that a more direct traceability system would ensure that the premiums paid for CMIA-certified Cotton directly reach the farmers who implement sustainable practices [61].

4.4 Responsible Environment Enhanced Livelihoods (REEL) Cotton

REEL focuses on improving the environmental and social performance of cotton production. It promotes sustainable practices like water conservation, integrated pest management, and improved soil health, aiming to reduce the negative impacts of conventional cotton farming. REEL also emphasises fair labour practices and community development, ensuring the farmers' well-being.

While REEL offers a promising approach to sustainable cotton production, its reach is limited to India. Expanding this model to other cotton-producing regions could significantly improve the industry's sustainability. Increased collaboration between REEL and similar initiatives across the globe would help foster wider adoption of its beneficial practices [62].

5 Conclusion

United Nations' Sustainable Development Goal (SDG) 12 asks for sustainable consumption and production of all consumable materials. The biomaterials mentioned and discussed in this chapter are exciting candidates as sustainable alternatives for the fashion industry. The scaling up of the production of various bio-fibres derived from natural, plant-based, microbe-based, and animal cell sources has the potential to reduce the environmental impact of the fashion textile industry.

References

1. M. Ramesh, C. Deepa, L.R. Kumar, M.R. Sanjay, S. Siengchin, Life-cycle and environmental impact assessments on processing of plant fibres and its bio-composites: a critical review. J. Ind. Text. **51**(4), 5518S-5542S (2022). https://doi.org/10.1177/1528083720924730

2. R. Blok, J. Smits, R. Gkaidatzis, P. Teuffel, Bio-based composite footbridge: design, production and in situ monitoring. Struct. Eng. Int. **29**(3), 453–465 (2019). https://doi.org/10.1080/10168664.2019.1608137
3. P.P. Das, V. Chaudhary, Environmental impact and effect of chemical treatment on bio fiber based polymer composites. Mater. Today Proc. **49**(xxxx), 3418–3422 (2020). https://doi.org/10.1016/j.matpr.2021.03.097
4. I. D'Adamo, A. Colasante, Survey data to assess consumers' attitudes towards circular economy and bioeconomy practices: a focus on the fashion industry. Data Br. **43** (2022). https://doi.org/10.1016/j.dib.2022.108385
5. T. Wen, W. Luo, Y. Li, Z. Lin, Advances and new insights in naturally colored cotton breeding and research. Indus. Crops Products **211**(118252) (2024). ISSN 0926-6690.https://doi.org/10.1016/j.indcrop.2024.118252
6. M. Bayraktar et al., Journey of naturally colored cottons: from field to final products and their product ranges. Deu Muhendis. Fak. Fen ve Muhendis. **18**(52), 108–108 (2016). https://doi.org/10.21205/deufmd.20165217551
7. G.K. Günaydin, O. Avinc, S. Palamutcu, A. Yavas, A.S. Soydan, *Naturally Colored Organic Cotton and Naturally Colored Cotton Fiber Production* (Springer Singapore, 2019)
8. M. Naoumkina, D.J. Hinchliffe, G.N. Thyssen, Naturally colored cotton for wearable applications. Front. Plant Sci. **15**(March), 1–7 (2024). https://doi.org/10.3389/fpls.2024.1350405
9. R.H. Blas-Sevillano et al., Physicochemical characterisation of several types of naturally colored cotton fibers from Peru. Carbohydr. Polym. **197**, 246–252 (2018). https://doi.org/10.1016/j.carbpol.2018.06.006
10. Brtish Wool, *Naturally coloured wool types*. https://shop.britishwool.org.uk/sheep/naturally-coloured/
11. Wool Colours in Shetland Sheep, https://www.shetland-sheep.org.uk/shetland-sheep-breed/colours-markings
12. M. Ma, M. Hussain, S. Dong, W. Zhou, Characterization of the pigment in naturally yellow-colored domestic silk. Dyes Pigment. **124**, 6–11 (2016). ISSN 0143–7208. https://doi.org/10.1016/j.dyepig.2015.08.003
13. A. Sailaubayeva, *Adopting Sustainable Materials in the Fashion Industry : A Multiple Case Study Research* (2024)
14. It's called Infinna™!-Infinited Fiber. https://infinitedfiber.com/blog/2021/01/20/its-called-infinna/
15. Agraloop—Circular Systems, *Regenerative impact*. https://circularsystems.com/agraloop/
16. Bananatexx-Stickerei, Textildruck für Beruf und Freizeit. https://www.bananatexx.de/
17. Polybion. https://www.polybion.bio/
18. BUCHA BIO, *Accelerates manufacturing plans for biobased materials*. https://www.greener-manufacturing.com/news/bucha-bio-accelerates-manufacturing-plans-biobased-materials
19. Lab-Grown Leather, VitroLabs Inc. https://www.vitrolabsinc.com/
20. Qorium, *Real leather, reinvented*. https://www.qorium.com/
21. F. Liu et al., Biobased fibers from natural to synthetic: processing, manufacturing, and application. Matter (2024)
22. R. Sukanyadevi, V. Muralidharan, C. Kavya, Bio-polymers derived from renewable plant sources and application in apparels, in *Sustainability in Fashion and Apparels* (WPI Publishing, 2018), pp. 40–45
23. J.Y. Chen, K. Polston, E. Nicols, B. Phung, A study on textile recycling in college student residence areas. J. Text. Inst. **113**(9), 1854–1861 (2022)
24. W. Zhou, P. Zhou, Y. Tang, Z. Zhou, The properties and application progress of sorona fiber. J. Phys. Conf. Ser. **2079**(1), 12019 (2021)
25. J.V. Kurian, A new polymer platform for the future-Sorona® from corn derived 1,3-propanediol. J. Polym. Environ. **13**(2), 159–167 (2005). https://doi.org/10.1007/s10924-005-2947-7
26. W. Zhou, P. Zhou, Y. Tang, Z. Zhou, The properties and application progress of sorona fiber. J. Phys. Conf. Ser. **2079**(1), 012019 (2021). https://doi.org/10.1088/1742-6596/2079/1/012019

27. MIRUM. https://nfw.earth/mirum
28. A. Matthes, K. Beyer, H. Cebulla, M.G. Arnold, A. Schumann, *Sustainable Textile and Fashion Value Chains: Drivers, Concepts, Theories and Solutions* (2020)
29. MIRUM®, *A biobased leather-like material.* https://circularmateriallibrary.org/material/mirum/
30. *What is MIRUM? The Plastic-Free Leather Alternative and Where to Get It—Sustainably Chic.* https://www.sustainably-chic.com/blog/what-is-mirum
31. T. Design, A. Auranen, *The New Cotton Project* (2022)
32. M. Ribul, Regenerative textiles: a framework for future materials circularity in the textile value chain. *Sustainability* **13**(24) (2021). https://doi.org/10.3390/su132413910
33. S. Elanthikkal, U. Gopalakrishnapanicker, S. Varghese, J.T. Guthrie, Cellulose microfibres produced from banana plant wastes: isolation and characterisation. Carbohydr. Polym. **80**(3), 852–859 (2010)
34. Orange Fiber, *Sustainable fabrics from citrus juice by-products.* https://orangefiber.it/
35. M. Agassi, The advantages of implementation of water conservation practices in arid, semiarid regions. J. Sustain. Agric. **18**(2–3), 63–69 (2001)
36. Home. https://www.appleskin.com/appleskin
37. Y.H. Arifin, Y. Yusuf, Mycelium fibers as new resource for environmental sustainability. Procedia Eng. **53**, 504–508 (2013)
38. C. Madusanka et al., A review of recent advances in fungal mycelium based composites. Discov. Mater. **4**(1), 13 (2024)
39. Bolt Threads–Mylo. https://boltthreads.com/technology/mylo/
40. The Reishi™ Book - Touch, Explore, Imagine, See. https://www.mycoworks.com/reishi-book
41. M. Gama, P. Gatenholm, D. Klemm, *Bacterial NanoCellulose* (2016)
42. X. Zhao et al., Sustainable bioplastics derived from renewable natural resources for food packaging. Matter **6**(1), 97–127 (2023). https://doi.org/10.1016/j.matt.2022.11.006
43. AIRCARBON, *Newlight.* https://www.newlight.com/aircarbon
44. Bolt Threads–Microsilk. https://boltthreads.com/technology/microsilk/
45. QMILK-The material of the future-www.qmilkfiber.eu. https://www.qmilkfiber.eu/?lang=en
46. Faircraft-Real leather, grown in a lab. https://www.faircraft.bio/
47. Modern Meadow, *Biomaterials.* https://modernmeadow.com/biomaterials
48. CSIRO scientists discover how to grow coloured cotton, removing need for harmful chemical dyes. https://www.abc.net.au/news/2020-06-27/csiro-scientists-grow-coloured-cotton/12395250. Accessed 13 April 2025
49. J. Militký, D. Křemenáková, M. Venkataraman, A.P. Aneja, Towards sustainable textiles for a safer planet: main topics. Sustainability **16**(13), 5344 (2024)
50. J. Sawant, R. Guru, D. Grewal, S.C. Talekar, S.P. Kulkarni, Sustainability in textiles: a critical review of eco–friendly practices and materials. ShodhKosh J. Vis. Perform. Arts **5**(2), 85–97 (2024). https://doi.org/10.29121/shodhkosh.v5.i2.2024.891
51. F. Stella, S.F. Garofalo, N. Cavallini, D. Fino, F.A. Deorsola, Closing the loop: analysis of biotechnological processes for sustainable valorisation of textile waste from the fast fashion industry. Sustain. Chem. Pharm. **38**, 101481 (2024)
52. S. Siddhartha, Sustainable fashion: exploring the concept of greenwashing and new trends in the fashion industry. Int. J. Innov. Sci. Res. Technol. **9**(7), 1264–1272 (2024). https://doi.org/10.38124/ijisrt/ijisrt24jul957
53. N. Varma, H. Wadatkar, R. Salve, T.V. Kumar, Advancing sustainable agriculture: a comprehensive review of organic farming practices and environmental impact. J. Exp. Agric. Int. **46**(7), 695–703 (2024)
54. S. Rani, V. Johar, Organic farming: a way forward for sustainable development. Futur. Trends Agric. Eng. Food Sci. **3**(1), 66–81 (2024). https://doi.org/10.58532/v3bcagp1ch4
55. What is organic cotton?, https://www.soilassociation.org/take-action/organic-living/fashion-textiles/organic-cotton/
56. N.N. Bhutto, A.M. Jiskani, G.M. Nizamani, Better cotton: an approach to sustainable agriculture. J. Appl. Agric. Sci. Technol. **6**(1), 85–89 (2022)

57. Mass Balance Chain of Custody Model - Better Cotton. https://bettercotton.org/what-we-do/connecting-supply-demand-chain-of-custody/mass-balance-chain-of-custody-model/.
58. *Exec sum: the false promise of certification: textiles, palm oil and fisheries* (2018)
59. The Better Cotton Initiative, https://bettercotton.org/what-we-do/
60. Home, *Cotton made in Africa | CmiA*. https://cottonmadeinafrica.org/en/
61. V. Bitzer, P. Glasbergen, Partnerships for sustainable change in cotton: an institutional analysis of African cases. J. Bus. Ethics **93**, 223–240 (2010)
62. The REEL Cotton Programme—CottonConnect. https://www.cottonconnect.org/resources-hub/the-reel-cotton-programme

Chapter 4
Advances in Bio-Based Regenerated and Synthetic Fibres and Their Sustainability Issues

Dewan Murshed Ahmed, Mir Abdul Gaffar, Muntaha Mahin, Borhan Uddin Khan, Rishad Rayyaan, and Afsana Mobin

Abstract Textile industry, despite being a front-line contributor to the global economy, is known to have significant impacts on the environment and society. This natural resource-demanding industry is associated with high consumption and substantial waste generation that lead to significant degradation of the environment. This chapter focuses on providing details on sustainable alternative approaches to the conventional consumption pattern of the natural resources with a view to paving a way for reducing the environmental footprint left by the world textile industry. It sheds light on the potentials of regenerated bio-based fibres, such as viscose, lyocell, modal, polylactic acid (PLA), polyhydroxyalkanoates (PHA), etc. in reducing the overall environmental impacts compared to other conventional fibres. The chapter presents the advancements made in the fields of regenerated bio-based fibres, the emerging technologies, sustainability metrices, the regulatory policies and

D. M. Ahmed (✉) · A. Mobin
Department of Fabric Engineering, Bangladesh University of Textiles, Dhaka, Bangladesh
e-mail: murshed@fe.butex.edu.bd; dewanmurshed.ahmed@postgrad.manchester.ac.uk

A. Mobin
e-mail: afsanatatex080@gmail.com

D. M. Ahmed
Department of Materials, The University of Manchester, Manchester, United Kingdom

M. A. Gaffar
Department of Wet Process Engineering, Bangladesh University of Textiles, Dhaka, Bangladesh
e-mail: mirabdulgaffer11914037@gmail.com

M. Mahin
Department of Yarn Engineering, Bangladesh University of Textiles, Dhaka, Bangladesh
e-mail: muntaha25dec@gmail.com

B. U. Khan
Department of Apparel Engineering, Bangladesh University of Textiles, Dhaka, Bangladesh
e-mail: borhan.uddin.khan22@gmail.com

R. Rayyaan
School of Design, University of Leeds, Leeds, England
e-mail: r.rayyaan@leeds.ac.uk

A. S. Muhammad Sayem and P. Goswami (eds.), *SDG 12 and Global Fashion Textiles Production*, SDGs and Textiles, https://doi.org/10.1007/978-981-96-9051-0_4

the measures adopted by the global community with a view to implementing those policies.

Keywords Textile fibres · Regenerated bio-based fibres · Environmental sustainability · Policy

1 Introduction

The textile industry is a significant contributor to the global economy, and hinges on two primary categories of materials: natural fibres (NF) and man-made fibres (MMF). MMFs encompass regenerated cellulosic fibres and synthetics, both of which undergo substantial chemical and structural transformations during production [1]. This transformation is pivotal in understanding the environmental implications associated with textile manufacturing. Textile Exchange's 2022 report highlights a record fibre output of 116 million tonnes and is projected to rise to 147 million tonnes by 2030 [2]. The ecological footprint of the textile industry is significant as it has a greater impact on the environment. Synthetic fibres such as polyester production is responsible for over 706 million tons of greenhouse gas emissions which is equivalent to the emissions from 185 coal-fired power plants or the energy use of 64 million American homes [3]. With a projected increase in polyester production, CO_2 emissions could rise up to 1.5 billion tons by 2030 [4].

While synthetic fibres have a considerable environmental impact, natural fibres also pose sustainability challenges. Conventional cotton has a significant carbon footprint with a global warming potential of 18.66 kg CO_2 equivalent per kg compared to just 0.47 kg CO_2 equivalent for per kg of organic cotton. Organic cotton is a more sustainable choice but it still affects freshwater eutrophication. Silk production has a similar global warming potential to cotton and conveys risks such as high fossil resource use, fine particulate matter formation, and freshwater ecotoxicity [5]. Because of these pressing issues, the importance of sustainable practices in the textile and material industries has become crucial now. Awareness of climate change, resource scarcity, and environmental regulations have converged to drive a global movement toward sustainability.

The United Nations World Commission on Environment and Development defines sustainability as "*Ongoing progress that fulfils current requirements while ensuring that future generations can also meet their own needs*". Sustainable development demands the minimization of detrimental effects left on air, water, and other natural resources and proper maintenance of the overall integrity of ecosystems [6–8]. As the textile industry navigates this shift towards sustainability, the integration of regenerated bio-based fibres represents a promising path forward.

This chapter aims to provide insights regarding regenerated bio-based fibres and bio-based synthetic fibres for enhancing sustainability in world textile sector. In addition, the chapter will define sustainability, analyse the environmental impact of traditional materials, and highlight the need for sustainable alternatives driven by consumer demand and technological innovations. Key topics include, but not limited to: biodegradable fibres, their recycling, sustainability metrics, and case studies of successful practices. Additionally, it will discuss challenges, policy considerations, and the importance of ongoing research in promoting a sustainable future for textiles.

2 Need for Sustainable Alternatives

2.1 Drivers of Change

The importance of more sustainable bio-based alternatives that hold the potential to reduce the overall environmental impact is becoming more prominent day by day. Factors like consumer demands, technological advancements, and policies are driving the textile industry to incorporate sustainable alternative materials in its supply chain. The consumers have been appearing to be more inclined towards sustainable clothing solutions over the last decade which ultimately forced the fashion and the textile industry to come up with such alternative solutions [9, 10]. A recent study conducted by Sigaard and Laitala [11] displayed that consumers are likely to choose fibres or clothing based on sustainability thinking [11]. Especially younger generation consumers share a larger amount of interest in alternative sustainable choices than the older generations [12, 13].

Furthermore, bio-based regenerated and synthetic materials are suitable approaches to impart sustainability in textiles with biosafety, biocompatibility, and biodegradability [14]. Driven by technical advancements and environmental concerns, sustainable processing methods of bio-based fibres are becoming popular. Compared to conventional fibres, bio-based fibre production is more environmentally friendly as it reduces the use of toxic chemicals and significantly decreases water use and air pollution. This encourages the development and adaptation of bio-based materials which can significantly reduce the industry's environmental footprint [15].

2.2 Current Global Trends in Textiles

Textile and fashion industries have become among the largest revenue-generating industries but they are receiving criticism because of social and environmental concerns. This sector contributes up to 10% of global carbon dioxide emissions annually which is equivalent to over 1.7 billion tonnes [16–19]. They are also responsible for approximately 20% of industrial water pollution which roughly contributes to 35% of oceanic microplastic pollution. Moreover, approximately 92 million tonnes per year of textile waste is produced by the textile and fashion industry including landfills, unsold products, incinerations, and so on [20]. Traditionally, the textile industry tends to follow a linear system which wastes many economic opportunities, and its impact on the environment and society is also crucial [21]. Addressing these challenges, the textile and fashion industry is making transitions from a linear economy towards a circular economy [22]. The circular economy paradigm rejects the traditional end-of-life concept and prioritizes the reuse, recycling, and recovery of materials throughout the production and distribution processes [23]. This framework is mainly based on three main principles: reduce, reuse, and recycle which ensures long-time waste management strategies [24].

Since the textile and fashion industry is responsible for producing a significant amount of waste within its supply chain, an effective waste management system is a must. Textile industries are taking initiatives like constructing solid recycling infrastructure, using advanced technologies (near-infrared spectroscopy, optical sorting, and AI) for sorting and grading textile waste, imparting different types of recycling methods (mechanical, chemical, biological, etc.), promoting campaigns for consumers and customers to for improving the practice of circular economy system [25–27]. Figure 1 illustrates the idea of both linear economy and circular economy.

Regenerated bio-based materials can become a great addition to the circular economy system. Regenerated bio-based fibres are manufactured from polymers derived from natural sources, mostly from plants, that involve the chemical processing of natural resources, ensuring renewability and sustainability. Their biodegradable properties reduce landfill waste and pollution, while their versatility allows for diverse applications across various sectors. These textiles can easily be recycled or composted, minimizing their environmental footprint.

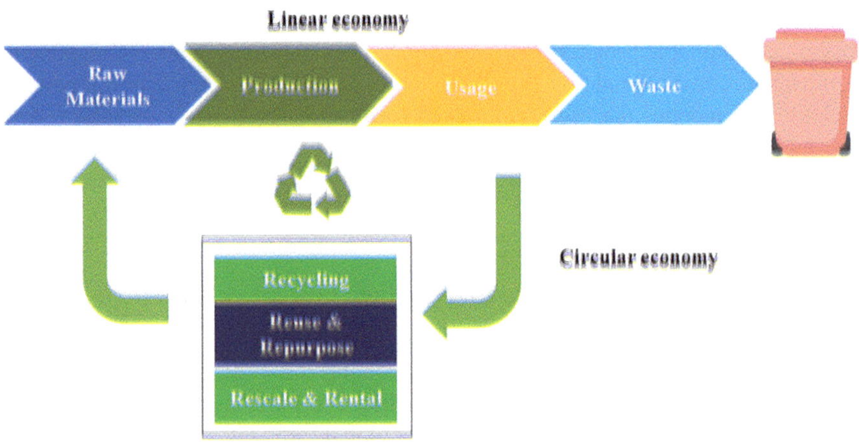

Fig. 1 Transition of a liner economy into a circular economy [28]

3 Regenerated Bio-Based Fibres

Regenerated bio-based fibres are made from renewable natural resources using different types of chemical and biological processes. These fibres can be very good alternative to the conventional fibres as they hold the potential to significantly reduce their environmental impacts. These fibres are mainly sourced from cellulose-based waste, alternative carbohydrates, and microbial biomass. The term "regenerated" refers to their potential to be generated from natural polymers as completely new fibre structures. In this regeneration process, natural polymers such as cellulose is converted into a soluble form and then the solution is re-formed to yield new fibre [29]. This process can give newer fibres having improved properties and biodegradability while ensuring sustainability by exploiting renewable resources as well [30]. A brief description of the regenerated bio-based fibres that are now being used commercially with their proven sustainability features is given below.

3.1 Viscose

Viscose, also called rayon, comes from cellulose found in wood pulp or cotton linters, usually from fast-growing plants like eucalyptus, beech, and spruce [31]. The production process of viscose starts with cellulose being treated with sodium hydroxide (NaOH) to form alkali cellulose. This substance is then pressed to remove excess liquid giving out a white, crumb-like product called sodium cellulose (Na-cellulose). Na-cellulose is then aged by exposure to air. The duration of this aging process affects the viscosity of the viscose solution; longer aging reduces its viscosity. After the ageing process, there is the xanthation phase. Here Na-cellulose reacts with carbon disulfide (CS_2) under controlled conditions to produce a compound named cellulose xanthate. This material has a similar viscosity to sodium cellulose and is dissolved in a dilute NaOH solution to create viscose. The viscose is then matured and extruded through a spinneret into a coagulating bath, where cellulose is regenerated into continuous filaments [32]. Technological advancements (see Table 1) made in the production of viscose have played a great role in improving emission control, especially in case of carbon disulfide recovery [33].

3.2 Lyocell

Lyocell fibres can be a more sustainable option as compared to viscose. Lyocell production involves a closed-loop solvent technology that recycles as much as 99% of the solvent (see Fig. 2). In the first step, cellulose is dissolved in a non-toxic solvent called N-methyl morpholine N-oxide (NMMO). Next the cellulose solution is extruded through spinnerets to obtain lyocell fibres. Higher strength, superior

Table 1 Recent developments in viscose and lyocell production process

Fiber type	Process name	Process description	References
Viscose	Wheat gluten powder processing	– Mixed with alcohol (30–70 °C, 2–4 h), centrifuged, and evaporated to extract protein – Cotton pulp is treated with NaOH to prepare viscose spinning liquid – Protein and viscose liquids are blended, mixed, and spun into the final product	[34]
	Grinding treatment	– The wood pulp was treated with saturated steam at 1.55 MPa and 200 °C for 3 to 9 min – It was then immersed in a diluted aqueous sodium hydroxide (NaOH) solution to form a cellulose-alkali slurry – Agitation and mixing were carried out to achieve complete dissolution	[38]
	Fire-resistant cellulose viscose with multi-wall carbon nanotubes	Medicinal cotton was dissolved in 16–20% NaOH and alkalized at 35–45 °C for 1.5–3 h, followed by the addition of CS_2 Yellowing was carried out at 35–45 °C for 3–5 h Dissolution was performed using 4–5% NaOH solution, with the addition of carboxylated multi-wall carbon nanotubes (0.1–1 wt% of the medicinal cotton) to obtain multi-wall carbon nanotube-dispersed yellowing cellulose viscose	[39]
	Xanthation process for casting cellulose film	Shaped articles are prepared by steeping at 30 °C to about 50 °C Xanthation is performed using 30–40% carbon disulfide by weight of the cellulose present Casting cellulose film involves xanthation being carried out, at least partially, at 20 to 30 °C, and dissolution is done, at least partially, at 10 °C to 25 °C	[40]
Lyocell	Hemp fiber processing*(Lyohemp)*	– Hemp fibre tow treated with NaOH (160 °C, 1 h) – NaOH and surfactants used as cooking liquor – Cellulose pulp DP adjusted (450–700) using H_2O_2 and NaOCl – Pulp dissolved at 100 °C under vacuum with continuous mixing – Dry–wet spinning used for dope extrusion	[41]

(continued)

Table 1 (continued)

Fiber type	Process name	Process description	References
	Characterization of Cellulose Solutions and Spinning Dopes via Degree of Polymerization and Refractive Index	– The degree of polymerization (DP) in cuoxam solution (DPCuoxam) was determined using capillary viscometry at 23 °C – The micro and macroscopic structures of pulps and spinning dopes were characterized, along with DP – Spinning dopes were made by pre-dispersing wet pulp (30% solid content) in aqueous NMMO (60%) at 100 °C – The obtained cellulose solutions were evaluated by refractive index	[42]
	Preparation of Hydroxyethyl Chitosan (EHC) and EHC/Cellulose Polyblends	– 10 g of chitosan was added to 150 mL of 50 wt% NaOH solution with mixing – The mixture was stored at −18 °C for 48 h for alkalization – After thawing, it was combined with 20 mL of isopropyl alcohol and stirred for homogenization – 6 g of chloroethanol was added under continuous stirring at 80 °C for 12 h – The mixture was filtered and washed twice with alcohol, then vacuum-dried to obtain hydroxyethyl chitosan (EHC) – 10 g of EHC was dispersed in 90 g of NMMO/H_2O and stirred at 90 °C for 12 h – The EHC/cellulose polyblends were dissolved in 50 wt% aqueous NMMO	[43]
	NMMO/cellulose solution	– Two separate series of simulations were conducted to study the effect of water content on NMMO/cellulose solutions, focusing on thermodynamics, structure, and hydrogen bonding – First simulation: Conducted at 1 wt% cellulose with a single unrestrained cellulose strand – Second simulation: Conducted at 15 wt% cellulose with 16 strands fixed in either a crystalline or dissociated state – The system was gradually heated to 373 K at a rate of 1 K/ps in the NPT ensemble at 1 bar pressure, using a Nosé–Hoover thermostat and barostat with coupling constants of 100 fs and 1000 fs, respectively – The 1 wt% simulations were used to analyze hydrogen bonding between NMMO, water, and cellulose, based on a set of geometric criteria	[44]

(continued)

Table 1 (continued)

Fiber type	Process name	Process description	References
	Lyocell fibre with quaternary ammonium salts for antibacterial wet wipes	– Lyocell fibre contained organic salt in amounts ranging from 0.1 to 50 wt%, preferably 1 to 10 wt%, and most preferably 2 to 7 wt%, based on the dry fibre weight – The invention relates to the use of lyocell fibre in wet or moist wipes for hygiene products, particularly disinfecting and sanitizing wipes for home, food service, medical, and other markets – The quaternary ammonium salt used for making wet or moist wipes is preferably a quaternary ammonium chloride	[45]

absorbency, and biodegradability makes lyocell an exceptional regenerated bio-based fibre [34].

3.3 *Modal*

The production of modal fibres resembles that of the viscose process. The cellulose sheets are dissolved in caustic soda to form alkali cellulose which is then treated with carbon disulfide that yields cellulose xanthate. The solution is then extruded through spinnerets into a coagulating bath that yields continuous threads [35, 36]. In order to improve strength and flexibility, the fibres are then stretched further. Despite having likewise production process as viscose, modal production involves lesser chemical usage than viscose production and also has better recovery of chemicals. Modal is known for its exceptional softness, superior moisture absorbency, resistant to shrinking and discoloration properties all of which make it a highly performing regenerated bio-based fibre [37].

4 Bio-based Synthetic Fibres

Fibrous materials derived from natural polymers or biopolymers have gained significant attention for their reduced ecological footprint compared to conventional petroleum-based polymers. Bio-based man-made organic fibres (MMOFs) exhibit considerable potential in textiles, sportswear, healthcare, and protective equipment [46]. This section provides an overview of various bio-based fibers, including Polylactic Acid (PLA), Polyhydroxyalkanoates (PHA), Polybutylene Succinate (PBS), Polytrimethylene Terephthalate (PTT), Poly(ethylene furanoate) (PEF), bio-nylon,

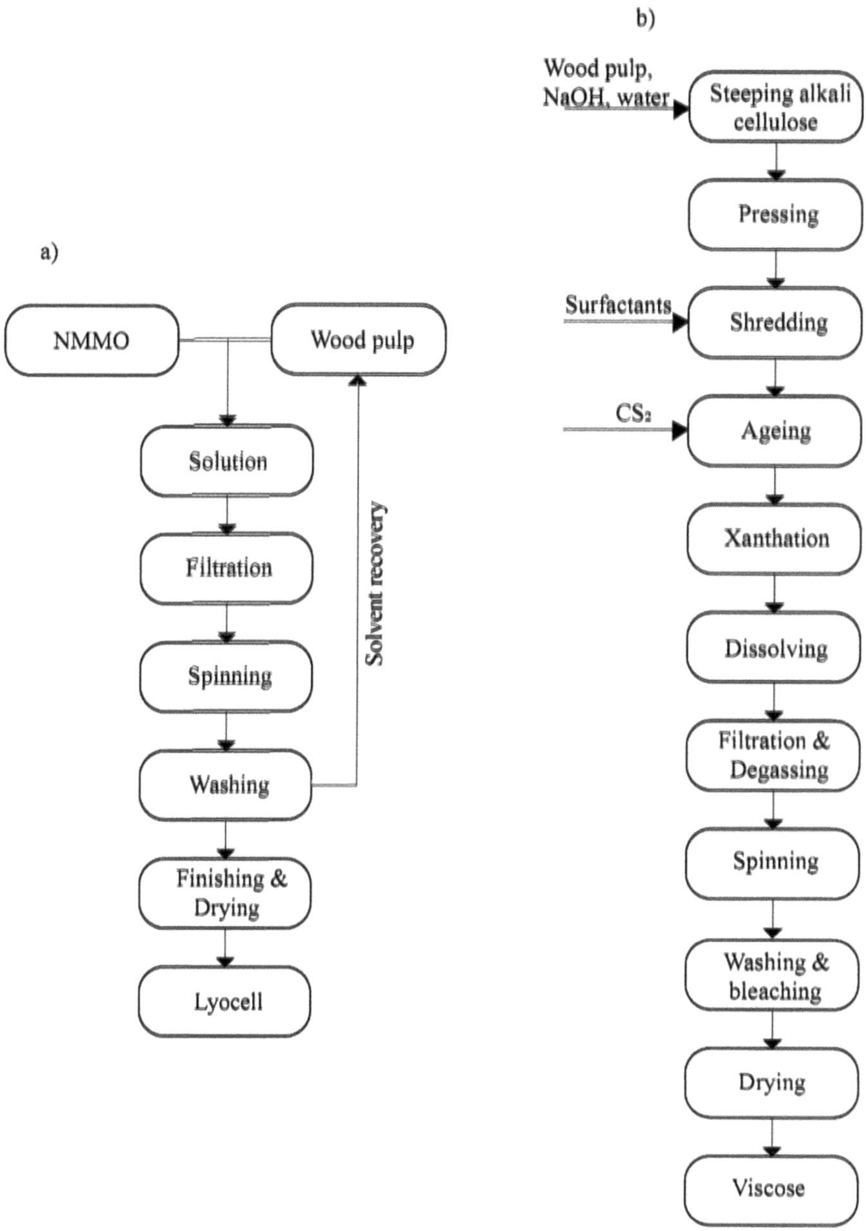

Fig. 2 Production process of **a** Lyocell and **b** Viscose

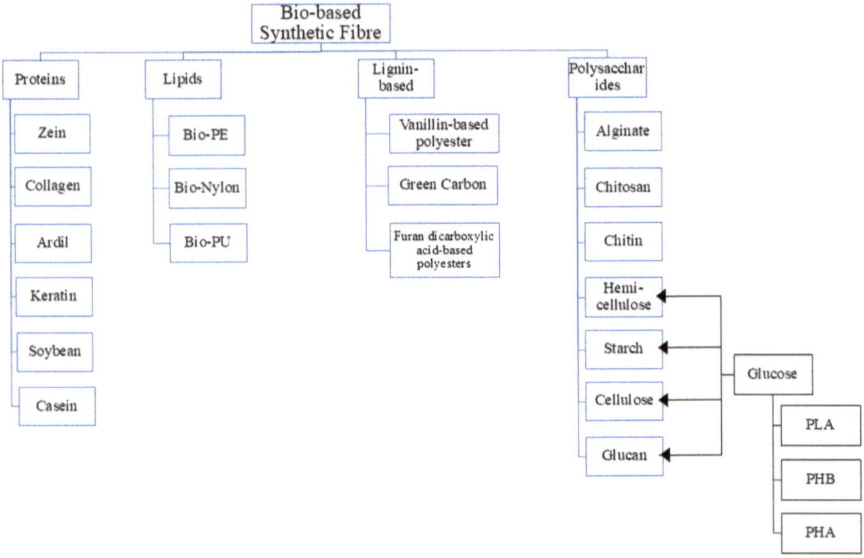

Fig. 3 Different types of Bio-based synthetic fibres [46]

bio-polyester, lignin-based fibres, starch-based fibres, chitin- and chitosan-based fibres, soy protein fibres, and algae-based fibres (Fig. 3).

4.1 *Polylactic Acid (PLA)*

Polylactic acid (PLA) is another biodegradable polyester fibre obtained from renewable resources such as corn and other polysaccharides. PLA production exploits plant-based feedstocks where the plants are hydrolysed in order to convert starches into sugars. The sugars then undergo a fermentation process and gets converted into lactic acid. Lactic acid then undergoes condensation or ring-opening polymerization which gives Polylactic acid which is processed using standard textile techniques like melt spinning or wet spinning to finally obtain PLA fibres. The softness, durability and moisture management properties of PLA fibres make them suitable as apparel wear, Medtech products, and packaging material [47]. Unlike other biodegradable polymers that degrade through surface enzymatic activity, PLA degradation in composting environments undergoes hydrolysis due to moisture and temperatures above 60 °C. The hydrolysis process reduces its molecular weight so that microorganisms can metabolize the fragments. Under industrial composting conditions, PLA fully biodegrades in a few months but degrades more slowly under mesophilic or ambient conditions [48].

4.2 Polyhydroxyalkanoates (PHA)

Polyhydroxyalkanoate (PHA) is a type of biodegradable polyester which is synthesized by utilizing microorganisms such as *Cupriavidus necator* and *Pseudomonas putida* under nutrient-limited conditions. Bacteria store carbon in their cells as PHA granules, using sustainable sources like plant-based carbohydrates and organic waste [49]. Once enough PHA has built up inside the cells, processes like solvent extraction, mechanical disruption, or filtration are used to extract them. PHAs break down quickly in composting environments by dint of microbial enzymes, turning into carbon dioxide and water. PHA degradation usually takes from a few weeks to months if is exposed to appropriate environmental conditions [50, 51].

4.3 Polybutylene Succinate (PBS)

PBS is an aliphatic polyester made from corn, cassava, sugarcane and such other renewable sources. Polycondensation of succinic acid and butanediol which are contained in the above-mentioned sources are mainly responsible for the production of PBS. PBS breaks down in different environments, including soil and composting conditions. Studies show that PBS loses between 0.2 and 0.5% of its weight in soil degradation tests over 30 days [52–54]. Unlike traditional synthetic polymers like PET, which resist breaking down due to their strong bonds and water-resistant properties, bio-based polyesters like PLA, PBS, and PHA are designed to degrade more easily making them suitable alternatives for more sustainable textile manufacturing [51].

4.4 Polytrimethylene Terephthalate (PTT)

Polytrimethylene Terephthalate (PTT) are bio-based fibres derived from non-grain sources such as starch and glycerine. Its main component is 1,3-propanediol (PDO). PTT is synthesized via direct esterification or transesterification involving terephthalic acid and PDO [51, 55]. PTT fibres demonstrate enhanced resilience, a higher tensile strength and a greater elongation at break relative to other polyester fibres [56, 57]. Bio-based PTT fibres are structurally similar to conventional polyester fibres which make them non-biodegradable just like conventional polyester [58].

4.5 Poly (Ethylene Furanoate) (PEF)

Bio-based Poly (ethylene furanoate) (PEF) is another type of polyester that is originated from furan dicarboxylic acid (FDCA). PEF can be produced using traditional polymerization techniques, such as polycondensation, transesterification, direct esterification etc. It has superior gas barrier properties, better mechanical strength and thermal properties making it suitable for sustainable packaging material application [57]. PEF films fully degrades within 72 h at 65 °C when treated with the enzyme *Humicola insolens cutinase* (HiC), resulting in significant weight loss and over 95% conversion into monomers [59]. However, to make fibre-grade PEF more susceptible to biological degradation in waste management systems, methods for de-crystallization and reducing particle size are necessary [48].

4.6 Bio-Nylon

Bio-Polyamide (Bio-Nylon or Bio-PA) known as N510 (nylon 510), is primarily derived from castor oil. This fibre is produced from sebacic acid and penta-methylene diamine. It has strength and thermal resistance similar to nylon 6, along with superior dimensional stability in humid conditions. To synthesize bio-based polyamides, bio-based 1,5-pentanediamine (DN5) undergoes polymerization with different long-chain dicarboxylic acids (LCDA). This process thus substantially diminishes dependence on petroleum and other fossil-derived raw materials. This fibre exhibits excellent properties as a clothing material such as low-temperature dyeability, softness, skin compatibility, wear resistance, etc. [51].

4.7 Bio-PE

Bio-PE (Bio-Polyethylene) is chemically identical to traditional polyethylene but it is sourced from renewable materials, i.e. sugarcane [60]. The synthesis of bio-polyethylene (Bio-PE) begins with fermentation to produce bio-ethanol. This bio-ethanol is then dehydrated to produce bio-ethylene which is the monomer for polyethylene synthesis. The bio-ethylene is polymerized through either free radical or coordination method, depending on the type of polyethylene needed, such as LDPE or HDPE. Bio-PE offers mechanical properties similar to conventional PE [57, 61]. When natural fibres are used as fillers, the stiffness and strength of Bio-PE composites improve, with Young's modulus reaching up to 5.81 GPa in bio-composites reinforced with fibres like corn and abaca [62]. Bio-PE can potentially reduce greenhouse gas emissions by 70–75% compared to traditional petroleum-based polyethylene due to the CO_2 absorption by biomass during cultivation [63].

4.8 Algae-Based Fibres

Algae-based fibres are spun from cellulose extracted from algae. Red algal waste from agar production is a raw material for bio-composite fibres. These algae-based fibres are often reinforced with polylactic acid (PLA) in order to obtain favourable mechanical properties and ecological benefits [64]. The cultivation of algae can be conducted in coastal waters which helps minimizing land usage and reduction in the environmental impact of fibre production [65]. Recent research has explored upcycling end-of-life textiles by incorporating microalgae, specifically *Chlorella vulgaris*, to create "living" bio-composites capable of atmospheric carbon dioxide (CO_2) removal [66].

4.9 Chitin and Chitosan Fibres

Chitosan is a polysaccharide-based biopolymer derived from chitin. Chitin is a linear and semi-crystalline polymer consisting of N-acetylglucosamine units connected by β (1 → 4) glycosidic linkages. Chitin and chitosan are found in exoskeletons of crustaceans like shrimps, lobsters, etc., mollusks like octopus, snails, etc., algae like brown algae, green algae, etc., insects like beetles, ants, spiders, cockroaches, scorpions, etc., and fungi cell walls. Chemical treatments, biological approaches, and microwave-assisted procedures are the most common methods used for chitin extraction. The conversion of chitin to chitosan occurs by deacetylation [67]. The traditional approach involves exposing chitin to strong alkaline solutions, primarily sodium hydroxide, at temperatures more than 100 °C. The biological deacetylation method uses an enzyme called deacetylase which is derived from fungi. The activity of chitin deacetylase can be augmented through the application of certain ionic solutions. Chitosan can be readily transformed into many forms, including films, fibres (nano and micro), gels, and membranes, through chemical modification. Chitosan is highly applicable in antimicrobial textiles and wound dressings [68].

4.10 Soy Protein Fibres

Soy Protein Fibre (SPF) is a biodegradable regenerated protein fibre which is extracted from renewable source like soybeans. The primary extraction method for soy protein is alkali solution–acid precipitation. However, there are other extraction methods such enzyme-assisted extraction, reverse micelle extraction, and ultrafiltration membrane extraction. The extraction of protein from soybeans begins with a degumming process eliminating lipids and oils, resulting in soy protein isolate. The isolate is subsequently dissolved in an alkaline solution, and continuously treated with chemicals such as polyvinyl alcohol (PVA) to boost spinnability and optimize fibre production. In the final stage, the solution is wet spun through spinnerets into an acidic coagulation bath, where protein molecules cross-link and form fibres [69]. Soy protein fibres offer exceptional mechanical properties and are suited for textile applications due to their breathability, moisture management, and softness. However, their lower durability and resistance in humid environments limit their broader applications [70].

4.11 Lignin-Based Fibres

Lignin-based fibres are a viable alternative to traditional carbon-based materials having excellent mechanical strength and thermal stability. Lignin is initially extracted as a residual from processes such as Kraft pulping or organosolv pulping of biomass. Then the residual is transformed into fibres using melt spinning or solution spinning. After spinning, the fibres undergo post-processing operations, such as stretching and drying, to further improve their mechanical strength [71].

4.12 Starch-Based Fibres

Starch, a naturally occurring polymer, is derived from plants like maize, rice, and potatoes. It is popular because of its cost-effectiveness, accessibility, and biodegradability. Starch can be exposed to enzymatic or chemical modifications to give different derivatives with specific intended properties. The synthesis of starch-based fibres begins with the modification of starch to improve its thermoplastic properties by the addition of plasticizers like glycerol. The resultant starch-polymer composite is then melted and extruded through spinnerets to produce fibres. Starch-based fibres are highly exceptional because of their excellent biodegradability and economic efficiency. However, the relatively low tensile strength limits their application in case of high-performance textiles. These fibres are mostly used in the production of packaging and disposable textiles where biodegradability is an important parameter [68].

4.13 Performance of Bio-Based Synthetic Fibres

Recent advances in sustainable bio-based fibres focus on achieving an optimum balance between durability, functionality, and biodegradability. The performance of these new sustainable fibres, in comparison to the traditional synthetics, in terms of durability, functionality, and biodegradability is an important area of research indeed (see Tables 2 and 3).

Table 2 Performance characteristics of bio-based regenerated and synthetic fibres

Fiber type	Strength	Moisture absorption (%)	Biodegradability	Functionality	References
Viscose	Dry: 22–26 (cN/tex), Wet: 10–15 (cN/tex)	11–14	Highly biodegradable	Highly absorbent, luster breathable, soft, and can be blended with other fibers for better durability	[84]
Lyocell	Dry: 38–42 (cN/tex), Wet: 34–38 (cN/tex)	11	Highly biodegradable	Softness, drapability, dimensional stability, dye uptake, and colorfastness	[84, 85]
Modal	1.3 dtex	11	Highly biodegradable	Softness, breathability, durability, color retention, pilling resistant, shrinkage resistant	[86]
PLA	53–60 MPa	<1	10–22 months	Easy melt processing, unique properties, renewable source, compostable, recyclable	[51, 87]
PHA	15–40 MPa	<5	6–10 months	Biodegradable, renewable, strong, moisture-resistant, thermally processable, biocompatible, barrier properties	[88, 89]
PBS	17–19 MPa	Low moisture absorption, at even elevated temperatures	Biodegradable	Biodegradable, renewable, flexible, durable, heat-resistant, compostable, processable, moisture-resistant	[90, 91]
PTT	58–67 MPa	0.0982	Non-biodegradable	Excellent resilience, wearability, dyeability, static resistance, and chemical resistance, making it suitable for high-end garments, home textiles, and industrial textiles	[92, 93]

Table 3 Environmental impact comparison of different textile fibers

Impact category	Energy consumption	Water usage	Carbon and GHG (green-house gas) Emissions (kg CO_2-eq./kg)	Chemical usage	Land use and deforestation	Waste and end-of-life impact	Refs
Cotton	Moderate to high	5732 m^3 per tonne of fibre	2.0	High (pesticides, fertilizers)	High (requires agricultural land, water-intensive)	Biodegradable but contributes to landfills if not composted	[84, 94]
Polyester	High (energy-intensive synthesis), energy usage:125 MJ/kg	130 m^3 (5 m^3 process water + 125 m^3 cooling water)	4.1	High (dyeing, finishing)	High (fossil fuel extraction impacts)	Non-biodegradable, persists in environment	[95–97]
Nylon	High (polymerization and processing)	Low	0.4507	High (dyeing, finishing, polymerization)	High (fossil fuel extraction impacts)	Non-biodegradable, persists in environment	[97, 98]
Regenerated fibres							
Viscose (Rayon)	High	319 m^3 per tonne of fibre	3.8	High (sodium hydroxide, carbon disulfide, acid treatments)	High (often sourced from non-sustainable forestry)	Biodegradable but contributes to landfill waste	[84, 94, 95]
Modal	Moderate (slightly more efficient than viscose)	472 m^3 per tonne of fiber	0.03	Moderate (fewer chemicals than viscose)	Moderate (sustainably managed beechwood)	Biodegradable with less chemical waste than viscose	[84, 94, 95]

(continued)

Table 3 (continued)

Impact category	Energy consumption	Water usage	Carbon and GHG (green-house gas) Emissions (kg CO_2-eq./kg)	Chemical usage	Land use and deforestation	Waste and end-of-life impact	Refs
Lyocell	Moderate to low	263 m^3 per tonne of fibre	1.1	Low (NMMO solvent, non-toxic and recyclable)	Low	Biodegradable with minimal processing waste	[84, 88, 94, 99, 100]
Bio-based synthetic fibres							
PLA	Renewable energy use: 60.4GJ, non-renewable energy use:28.8GJ	786 m^3 per tonne of fibre	501 kg CO_2 eq/ton	Low	Moderate, requires 1,775 m^2 of land per ton of PLA	Biodegradable in specific conditions	[101, 102]
PHA	High (fermentation process), energy demand:106–158 MJ/Kg	Low (microbial feedstock does not require land)	2–4.3	Low (enzymes, microbes)	Low (microbial feedstock does not require land)	Decomposes naturally without residual waste	[49, 103, 104]

(continued)

Table 3 (continued)

Impact category	Energy consumption	Water usage	Carbon and GHG (green-house gas) Emissions (kg CO_2-eq./kg)	Chemical usage	Land use and deforestation	Waste and end-of-life impact	Refs
PBS	Moderate	PBS from sugar beet pulp has lower energy consumption than fossil-based plastics	5.88	Moderate (catalysts, additives)	Requires land for biomass feedstock but can be optimized with efficient biorefinery processes and Lower risk of deforestation compared to Bio-PE if non-food feedstocks are used	Biodegradable (industrial composting needed)	[105, 106]
PTT	Moderate (less than PET due to partial renewable feedstock)	Low water usage ~ 3–7 L/kg	1.8–2.5	Moderate	Moderate (due to agricultural feedstocks)	Non-biodegradable	[107]
Bio-Nylon	High	Bio-nylon production from lignocellulosic biomass requires water but is more efficient	–	Low	Moderate	Non-biodegradable, but recyclable under specific systems	[108]

(continued)

Table 3 (continued)

Impact category	Energy consumption	Water usage	Carbon and GHG (green-house gas) Emissions (kg CO_2-eq./kg)	Chemical usage	Land use and deforestation	Waste and end-of-life impact	Refs
Bio-PE	Bio-PE from sugarcane uses less fossil energy but still requires substantial energy for cultivation and processing	Sugarcane-derived bio-PE has high water demands, leading to concerns about water depletion	Reduces fossil CO_2 emissions but may contribute to indirect emissions from land-use changes	Low	Sugarcane cultivation for bio-PE leads to significant land-use expansion and expansion of sugarcane plantations for Bio-PE can drive deforestation in tropical regions	Non-biodegradable	[63, 109, 110]
Alginate fibres	Low (simple extraction)	Low (marine-based raw material)	Low	Minimal (alkaline-based extraction)	Very low (grown in oceans, no deforestation)	Dissolves in water, minimal waste	[111, 112]

Durability refers to a material's ability to resist wear, tear, and other environmental factors while maintaining its functional properties over its lifetime. Synthetic fibres are widely known for their durability, with applications across the textile industry. Bio-based synthetic fibres such as polylactic acid (PLA) and polybutylene succinate (PBS) provide durable mechanical properties while also offering ecological benefits by degrading in the environment over time. This enhances the overall environmental sustainability. Various modifications, such as fibre treatments, nanocomposite integration, and fibre blending, have been explored to enhance their mechanical strength and durability, making them comparable to traditional synthetic fibres [72].

The functionality of fibres includes properties like moisture management, thermal regulation, and elasticity. Synthetic fibres such as nylon and polyester are often exposed to vigorous chemical treatment or modification to enhance their performance properties [73, 74]. On the other hand, bio-based synthetics are developed with improved functionality through more environmentally friendly approaches without the employment of harmful chemical treatments processes [75]. When it comes to thermal stability, bio-based synthetic like PBS has a relatively low melting point while still exhibiting superior flexibility and toughness [76]. However, PLA's thermal performance is limited by its low melting point and glass transition temperature. However, blending PLA with PBS and polyhydroxyalkanoates (PHAs) significantly enhances its thermal stability improving their usability in high-temperature conditions [77].

Biodegradability of synthetic fibres is a critical aspect of sustainability. Conventional synthetic fibres are non-biodegradable leaving detrimental effects on landfills and aquatic ecosystems. These materials fragment into microplastics which penetrate into the aquatic habitats and may eventually enter the human food chain [78]. Compared to these conventional synthetics, bio-based synthetic materials demonstrate better biodegradability [79]. For instance, PLA and PBS are found to degrade under industrial composting conditions while PHA can decompose within a few months under composting conditions [80, 81]. Although bio-based polymers degrade more easily than traditional synthetics, factors such as polymer structure and environmental conditions significantly affect their rates of degradation [82, 83].

5 Advancements in Sourcing of Raw Materials

The constant high demand for eco-friendly textile products has led to significant advancements in the sourcing of renewable raw materials for regenerated bio-based textiles (see Table 4). Renewable resources such as wood pulp, agricultural waste, and bacterial cellulose provide sustainable alternatives that minimize environmental impacts [113].

Table 4 Advancements in raw material sourcing

Raw material	Description	Reference(s)
Wood pulp	Sourced from rapidly growing species such as eucalyptus, spruce, and pine. The main pulping methods are the Kraft process (wood chips cooked with sodium hydroxide and sodium sulphide to remove lignin) and the sulphite process. Both methods break down lignin and hemicellulose to yield dissolving pulp, which contains over 90% cellulose. This pulp is cleaned, bleached, and treated to remove contaminants. Wood pulp reduces dependence on petroleum-derived, non-biodegradable fibres like polyester, a leading cause of plastic pollution	[32]
Agricultural waste	Agricultural waste, such as banana fibres, areca nut fibres, and other lignocellulosic materials, is used to reduce raw material costs and bolster rural economies by creating new markets for agricultural residues. Banana fibres are degummed and made into weft yarn, which can be blended with cotton or silk for enhanced durability. Areca nut fibres are strong, flexible, and soft, making them suitable for producing yarn. These fibres exhibit good strength and flexibility, making them viable alternatives for sustainable textile manufacturing	[114]
Bacterial cellulose	Produced through microbial fermentation, typically using Acetobacter xylinum, which produces cellulose as a by-product of glucose metabolism. Bacterial cellulose (BC) has superior mechanical strength, high crystallinity, and a three-dimensional structure. It can be moulded into customized garment forms, enabling a zero-waste strategy by eliminating fabric waste. Ongoing research focuses on chemically modifying BC to reduce its hydrophilicity, enabling the production of water-repellent textiles with desired comfort properties	[115, 116]

6 Recycling of Regenerated Bio-Based Fibres

Recycling of fibres are of following types- primary, secondary, tertiary, and quaternary [8]. Primary recycling refers to the reuse of manufacturing waste, secondary recycling includes mechanical processes that transform textiles through cutting and shredding, and tertiary recycling refers to various chemical processes, such as pyrolysis, which breaks down materials into their original monomers. Quaternary recycling is focused on the recovery of energy through waste incineration.

6.1 Mechanical Recycling

Mechanical recycling process reduces the materials' size by grinding or shredding and ultimately reshaping the material in its molten state (see Fig. 4). Primary mechanical recycling method mainly works with post-industrial waste and significantly reduces costing by recovering materials that would otherwise be discarded [117].

This method can be best utilized if the recycling process gets implemented directly during the production process. Although it can also be applied to post-consumer waste, primary recycling is less frequently used in these cases due to the additional complex steps required for material purity which gives a rise in overall costing. Therefore, in case of post-consumer (PC) or end-of-life (EOL) waste having uncertain composition and purity, secondary mechanical recycling is employed [117]. Secondary mechanical recycling includes additional processes like material separation and purification. Although secondary recycling does not cause any kind of chemical alteration of the polymer, the molecular weight can still be reduced due to chain scissions as a consequence of exposure to water or trace acids, which can eventually cause the materials' mechanical properties to be degraded. However, this degradation can be mitigated by methods such as intensive drying, vacuum degassing, and using stabilizing additives [118]. Another issue is contamination with incompatible polymers. Most polymers have poor compatibility with one another, resulting in subpar mechanical properties when blended. For example, impurities like PET in PVC can form solid lumps, degrading the quality of recycled PVC [119].

6.2 Chemical Recycling

Chemical recycling implies the transformation of polymeric waste by modification of its chemical structure, converting it into monomers that can be utilized as raw materials for the synthesis of new polymers. This procedure facilitates the decomposition of fibres into monomers or fundamental constituents, which can subsequently be re-polymerized to produce new fibres of similar or superior quality. Figure 5 shows the chemical recycling processes of cellulosic textile wastes. Despite its advantages, chemical recycling poses the risk of retaining or dispersing hazardous substances present in dyes, anti-wrinkle agents, softeners, and other commonly used additives [25]. Chemical recycling can be carried out through various methods, including glycolysis, methanolysis, hydrolysis, aminolysis, ammonolysis, pyrolysis, and catalytic depolymerization.

7 Environmental and Economic Impacts

Life cycle assessment (LCA) is a systematic approach used in order to evaluate the entire environmental impacts of a certain product "from origin to disposal". It examines each stage of the goods' life cycle in a following a systematic method. The assessment begins with the extraction of raw materials and ends with the yield of the final product. Life Cycle Assessment evaluates the environmental impacts that are associated with a product at every phase. This evaluation examines elements such as energy consumption, water usage, CO_2 emissions, and waste generation [122, 123]. In case of bio-based textiles and recycled synthetics, LCA framework evaluates

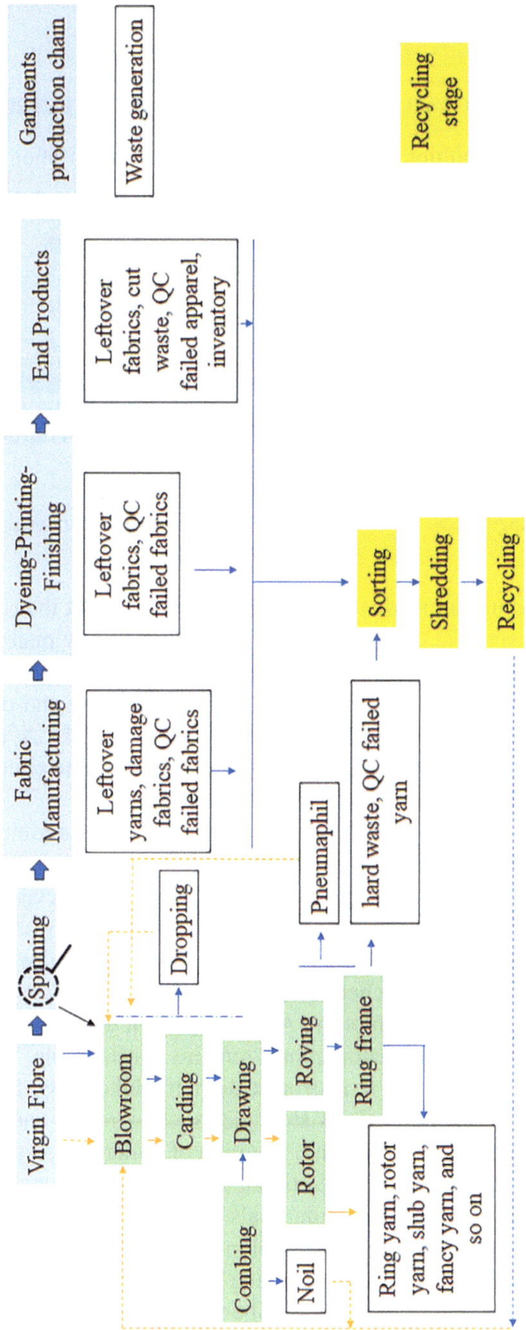

Fig. 4 Flowchart of mechanical recycling of waste materials produced in the textile industries [28]

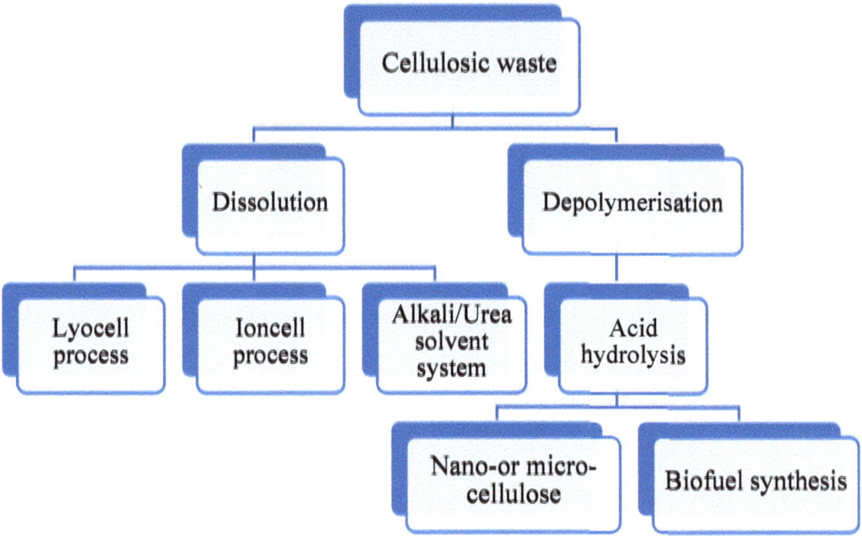

Fig. 5 Chemical recycling process of cellulosic waste produced in textile industries [28]

the overall environmental impacts of these materials throughout their lifetime. Key aspects include assessing the environmental implications of raw material sourcing, analysing energy use and emissions during production, evaluating the performance and durability of the textiles during their usage, and considering end-of-life options such as biodegradability and recyclability [122]. By employing LCA, designers can identify improvements, reduce ecological footprints, and promote circular economy principles in the development of sustainable textiles.

Shen et al. [124] conducted a LCA (life cycle assessment) examining three man-made cellulose fibres (Modal, Tencel, and Viscose) alongside traditional textile fibres like cotton, polyester (PET), and polypropylene (PP). Their study revealed that modal and viscose offer significant environmental advantages due to lower fossil energy consumption during fibre and pulp production. Specifically, Tencel was noted for its overall reductions in energy use, chemical inputs, and emissions of CO_2 and SO_2, along with decreased water consumption. Although cotton does not require energy-intensive processing, its environmental impact remains considerable. This is because of the high freshwater demand and terrestrial ecotoxicity linked to cotton production. To mitigate these environmental effects strategies such as reuse, and recycling are favoured over incineration and landfilling. Reusage of material is especially advantageous as it minimizes the need for additional processing. However, textile waste consists of both natural and synthetic materials. This leads to more challenges to recycling when different components are combined. As a result, there is a reduction in product quality [120]. This is why recycling requires proper separation during production. Additionally, the LCA study by Shen et al. [121] examined greenhouse gas emissions and non-renewable energy consumption for various plastic materials, including partially bio-based PET, recycled PET, partially recycled bio-based PET, and petrochemical PET. Their comparison with polylactic acid (PLA) and man-made cellulose fibres like Tencel, Viscose, and Modal indicated that both bio-based and recycled polymers have a lower environmental impact than single-use petrochemical PET. Therefore, emphasizing the significance of implementing sustainable practices in textile manufacturing.

8 Policy and Regulatory Considerations

In order to ensure the advancement of sustainable practices in the field of textiles, it is necessary to come up with concrete policy frameworks and regulatory measures at both local and global levels, which can impact and reshape the existing supply chain in a environmentally sustainable way. Effective implementation of such policies is crucial for advocating for the sustainable practices and also for encouraging the industry to shift towards regenerated bio-based materials.

8.1 Government Initiatives

There is no Government that does not acknowledge the threats that the traditional textile manufacturing processes pose to the health and the environment. Excessive energy consumption, soil and water pollution, high carbon emissions are some of the major threats to be listed [125, 126]. Various proposals have been put forth in order mitigating such threats and promoting sustainability in the textile industry. For instance, the European Union's Circular Economy Action Plan aims at reducing textile waste and increasing recycling approaches by establishing strict production requirements [127]. Other measures such as Waste Framework Directive and Extended Producer Responsibility (EPR), make it a requirement for the manufacturers to manage the entire lifecycle of textile products, from production to disposal [128]. The USA offers incentives such as tax breaks for companies that adopt bio-based alternative practices through their Sustainable Textiles Act [129]. These initiatives indicate the increasing global effort to attain Sustainable Development Goal (SDG) 12 by improving sustainable production practices across diverse economies.

In other developing countries of Southeast Asia, such regulations on sustainability are basically proposed and also imposed by international trade agreements and buyer demands. For example, in Vietnam and Bangladesh, focus have been shifted towards reducing water consumption and aligning local practice with global norms as a gesture of embracing sustainable practices [130–134].

8.2 Global Policies and Trade Impacts

Global trade policies also play a significant role in establishing sustainable practices by re-shaping production and distribution networks. The United Nations Sustainable Development Goals (SDGs) advocates for responsible production and consumption and production in the fields of textiles [135]. International agreements like the Paris Agreement aims at reducing carbon emissions which has influenced regulations around textile production [136]. The Basel Convention addresses the disposal of hazardous chemicals used in traditional textile manufacturing [137]. Another notable policy has been adopted by The European Union. Their trade regulations impose tariffs on countries that fail to adhere to the sustainable manufacturing practices [138].

The European Commission has launched various bioeconomy initiatives to secure worldwide dominance in the knowledge-driven economy to promote growth and competitiveness since 1982. These include the Strategy on Biotechnology (2002), the Lisbon Agenda (2000), and the EC White Paper (1993). They are all aimed at advancing industrial biotechnology to lower environmental impacts while maintaining production levels [139, 140]. Bioeconomy, also known as the knowledge-based or bio-based bio-economy, is an economic system that generates chemicals, energy, and materials, from renewable biological sources, encompassing plant and animal life [141]. The EU's Bioeconomy Action Plan (2018) outlines actions to enhance competitiveness, develop new technologies, and foster collaboration for ecological well-being [142].

Modernizing the industrial base through new value chains is vital. Thus proper research is needed for sustainable bio-based products like bio-fuels and bio-chemicals to replace fossil materials across various sectors [143, 144]. In 2017, Europe's bioeconomy generated over 2.4 trillion Euros in total turnover, reflecting a 25% increase since 2008 [145]. The implementation of the EU Bioeconomy Strategy relies on the European industry's ability to develop new bio-based products like bio-textiles. This can reduce fossil fuel dependency and diversify the wood industry. A key factor for success is understanding new markets, with consumer demand playing a pivotal role.

A recent study revealed that a significant number of consumers are prepared to pay a premium for environmentally sustainable products, with some willing to pay 64 to 128% more [146]. Marketing campaigns promoting eco-friendly products could raise consumer awareness and increase environment-friendly purchases. Price remains a significant barrier, but high-quality bio-textiles could help overcome this, especially in the luxury market. Gender has no significant effect on willingness to pay, suggesting a market for both men's and women's clothing. This study highlights the potential for bio-textiles across Europe which can align with the EU's Circular Economy Action Plan to promote sustainable products. Another research reveals that consumers in the Netherlands and Ireland prefer bio-based products (BBPs) over fossil-based alternatives. Approximately 50% of respondents in both countries are unwilling to pay a premium for BBPs. However, 50% of Irish consumers indicated a willingness to pay a premium of 25 to 50% for specific categories like disposable products and cosmetics. Notably, 61% of respondents in Ireland believe their choices positively impact the environment, compared to 50% in the Netherlands. Irish females are most inclined to purchase BBPs, while Dutch males favour bio-based construction materials. Price remains a critical factor, with 27% of consumers identifying it as the primary motivator. The study emphasizes the need for improved consumer education on the sustainability and functionalities of BBPs to enhance market acceptance and growth [147]. Despite limited research on bio-based apparel consumption, an online survey was done with 1,673 German participants over 16 years old. The survey incorporated a choice-based consumer experiment on a bio-based rain jacket [148]. The study identified two consumer segments: 39% showed acceptance of moderate prices for bio-based products, while 61% preferred the lowest prices. Innovativeness and altruism had moderate impacts on preferences while socio-demographic factors did not significantly influence consumer segmentation. The findings underscore the diversity of customer preferences about bio-based garment pricing. Which leads to a necessity for specialized pricing methods in supply chain development.

9 Challenges of Bio-Based Materials and Future Directions

Regenerated bio-based textiles and synthetic materials provide sustainable alternatives to traditional textile fibres. However, there are several challenges they are facing such as issues with scalable production, high cost, and supply chain issues. Bio-based textiles have limited production capacity compared to fossil fuel-driven synthetic fibres. Moreover, bio-based regenerated fibres (Lyocell, Modal) and synthetic materials (PLA) are typically manufactured from crops, wood, and bamboo sources [149, 150]. As a result, converting these materials into textile fibres is energy-intensive and also requires specialized infrastructure. However, sourcing sustainable wood pulp also leads to issues of deforestation [151]. These challenges are impacting the scalable production of sustainable textile materials. Moreover, implementing sustainability initiatives is associated with cost. Despite the long-term cost savings, the implementation of sustainability initiatives often involves a significant amount of initial investment which may pose a financial burden for smaller businesses and those with limited resources. The unavailability of government funding, lack of basic knowledge about the potential benefits, and regulatory challenges can hinder the adaptation of sustainability practices These challenges limit the ability to meet the growing demand for sustainable textiles.

The utilization of local feedstock along with alternative sources of raw materials can partially solve this problem. Recently, lyocell has been manufactured with the help of industrial hemp which works as an alternative source of cellulose. Since industrial hemp is a fast-growing plant, it requires less water and fertilizers than other crops. Moreover, the risk of soil erosion, cost, and waste issues within manufacturing becomes less than before [152]. The implementation of a circular textile industry model can significantly improve sustainability. This concept encompasses the comprehensive evaluation of the entire product life cycle, incorporating both technical and regenerative aspects, to ensure that the product fulfils its intended function while also allowing for potential reuse or recycling.

Therefore, to achieve scalable and sustainable production of bio-based regenerated and synthetic materials, a collaboration between specialized process techniques, product development, market applications, and advanced manufacturing equipment is a must. To impart this system within the supply chain, textile industries need to establish a robust supply of raw materials, develop system-oriented production technologies, and improve other necessary processes.

10 Conclusion

The advancements in developing regenerated bio-based textiles play a significant role in guiding the textile industry towards sustainability by reducing the textile industry's environmental footprint. The innovations made in raw material sourcing, recycling, production technologies and the emphasis put on circular economy practices opened up new dimensions towards achieving an economically and environmentally sustainable industrial practice. However, despite the advancements made, challenges remain in the widespread use of these materials such as scalability, cost considerations, trading off performance, supply chain integration and dealing with infrastructural limitations. Sustained exploration and funding for ecological innovations alongside policy setups are essential in addressing these challenges. In order to successfully retain the advancements made towards achieving this sustainable approach, confronted effort from the industry stakeholders, the policymakers and the researchers must be put forth. Conversely, companies should restructure their capital, assets, and resources using circular economy ideas at their core guiding force. This guarantees the direction of the business and the accomplishment of Sustainable Development Objectives. On the other hand, emphasis must be put on creating increased consumer awareness with a view to enhancing the demand for such sustainable products and services. SDG 12 guides the 2030 Agenda by necessitating collaborative and regenerative initiatives that emphasise environmental preservation and social capital. So, contemporary company models, mostly centred on profit maximisation, must transition towards sustainability to guarantee long-term industrial survival. Continuous research into utilizing the full potential of regenerated bio-based fibres should get patronized worldwide in achieving the intended sustainability.

References

1. Man-made fibers continue to grow (2024). Accessed 05 Oct 2024. https://www.textileworld.com/textile-world/fiber-world/2015/02/man-made-fibers-continue-to-grow/

2. Textile Exchange's annual materials market report shows further growth in the overall production of new materials, including fossil-based synthetic fibers. Textile exchange. Accessed 05 Oct 2024. https://textileexchange.org/news/textile-exchanges-annual-materials-market-report-shows-further-growth-in-the-overall-production-of-new-materials-including-fossil-based-synthetic-fibers/
3. D. Drew, G. Yehounme, The apparel industry's environmental impact in 6 graphics (2017). Accessed 02 Feb 2025. https://www.wri.org/insights/apparel-industrys-environmental-impact-6-graphics
4. S. Greene, R. Kirchain, T.R. Miller, E. Olivetti, Sustainable apparel materials. Accessed 05 Oct 2024. https://search.issuelab.org/resource/sustainable-apparel-materials.html
5. V. Gonzalez, X. Lou, T. Chi, Evaluating environmental impact of natural and synthetic fibers: a life cycle assessment approach. Sustainability **15**(9), art no. 9 (2023). https://doi.org/10.3390/su15097670
6. G. Hariem Brundtland, World commission on environment and development. Environ. Policy Law **14**(1), 26–30 (1985). https://doi.org/10.1016/S0378-777X(85)80040-8
7. A. Joy, J.F. Sherry Jr., A. Venkatesh, J. Wang, R. Chan, Fast fashion, sustainability, and the ethical appeal of luxury brands. Fash. Theory **16**(3), 273–295 (2012). https://doi.org/10.2752/175174112X13340749707123
8. S.S. Muthu, *Textiles and Clothing Sustainability: Recycled and Upcycled Textiles and Fashion*. Springer (2016)
9. A. Pérez, J. Collado, M.T. Liu, Social and environmental concerns within ethical fashion: general consumer cognitions, attitudes and behaviours. J. Fash. Mark. Manag. Int. J. **26**(5), 792–812 (2022). https://doi.org/10.1108/JFMM-04-2021-0088
10. E.L. Ritch, Consumers interpreting sustainability: moving beyond food to fashion. Int. J. Retail Distrib. Manag. **43**(12), 1162–1181 (2015). https://doi.org/10.1108/IJRDM-04-2014-0042
11. A.S. Sigaard, K. Laitala, Natural and Sustainable? Consumers Textile Fiber Preferences, Fibers **11**(2), 12 (2023). https://doi.org/10.3390/fib11020012
12. A. Diamantopoulos, B.B. Schlegelmilch, R.R. Sinkovics, G.M. Bohlen, Can socio-demographics still play a role in profiling green consumers? A review of the evidence and an empirical investigation. J. Bus. Res. **56**(6), 465–480 (2003). https://doi.org/10.1016/S0148-2963(01)00241-7
13. E.W. Johnson, P. Schwadel, It Is Not a Cohort Thing: Interrogating the Relationship Between Age, Cohort, and Support for the Environment. Environ. Behav. **51**(7), 879–901 (2019). https://doi.org/10.1177/0013916518780483
14. W. Tian et al., Recent progress in biobased synthetic textile fibers. Front. Mater. **9**, 1098590 (2022). https://doi.org/10.3389/fmats.2022.1098590
15. J. Deckers, S. Manshoven, L.F. Mortensen, The role of bio-based textile fibres in a circular and sustainable textiles system (2023)
16. A. Andooz, M. Eqbalpour, E. Kowsari, S. Ramakrishna, Z. Ansari Cheshmeh, A comprehensive review on pyrolysis from the circular economy point of view and its environmental and social effects. J. Clean. Prod. **388**, 136021 (2023). https://doi.org/10.1016/j.jclepro.2023.136021
17. G. Gereffi S. Frederick, The global apparel value chain, trade and the crisis: challenges and opportunities for developing countries. World Bank Policy Res. Work. Pap. Ser. (2010)
18. F. Jia, S. Yin, L. Chen, X. Chen, The circular economy in the textile and apparel industry: a systematic literature review. J. Clean. Prod. **259**, 120728 (2020). https://doi.org/10.1016/j.jclepro.2020.120728
19. K. Shirvanimoghaddam, B. Motamed, S. Ramakrishna, M. Naebe, Death by waste: fashion and textile circular economy case. Sci. Total. Environ. **718**, 137317 (2020). https://doi.org/10.1016/j.scitotenv.2020.137317
20. H. Dahlbo, K. Aalto, H. Eskelinen, H. Salmenperä, Increasing textile circulation—consequences and requirements. Sustain. Prod. Consum. **9**, 44–57 (2017). https://doi.org/10.1016/j.spc.2016.06.005

21. B. Piribauer, A. Bartl, Textile recycling processes, state of the art and current developments: a mini review. Waste Manag. Res. **37**(2), 112–119 (2019). https://doi.org/10.1177/0734242X18819277
22. M. Gardetti, Introduction and the concept of circular economy, pp. 1–11 (2019). https://doi.org/10.1016/B978-0-08-102630-4.00001-7
23. J. Kirchherr, D. Reike, M. Hekkert, Conceptualizing the circular economy: an analysis of 114 definitions. Resour. Conserv. Recycl. **127**, 221–232 (2017). https://doi.org/10.1016/j.resconrec.2017.09.005
24. P. Manickam G. Duraisamy, 4–3Rs and circular economy, in *Circular Economy in Textiles and Apparel*, S.S. Muthu, Ed., in The Textile Institute Book Series., Woodhead Publishing, pp. 77–93 (2019). https://doi.org/10.1016/B978-0-08-102630-4.00004-2
25. R.B. Baloyi, O.J. Gbadeyan, B. Sithole, V. Chunilall, Recent advances in recycling technologies for waste textile fabrics: a review. Text. Res. J. **94**(3–4), 508–529 (2024). https://doi.org/10.1177/00405175231210239
26. M. Koszewska, Circular economy in textiles and fashion–the role of a consumer, pp. 183–206 (2019). https://doi.org/10.1016/B978-0-08-102630-4.00009-1
27. I. Wojnowska-Baryła, K. Bernat, M. Zaborowska, D. Kulikowska, The growing problem of textile waste generation—the current state of textile waste management. Energies. **17**(7), art no. 7 (2024) https://doi.org/10.3390/en17071528
28. A.K. Das, Md.F. Hossain, B.U. Khan, Md.M. Rahman, M.A.Z. Asad, M. Akter, Circular economy: a sustainable model for waste reduction and wealth creation in the textile supply chain. SPE Polym. **6**(1), e10171 (2025). https://doi.org/10.1002/pls2.10171
29. T. Kim, D. Kim, Y. Park, Recent progress in regenerated fibers for "green" textile products. J. Clean. Prod. **376**, 134226 (2022). https://doi.org/10.1016/J.JCLEPRO.2022.134226
30. M. Ribul, Regenerative textiles: a framework for future materials circularity in the textile value chain. Sustain. Switz. **13**(24) (2021). https://doi.org/10.3390/su132413910
31. M. Shabbir, F. Mohammad, Sustainable production of regenerated cellulosic fibres. Elsevier Ltd (2017). https://doi.org/10.1016/B978-0-08-102041-8.00007-X
32. A. Matthes, K. Beyer, H. Cebulla, M.G. Arnold, A. Schumann, Sustainable textile and fashion value chains: drivers, concepts. Theories and Solutions. (2020). https://doi.org/10.1007/978-3-030-22018-1
33. I.S.F. Mendes, A. Prates, D.V. Evtuguin, Production of rayon fibres from cellulosic pulps: state of the art and current developments. Carbohydr. Polym **273** (2021). https://doi.org/10.1016/j.carbpol.2021.118466
34. A.J. Sayyed, N.A. Deshmukh, D.V. Pinjari, A critical review of manufacturing processes used in regenerated cellulosic fibres: viscose, cellulose acetate, cuprammonium, LiCl/DMAc, ionic liquids, and NMMO based lyocell. Cellulose **26**(5), 2913–2940 (2019). https://doi.org/10.1007/s10570-019-02318-y
35. P. Parajuli, S. Acharya, S.S. Rumi, Md. T. Hossain, N. Abidi, Regenerated cellulose in textiles: rayon, lyocell, modal and other fibres, in *Fundamentals of Natural Fibres and Textiles*, Elsevier, pp. 87–110 (2021). https://doi.org/10.1016/B978-0-12-821483-1.00015-2
36. C. Vigneswaran, M. Ananthasubramanian, P. Kandhavadivu, Bioprocessing of synthetic fibres, in *Bioprocessing of Textiles*, Elsevier, pp. 189–250 (2014). https://doi.org/10.1016/B978-93-80308-42-5.50004-4
37. H.B. Öztürk, T. Bechtold, Splitting tendency of cellulosic fibers. Part 3: splitting tendency of viscose and modal fibers. Cellulose **15**(1), 101–109 (2008). https://doi.org/10.1007/s10570-007-9149-5
38. C. Yamane, K. Abe, M. Satho, H. Miyamoto, Dissolution of cellulose nanofibers in aqueous sodium hydroxide solution. Nordic Pulp Pap. Res. J. **30**(1), 92–98 (2015). https://doi.org/10.3183/npprj-2015-30-01-p092-098
39. X. Wang, Q. Li, Y. Di, G. Xing, Preparation and properties of flame-retardant viscose fiber containing phosphazene derivative. Fibers Polym. **13**(6), 718–723 (2012). https://doi.org/10.1007/s12221-012-0718-3

40. M.R. Cockroft, L. Fisher, process for processing cellulose films or shaped articles, WO2012160362A1. Accessed 07 Feb 2025. https://patents.google.com/patent/WO2012160362A1/en
41. J. Paulitz, I. Sigmund, B. Kosan, F. Meister, Lyocell fibers for textile processing derived from organically grown hemp. Procedia Eng. **200**, 260–268 (2017). https://doi.org/10.1016/j.proeng.2017.07.037
42. F. Meister, B. Kosan, A tool box for characterization of pulps and cellulose dopes in Lyocell technology. Nord. Pulp Pap. Res. J. **30**(1), 112–120 (2015). https://doi.org/10.3183/npprj-2015-30-01-p112-120
43. Y. Li, X. Liu, X. Zhuang, X. Jin, Q. Liu, Rheological behavior and spinnability of ethylamine hydroxyethyl chitosan/cellulose co-solution in N-methylmorpholine-N-oxide system. Fibers Polym. **17**(5), 778–788 (2016). https://doi.org/10.1007/s12221-016-5578-9
44. B.D. Rabideau, A.E. Ismail, Effect of water content in N-Methylmorpholine N-Oxide/Cellulose solutions on thermodynamics, structure, and hydrogen bonding. J. Phys. Chem. B **119**(48), 15014–15022 (2015). https://doi.org/10.1021/acs.jpcb.5b07500
45. M. Opietnik, G. Goldhalm, H. Firgo, Use of a lyocell fiber, US20180258375A1 (2018). Accessed 07 Feb 2025. https://patents.google.com/patent/US20180258375A1/da
46. A.C. Vale, L. Leite, V. Pais, J. Bessa, F. Cunha, R. Fangueiro, Extraction of natural-based raw materials towards the production of sustainable man-made organic fibres. Polymers **16**(24), art. no. 24 (2024). https://doi.org/10.3390/polym16243602
47. S.O. Lakshmanan, G. Raghavendran, *Regenerated Sustainable Fibres*. (Springer Singapore, 2018). https://doi.org/10.1007/978-981-10-8578-9_2
48. J. Egan, S. Salmon, *Strategies and Progress In Synthetic Textile Fiber Biodegradability*, vol. 4, no. 1. Springer International Publishing (2022). https://doi.org/10.1007/s42452-021-04851-7
49. M.K. Uddin, L. Novembre, A. Greco, A. Sannino, Polyhydroxyalkanoates, a prospective solution in the textile industry–a review. Polym. Degrad. Stab. **219**, 110619 (2024). https://doi.org/10.1016/J.POLYMDEGRADSTAB.2023.110619
50. P. Borelbach, R. Kopitzky, J. Dahringer, P. Gutmann, Degradation behavior of Biodegradable Man-Made Fibers in natural soil and in compost. Polymers **15**(13), 1–21 (2023). https://doi.org/10.3390/polym15132959
51. W. Tian et al., Recent progress in biobased synthetic textile fibers. Front. Mater. **9**(December), 1–12 (2022). https://doi.org/10.3389/fmats.2022.1098590
52. L. Aliotta, M. Seggiani, A. Lazzeri, V. Gigante, and P. Cinelli, A brief review of Poly (Butylene Succinate) (PBS) and Its main copolymers: synthesis, blends, composites, biodegradability, and applications. Polymers **14**(4) (2022). https://doi.org/10.3390/polym14040844
53. L. Liu, J. Yu, L. Cheng, X. Yang, Biodegradability of poly(butylene succinate) (PBS) composite reinforced with jute fibre. Polym. Degrad. Stab. **94**(1), 90–94 (2009). https://doi.org/10.1016/J.POLYMDEGRADSTAB.2008.10.013
54. V. Tserki, P. Matzinos, E. Pavlidou, C. Panayiotou, Biodegradable aliphatic polyesters. Part II. Synthesis and characterization of chain extended poly(butylene succinate-co-butylene adipate). Polym. Degrad. Stab. **91**(2), 377–384 (2006). https://doi.org/10.1016/J.POLYMDEGRADSTAB.2005.04.036
55. M. Shen et al., Are biodegradable plastics a promising solution to solve the global plastic pollution? Environ. Pollut. **263**, 114469 (2020). https://doi.org/10.1016/j.envpol.2020.114469
56. M.Y. Khalid, Z.U. Arif, Novel biopolymer-based sustainable composites for food packaging applications: a narrative review. Food Packag. Shelf Life **33**, 100892 (2022). https://doi.org/10.1016/J.FPSL.2022.100892
57. Q. Zhang, M. Song, Y. Xu, W. Wang, Z. Wang, L. Zhang, Bio-based polyesters: Recent progress and future prospects. Prog. Polym. Sci. **120**, 101430 (2021). https://doi.org/10.1016/j.progpolymsci.2021.101430
58. M.H. Rahman, P.R. Bhoi, An overview of non-biodegradable bioplastics. J. Clean. Prod. **294** (2021). https://doi.org/10.1016/j.jclepro.2021.126218

59. S. Weinberger et al., Enzymatic degradation of poly(Ethylene 2,5-furanoate) powders and amorphous films. Catalysts **7**(11), 1–11 (2017). https://doi.org/10.3390/catal7110318
60. V. Siracusa, Polymers analogous to petroleum-derived ones for packaging and engineering applications. Polymers **12**, 1641 (2020)
61. T. Mekonnen, P. Mussone, H. Khalil, D. Bressler, Progress in bio-based plastics and plasticizing modifications. J. Mater. Chem. A **1**(43), 13379–13398 (2013). https://doi.org/10.1039/c3ta12555f
62. X.Y.D. Soo et al., Bio-polyethylene and polyethylene biocomposites: an alternative toward a sustainable future. Macromol. Rapid Commun. **45**(14), 2400064 (2024). https://doi.org/10.1002/MARC.202400064
63. R. Kumar et al., An investigation of the environmental implications of bioplastics: recent advancements on the development of environmentally friendly bioplastics solutions. Environ. Res. **244**, 117707 (2024). https://doi.org/10.1016/J.ENVRES.2023.117707
64. F.N.M. Rosdi, N. Salim, R. Roslan, N.H.A. Bakar, S.N. Sarmin, Potential red algae fibre waste as a raw material for biocomposite. J. Adv. Res. Appl. Sci. Eng. Technol. **30**(1), 303–310 (2023). https://doi.org/10.37934/ARASET.30.1.303310
65. A. Constante, S. Pillay, H. Ning, U.K. Vaidya, Utilization of algae blooms as a source of natural fibers for biocomposite materials: Study of morphology and mechanical performance of Lyngbya fibers. Algal Res. **12**, 412–420 (2015). https://doi.org/10.1016/J.ALGAL.2015.10.005
66. P. In-na, J. Lee, G. Caldwell, Living textile biocomposites deliver enhanced carbon dioxide capture. J. Ind. Text. **51**(4), 5683S-5707S (2022). https://doi.org/10.1177/15280837211025725
67. D. Elieh-Ali-Komi, M.R. Hamblin, E.-A.-K. Daniel, Chitin and Chitosan: production and application of versatile biomedical nanomaterials. Int. J. Adv. Res. **4**(3), 411 (2016)
68. Roy, P.P. Chavan, Y. Rajeev, T. Praveenraj, P. Kolazhi, *Sustainable Manufacturing Practices in Textiles and Fashion* (2024). https://doi.org/10.1007/978-3-031-51362-6_1
69. Y. Zhao, R. Tian, Z. Xu, L. Jiang, X. Sui, Recent advances in soy protein extraction technology. J. Am. Oil Chem. Soc. **100**(3), 187–195 (2023). https://doi.org/10.1002/AOCS.12676
70. T. Rijavec, Z. Zupi, Soybean protein fibres (SPF), recent trends enhancing divers. Qual. Soybean Prod. (2011). https://doi.org/10.5772/19614
71. S. Wang et al., Lignin-based carbon fibers: Formation, modification and potential applications. Green Energy Environ. **7**(4), 578–605 (2022). https://doi.org/10.1016/j.gee.2021.04.006
72. T. Biswal, Biopolymeric conjugation with synthetic fibers and applications. Phys. Sci. Rev. **9**(9), 2827–2849 (2023). https://doi.org/10.1515/PSR-2022-0183/MACHINEREADABLECITATION/RIS
73. M. Jaffe, A.J. Easts, X. Feng, Polyester fibers, in *Thermal Analysis of Textiles and Fibers*, Elsevier, pp. 133–149 (2020). https://doi.org/10.1016/B978-0-08-100572-9.00008-2
74. M. Najafi, L. Nasri, R. Kotek, High-Perform Nylon Fibers **118**(2) (2017). https://doi.org/10.1016/B978-0-08-100550-7.00009-7
75. N. Abdouramane, D. Sibiescu, N.D. Efeze, P.M.A. Noah, F.E. Betene, H. Gomdje Valery, Analysis of bio-based antimicrobial textiles: a Review. Environ. Eng. Manag. J **22**(6), 1137–1143 (2023). https://doi.org/10.30638/eemj.2023.094
76. C.F.J. Kuo, W.L. Lan, S.H. Chen, C.Y. Chen, Property modification and process parameter optimization design of polylactic acid composite materials Part II: application of response surface methodology and multi-objective particle swarm optimization in the processing of polylactic acid composite fibe. Text. Res. J. **85**(7), 687–700 (2015). https://doi.org/10.1177/0040517514555793
77. Z. Fan et al., Highly enhanced mechanical, thermal, and crystallization performance of PLA/PBS composite by glass fiber coupling agent modification. Polymers **15**(15) (2023). https://doi.org/10.3390/polym15153164
78. R.C. Thompson, C.J. Moore, F.S.V. Saal, S.H. Swan, Plastics, the environment and human health: current consensus and future trends. Philos. Trans. R. Soc. B Biol. Sci. **364**(1526), 2153–2166 (2009). https://doi.org/10.1098/rstb.2009.0053

79. L.F. Vargas, B.A. Welt, A. Teixeira, P. Pullammanappallil, M. Balaban, C. Beatty, Biodegradation of treated Polylactic Acid (PLA) under anaerobic conditions. Trans. ASABE **52**(3), 1025–1030 (2009). https://doi.org/10.13031/2013.27371
80. S. Jha, B. Akula, H. Enyioma, M. Novak, V. Amin, H. Liang, Biodegradable biobased polymers: a review of the state of the art challenges, and future directions. Polymers **16**(16), 2262 (2024). https://doi.org/10.3390/polym16162262
81. D. Briassoulis, P. Tserotas, I. G. Athanasoulia, Alternative optimization routes for improving the performance of poly(3-hydroxybutyrate) (PHB) based plastics. J. Clean. Prod. **318**, 128555 (2021). https://doi.org/10.1016/j.jclepro.2021.128555
82. B.P. Chang, A.K. Mohanty, M. Misra, Studies on durability of sustainable biobased composites: a review. RSC Adv. **10**(31), 17955–17999 (2020). https://doi.org/10.1039/c9ra09554c
83. O. Okolie et al., Bio-based sustainable polymers and materials: from processing to biodegradation. J. Compos. Sci. **7**(6) (2023). https://doi.org/10.3390/jcs7060213
84. S. Rana, S. Pichandi, S. Parveen, R. Fangueiro, Regenerated cellulosic fibers and their implications on sustainability, in *Roadmap to Sustainable Textiles and Clothing: Eco-friendly Raw Materials, Technologies, and Processing Methods*, S.S. Muthu, Ed., (Singapore: Springer, 2014), pp. 239–276. https://doi.org/10.1007/978-981-287-065-0_8
85. J. Chen, Chapter 4–Synthetic textile fibers: regenerated cellulose fibers, in *Textiles and Fashion*, R. Sinclair, Ed., in Woodhead Publishing Series in Textiles, Woodhead Publishing, pp. 79–95 (2015). https://doi.org/10.1016/B978-1-84569-931-4.00004-0
86. W. Latif et al., Study of mechanical and comfort properties of modal with cotton and regenerated fibers blended woven fabrics. J. Nat. Fibers **16**(6), 836–845 (2019). https://doi.org/10.1080/15440478.2018.1441084
87. O. Avinc, A. Khoddami, Overview of Poly(lactic acid) (PLA) fibre. Fibre Chem. **41**(6), 391–401 (2009). https://doi.org/10.1007/s10692-010-9213-z
88. C. Var, S. Palamutcu, Man-made bio-based and biodegradable fibers for textile applications, in *Sustainable Manufacturing Practices in the Textiles and Fashion Sector*, S.S. Muthu, Ed., (Cham: Springer Nature Switzerland, 2024, pp. 229–280). https://doi.org/10.1007/978-3-031-51362-6_10
89. C.-S. Wu, D.-Y. Wu, S.-S. Wang, Preparation, characterization, and functionality of bio-based polyhydroxyalkanoate and renewable natural fiber with waste oyster shell composites. Polym. Bull. **78**(9), 4817–4834 (2021). https://doi.org/10.1007/s00289-020-03341-x
90. H.A. Saffian et al., Thermal, physical and mechanical properties of Poly(Butylene Succinate)/Kenaf core fibers composites reinforced with Esterified Lignin. Polymers **13**(14), art. no. 14 (2021). https://doi.org/10.3390/polym13142359
91. S.A. Rafiqah et al., A review on properties and application of bio-based Poly(Butylene Succinate). Polymers **13**(9), art. no. 9 (2021). https://doi.org/10.3390/polym13091436
92. J. Chen, Advances in bio-based fibers | International Fiber Journal. Accessed 08 Feb 2025. https://www.fiberjournal.com/advancesin-bio-based-fibers/
93. A.P. Rath, P.S.G. Krishnan, K. Kanny, Studies on (polytrimethylene terephthalate)/graphene oxide/f-MWCNT hybrid nanocomposites. Discov. Nano **19**(1), 21 (2024). https://doi.org/10.1186/s11671-024-03966-1
94. F.A.G.S. Silva, S. Branco, F. Dourado, B. Neto, M. Gama, Life cycle assessment of bacterial cellulose and comparison to other cellulosic sources. J. Clean. Prod. **493**, 144876 (2025). https://doi.org/10.1016/j.jclepro.2025.144876
95. L. Shen, E. Worrell, M. Patel, Environmental impact assessment of man-made cellulose fibres. Resour. Conserv. Recycl. **55**, 260–274 (2010). https://doi.org/10.1016/j.resconrec.2010.10.001
96. M. Jaffe, A.J. Easts, X. Feng, 8–Polyester fibers, in *Thermal Analysis of Textiles and Fibers*, M. Jaffe, J.D. Menczel, Eds., in The Textile Institute Book Series., Woodhead Publishing, pp. 133–149 (2020). https://doi.org/10.1016/B978-0-08-100572-9.00008-2
97. G. Kilgore, Carbon footprint of polyester vs cotton vs Wool vs Leather vs Nylon (Calculator), 8 billion trees: carbon offset projects & ecological footprint calculators. Accessed 09 Feb 2025. https://8billiontrees.com/carbon-offsets-credits/carbon-footprint-of-polyester/

98. N.M. Van Der Velden, M.K. Patel, J.G. Vogtländer, LCA benchmarking study on textiles made of cotton, polyester, nylon, acryl, or elastane. Int. J. Life Cycle Assess. **19**(2), 331–356 (2014). https://doi.org/10.1007/s11367-013-0626-9
99. Regenerated cellulose by the Lyocell process, a brief review of the process and properties: BioResources. Accessed 09 Feb 2025. https://bioresources.cnr.ncsu.edu/
100. C. Felgueiras, N.G. Azoia, C. Gonçalves, M. Gama, F. Dourado, Trends on the cellulose-based textiles: raw materials and technologies. Front. Bioeng. Biotechnol **9** (2021). https://doi.org/10.3389/fbioe.2021.608826
101. E. Rezvani Ghomi et al., The life cycle assessment for Polylactic Acid (PLA) to make it a low-carbon material. Polymers. **13**(11), 1854 (2021). https://doi.org/10.3390/polym13111854
102. A.Z. Naser, I. Deiab, B.M. Darras, Poly(lactic acid) (PLA) and polyhydroxyalkanoates (PHAs), green alternatives to petroleum-based plastics: a review. RSC Adv. **11**(28), 17151–17196 (2021). https://doi.org/10.1039/D1RA02390J
103. P. Borelbach, R. Kopitzky, J. Dahringer, P. Gutmann, (2023) Degradation behavior of biodegradable man-made fibers in natural soil and in compost. Polymers. **15**(13), art no. 13 (2023). https://doi.org/10.3390/polym15132959
104. L. Vogli et al., Life cycle assessment of a novel polyhydroxyalkanoates production process with mixed microbial cultures fed on winery waste. ITA (2017). Accessed 09 Feb 2025. https://cris.unibo.it/handle/11585/609772
105. L. Aliotta, M. Seggiani, A. Lazzeri, V. Gigante, P. Cinelli, A brief review of poly (butylene succinate) (PBS) and its main copolymers: synthesis, blends, composites, biodegradability, and applications. Polymers. **14**(4), art. no. 4 (2022). https://doi.org/10.3390/polym14040844
106. M.K. Patel et al., Second-generation bio-based plastics are becoming a reality—Non-renewable energy and greenhouse gas (GHG) balance of succinic acid-based plastic end products made from lignocellulosic biomass. Biofuels Bioprod. Biorefining **12**(3), 426–441 (2018). https://doi.org/10.1002/bbb.1849
107. M.H. Rahman, P.R. Bhoi, An overview of non-biodegradable bioplastics. J. Clean. Prod. **294**, 126218 (2021). https://doi.org/10.1016/j.jclepro.2021.126218
108. R. Aryapratama, M. Janssen, Prospective life cycle assessment of bio-based adipic acid production from forest residues. J. Clean. Prod. **164**, 434–443 (2017). https://doi.org/10.1016/j.jclepro.2017.06.222
109. Green PE (Bio-PE) for extrusion, injection & blow molding–FKuR (EN). Accessed 09 Feb 2025. https://fkur.com/en/bioplastics/im-green-polyethylene/
110. V. Siracusa, I. Blanco, Bio-Polyethylene (Bio-PE), Bio-Polypropylene (Bio-PP) and Bio-Poly(ethylene terephthalate) (Bio-PET): recent developments in bio-based polymers analogous to petroleum-derived ones for packaging and engineering applications. Polymers **12**(8), art no. 8 (2020). https://doi.org/10.3390/polym12081641
111. Md.M. Rahman et al., Sources, extractions, and applications of alginate: a review. Discov. Appl. Sci. **6**(8), 443 (2024). https://doi.org/10.1007/s42452-024-06151-2
112. R. Roy, P.P. Chavan, Y. Rajeev, T. Praveenraj, P. Kolazhi, Sustainable manufacturing practices in textiles and fashion, in *Sustainable Manufacturing Practices in the Textiles and Fashion Sector*, S.S. Muthu, Ed., (Cham: Springer Nature Switzerland, 2024), pp. 1–22. https://doi.org/10.1007/978-3-031-51362-6_1
113. S. Seisl, R. Hengstmann, Manmade cellulosic fibers (MMCF)—a historical introduction and existing solutions to a more sustainable production. Sustain. Text. Fash. Value Chains Driv. Concepts Theor. Solut. 3–22 (2021). https://doi.org/10.1007/978-3-030-22018-1_1
114. T.R. Mariamma, H.S. Nair, Popularization of agrowaste fibres—banana and areca nut fibre—a sustainable approach, pp. 81–99 (2022) https://doi.org/10.1007/978-981-19-0878-1_4
115. C.K. Chan, J. Shin, S.X.K. Jiang, Development of Tailor-shaped bacterial cellulose textile cultivation techniques for zero-waste design. Cloth. Text. Res. J. **36**(1), 33–44 (2018). https://doi.org/10.1177/0887302X17737177
116. A.P. Provin, V.O. dos Reis, S.E. Hilesheim, R.T. Bianchet, A.R. de Aguiar Dutra, A.L.V. Cubas, Use of bacterial cellulose in the textile industry and the wettability challenge—a review. Cellulose **28**(13), 8255–8274 (2021) https://doi.org/10.1007/s10570-021-04059-3

117. I.A. Ignatyev, W. Thielemans, B. Vander Beke, Recycling of polymers: a review. ChemSusChem. **7**(6), 1579–1593 (2014). https://doi.org/10.1002/cssc.201300898
118. D.S. Achilias, Material recycling–trends and perspectives (2012). https://doi.org/10.5772/2003
119. J. Hopewell, R. Dvorak, E. Kosior, Plastics recycling: challenges and opportunities. Philos. Trans. R. Soc. B Biol. Sci. **364**(1526), 2115–2126 (2009). https://doi.org/10.1098/rstb.2008.0311
120. Y. Wang, Fiber and textile waste utilization. Waste Biomass Valorization **1**(1), 135–143 (2010). https://doi.org/10.1007/s12649-009-9005-y
121. L. Shen, E. Worrell, M.K. Patel, Comparing life cycle energy and GHG emissions of bio-based PET, recycled PET, PLA, and man-made cellulosics. Biofuels Bioprod. Biorefining **6**(6), 625–639 (2012). https://doi.org/10.1002/bbb.1368
122. S.H. Eryuruk, Life cycle assessment method for environmental impact evaluation and certification systems for textiles and clothing, in *Handbook of Life Cycle Assessment (LCA) of Textiles and Clothing*, Elsevier, pp. 125–148 (2015). https://doi.org/10.1016/B978-0-08-100169-1.00007-1
123. L. Nguyen, G. Hsuan, S. Spatari, Life cycle economic and environmental implications of pristine high density polyethylene and alternative materials in drainage pipe applications. J. Polym. Environ. **25** (2017). https://doi.org/10.1007/s10924-016-0843-y
124. L. Shen, E. Worrell, M.K. Patel, Environmental impact assessment of man-made cellulose fibres, resources, conservation and recycling, vol. 55, no. 2, pp. 260–274 (2010). ISSN 0921–3449. https://doi.org/10.1016/j.resconrec.2010.10.001
125. European Parliament, The impact of textile production and waste on the environment. Accessed 12 Nov 2024. https://www.europarl.europa.eu/topics/en/article/20201208STO93327/the-impact-of-textile-production-and-waste-on-the-environment-infographics
126. A.K. Roy Choudhury, Environmental impacts of the textile industry and its assessment through life cycle assessment. In *Roadmap to Sustainable Textiles and Clothing: Environmental and Social Aspects of Textiles and Clothing Supply Chain*, S.S. Muthu, Ed. (Singapore: Springer, 2014), pp. 1–39. https://doi.org/10.1007/978-981-287-110-7_1
127. European Commission, Circular economy action plan–European Commission. Accessed 24 Sep 2024. https://environment.ec.europa.eu/strategy/circular-economy-action-plan_en
128. European Commission, Implementation of the waste framework directive. Accessed12 Nov 2024. https://environment.ec.europa.eu/topics/waste-and-recycling/implementation-waste-framework-directive_en
129. EPA. Accessed 12 Nov 2024. https://www.epa.gov/facts-and-figures-about-materials-waste-and-recycling/textiles-material-specific-data
130. S. Adnan, Md. N. Nabi, D. Hossain, Prospects & challenges of social sustainability of textile & RMG industry: evidence from Bangladesh, vol. 16 (2024)
131. C. Blanco González-Tejero, S. Ribeiro-Navarrete, R. Diaz, R. Furio-Piqueres, Textile industry and sustainability: an integrated examination. In *Textile Industry and the SDGs: Exploring Synergies for a Better Future*, J.F. Gallego-Nicholls, A. Carrilero-Castillo, Eds., (Singapore: Springer Nature), pp. 1–15 (2024). https://doi.org/10.1007/978-981-97-5073-3_1
132. Md.T. Islam et al., Sustainable textile industry: an overview. Non-Met. Mater. Sci. **4**(2), 15–32 (2022). https://doi.org/10.30564/nmms.v4i2.4707
133. V. Marinova, R. Radev, Sustainability of the European textile industry. E3S Web Conf. **452**, 05020 (2023). https://doi.org/10.1051/e3sconf/202345205020
134. Md.T. Rahaman, A.D. Pranta, Md.R. Repon, Md.S. Ahmed, T. Islam, Green production and consumption of textiles and apparel: importance, fabrication, challenges and future prospects. J. Open Innov. Technol. Mark. Complex. **10**(2), 100280 (2024). https://doi.org/10.1016/j.joitmc.2024.100280
135. UNDP, Sustainable development goals. Accessed 12 Nov 2024. https://www.undp.org/sustainable-development-goals
136. UNFCCC, The Paris Agreement. Accessed 12 Nov 2024. https://unfccc.int/process-and-meetings/the-paris-agreement

137. UNEP, Basel convention on the control of Transboundary Movements of Hazardous Wastes |. Accessed 12 Nov 2024. https://www.unep.org/resources/report/basel-convention-control-transboundary-movements-hazardous-wastes
138. European Commission, Sustainable development in EU trade agreements. Accessed 12 Nov 2024. https://policy.trade.ec.europa.eu/development-and-sustainability/sustainable-development/sustainable-development-eu-trade-agreements_en
139. K. McCormick, N. Kautto, The Bioeconomy in Europe: an overview. Sustainability **5**(6), art. No (2013). https://doi.org/10.3390/su5062589
140. C. Patermann, A. Aguilar, The origins of the bioeconomy in the European Union. New Biotechnol. **40**, 20–24 (2018). https://doi.org/10.1016/j.nbt.2017.04.002
141. OECD, The Bioeconomy to 2030. Accessed 24 Sep 2024. https://www.oecd.org/en/publications/2009/04/the-bioeconomy-to-2030_g1gha07e.html
142. L. Hetemäki, Future of the European forest-based sector: structural changes towards bioeconomy | European forest institute, Future of the European forest-based sector: structural changes towards bioeconomy. What Science Can Tell Us 6. European forest institute. Accessed 24 Sep 2024. https://efi.int/publications-bank/future-european-forest-based-sector-structural-changes-towards-bioeconomy
143. A. Paletto, S. Bernardi, E. Pieratti, F. Teston, M. Romagnoli, Assessment of environmental impact of biomass power plants to increase the social acceptance of renewable energy technologies. Heliyon **5**(7), e02070 (2019). https://doi.org/10.1016/j.heliyon.2019.e02070
144. A. Toppinen, A. Röhr, S. Pätäri, K. Lähtinen, R. Toivonen, The future of wooden multistory construction in the forest bioeconomy—A Delphi study from Finland and Sweden. J. For. Econ. **31**, 3–10 (2018). https://doi.org/10.1016/j.jfe.2017.05.001
145. Renewable carbon, Bioeconomy worth 2.4 trillion EUR to the European economy as bio-based industries mark sizeable jump in turnover and bio-based share of chemicals reaches record high of 15%. Accessed 24 Sep 2024. https://renewable-carbon.eu/news/bioeconomy-worth-2-4-trillion-eur-to-the-european-economy-as-bio-based-industries-mark-sizeable-jump-in-turnover-and-bio-based-share-of-chemicals-reaches-record-high-of-15/
146. N. Sandra, P. Alessandro, Consumers' preferences, attitudes and willingness to pay for bio-textile in wood fibers. J. Retail. Consum. Serv. **58**, 102304 (2021). https://doi.org/10.1016/j.jretconser.2020.102304
147. J. Gaffey, H. McMahon, E. Marsh, K. Vehmas, T. Kymäläinen, J. Vos, Understanding consumer perspectives of bio-based products—a comparative case study from Ireland and the Netherlands. Sustainability 13(11), art. no 11 (2021). https://doi.org/10.3390/su13116062
148. F. F. Stahl, A. Emberger-Klein, K. Menrad, Consumer preferences in Germany for bio-based apparel with low and moderate prices, and the influence of specific factors in distinguishing between these groups. Front. Sustain **2** (2021). https://doi.org/10.3389/frsus.2021.624913
149. T. Ivanović, R. Hischier, C. Som, Bio-based polyester fiber substitutes: from GWP to a more comprehensive environmental analysis. Appl. Sci. **11**(7), art. no 7 (2021). https://doi.org/10.3390/app11072993
150. L. Lawson, L.M. Degenstein, B. Bates, W. Chute, D. King, P.I. Dolez, Cellulose textiles from hemp biomass: opportunities and challenges. Sustainability **14**(22), art no 22 (2022). https://doi.org/10.3390/su142215337
151. L. Lawson, M. Ford, M.S. Hoque, W. Chute, D.C. Bressler, P.I. Dolez, Processes and challenges for the manufacturing of Lyocell fibres with alternative agricultural feedstocks. Appl. Sci. **13**(23), art no 23 (2023). https://doi.org/10.3390/app132312759
152. G. Kaur, R. Kander, The sustainability of industrial hemp: a literature review of its economic, environmental, and social sustainability. Sustainability. **15**(8), art no 8 (2023). https://doi.org/10.3390/su15086457

Chapter 5
Progress in Weaving Technology in Context of Sustainability

Kazi Sowrov, Md. Faisal Mahmud, Md. Abdullah Al Faruque, and Md. Abdullah Al. Mamun

Abstract The advances in weaving technology and their integration with sustainability practices along environmental, economic, and social dimensions are reviewed in this chapter. The basics of weaving are presented at the beginning of this chapter, including the complex processes of yarn preparation, the different types of looms that can be used, and the basic motions necessary to produce fabrics. The chapter also addresses diverse aspects of sustainability regarding weaving: energy and water consumption, carbon and water footprints, waste management, with the addition of new innovative technologies that could play a significant role in these fields advanced prototyping machines, automation for efficient production, and sustainable sizing processes. Relatively new near-net-shape weaving is discussed as a game-changing strategy for reducing textile production waste in applications like fashion, footwear, composites, and medical textiles. Issues of social sustainability, like fair labour practices and transparent supply chains, are emphasized as key to meeting the growing consumer demand for ethical products. Weaving technology development moves forward by the introduction of energy-efficient systems, recyclable materials, and automation for sustainable practice. Finally, the chapter concludes by addressing important challenges and needing an interdisciplinary approach toward reaching sustainability in weaving.

K. Sowrov · Md. F. Mahmud · Md. A. Al. Mamun (✉)
Department of Materials, Faculty of Science and Engineering, The University of Manchester, Manchester, England, UK
e-mail: mdabdullah.mamun@manchester.ac.uk

Md. F. Mahmud
Department of Textile Engineering, Ahsanullah University of Science and Technology, Dhaka, Bangladesh

Md. A. Al Faruque
Institute for Frontier Materials (IFM), Deakin University, Waurn Ponds, VIC, Australia

Md. A. Al. Mamun
Department of Fabric Engineering, Faculty of Textile Engineering, Bangladesh University of Textiles (BUTEX), Dhaka, Bangladesh

A. S. Muhammad Sayem and P. Goswami (eds.), *SDG 12 and Global Fashion Textiles Production*, SDGs and Textiles, https://doi.org/10.1007/978-981-96-9051-0_5

Keywords Weaving · Sustainability · Eco-friendly · Special weaving · Production parameters

1 Introduction

Weaving holds the distinction of being the oldest fabric manufacturing method, dating back to the introduction of human civilization [1]. It has been shown through historical evidence that the Egyptians utilized woven fabrics approximately 6000 years ago, while the Chinese utilized silk to create delicate garments more than 4000 years ago. From primitive to modern loom, weaving technology transformed from one phase to another phase by adopting various technological advancement including, energy source, process parameter, loom size, production rate etc. Sustainability is a growing concern of the modern world for living with survival development. Along with other industries textile industry will be a potential threat on the way of sustainable development goal if sustainability is not immediately introduced in all stages of the textile manufacturing chain. In textile industry, establishing sustainable practices among man, machine, materials are the key driving force to develop a clean sustainable production system. Sustainable weaving technology can optimize existing technologies to save natural resources by introducing innovative and eco-friendly weaving process that will pave the way to achieve sustainable goal [2]. By considering urgency of eco-friendly environment for greener world, sustainability inclusion in weaving industry will be an effective weapon to reduce usage of natural resources, cost of production and carbon footprint that minimized industries overall contribution to climate change [3].

Traditional weaving relied on handlooms, characterized by labour-intensive processes and limited scalability. The advent of power looms marked a paradigm shift, enabling mass production and enhancing efficiency. In recent decades, the focus has shifted towards integrating modern looms with sustainable practices. Shuttle-less looms, such as air-jet and rapier looms, have emerged, minimizing human intervention while optimizing speed and precision [4]. Despite these advancements, weaving faces significant sustainability challenges. The industry's reliance on energy-intensive machinery and the environmental impact of synthetic sizing agents underscores the urgent need for reform. Additionally, weaving processes generate waste, including leftover yarns and fabric scraps, which exacerbate environmental degradation [5]. Recognizing these challenges, stakeholders have prioritized sustainability as a core objective, prompting the development of energy-efficient looms, biodegradable materials, and water-saving techniques.

A notable innovation is near-net-shape weaving, which minimizes material waste by creating fabrics close to their final dimensions. This approach is particularly transformative for industries like aerospace, automotive, and fashion, where precision and efficiency are paramount [6]. Technological advancements such as 3D weaving further underscore the potential of this method to revolutionize textile production.

The chapter underscores the intertwined nature of technological innovation and sustainability in weaving. By embracing resource-efficient practices, integrating renewable energy, and leveraging automation, the weaving industry can meet the dual objectives of environmental stewardship and economic viability. Such advancements not only enhance productivity but also contribute to the broader goals of sustainable development.

2 Basic Principles of Weaving

Weaving is the process of making fabrics using two sets of threads named as warp yarn (also known as ends) and weft yarn (also known as picks/filling yarn) [7–9]. The warp yarns are placed vertically under a certain amount of tension whereas the weft yarns pass through the warp yarns using a medium (shuttle) or non-medium (shuttle-less) and thus the woven fabric is prepared due to their interlacement (warp and weft yarn) [8, 9].

Woven fabrics are made on a special type of machine called Loom [10]. Looms are classified in several ways. As for power operation, the loom is mainly classified as a handloom and a power loom [11–13]. Handloom is divided into five types, such as primitive or vertical loom, pit loom, frame loom, semi-automatic or Chittaranjan loom, and Hattersley loom [14]. Again, a power loom is distinguished as a water wheel loom, a conventional loom, and a modern loom. Based on the weft insertion mechanism, modern looms are classified as projectile, rapier, air-jet, and water-jet looms [13]. The modern looms are also known as shuttle-less looms [13, 14].

To enable the weaving process, in the case of shuttle looms, the weft yarns are accumulated onto a pirn or cop and placed into the shuttle to transfer the yarns between the shed (a gap between the warp yarns). Whereas, in the case of the shuttle-less or modern looms, weft yarns are collected onto a cone and directly used as received from the spinning industry [11, 13]. On the other hand, the warp yarns are first processed through the warping process. As per the pattern or design of the woven fabrics, the warp yarns are wrapped on a warping or weaving beam using either the direct/high-speed warping process or the indirect/sectional warping process. Thereafter, the warp yarns are passed through a special chemical treatment process, preferably known as the sizing process, where the warp yarns are treated with several sizing chemicals e.g., polyvinyl alcohols, starch, etc. to increase their strength, evenness, and reduce hairiness. Then, the warp yarns are placed into the heald frames and reed according to the design of the woven fabric, and this process is known as drawing and denting [12, 15]. Finally, this warping beam along with the heald frames and reed placed onto the loom using a special truck to enable the weaving process [11]. Figure 1 shows the yarn preparation method for the weaving technique [16].

All sorts of looms perform three basic mechanical motions to produce the woven fabrics, they are primary motions, secondary motions, and tertiary motions [7, 11]. Primary motions are shedding, picking, and beating whereas secondary motions are let-off and take-up. All other associated motions of a loom are known as tertiary

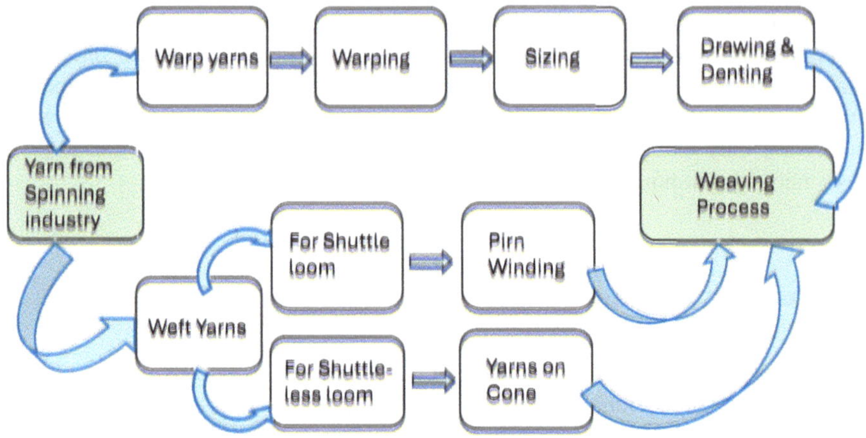

Fig. 1 Yarn preparation method for both warp and weft yarns [16]

motions such as warp stop motion, weft stop motion, and warp protector motion [7, 13]. It has been found that the weaving process is not possible without the primary and secondary motions whereas the tertiary motion is required to reduce the machine stoppage time and increase the productivity of the looms. Figure 2 demonstrates the primary and secondary motions of a loom [14].

Primary motions are shedding, picking, and beating or beat-up process. Shedding is defined as the separation of the warp yarns into two parts and the creation of an opening/tunnel for the weft yarns to be passed from one side of the loom to the other side. The opening or tunnel is called a "shed" and is controlled by the heald frames as of the design or pattern of the woven fabrics [12, 17]. The shedding mechanism is distinguished mainly into three parts such as tappet shedding, dobby shedding, and jacquard shedding [17]. These are responsible for controlling the lifting and lowering mechanism of the heald frames of a loom. Picking is termed as the transfer of weft yarns from one side of the loom to the other side. In conventional looms, this process was performed using a flying object named a shuttle that hold the pirn or cop filled with weft yarns into it [18]. However, the speed of the shuttle wasn't good enough to increase productivity, and as the ends of the shuttle were pointy, it increased the health-related risks of the weavers. Therefore, modern looms do not use the shuttle for the weft insertion but some other methods such as projectile, rapier, air-jet, and waterjet [16]. These weft insertion systems are safe and sound for the weavers and also enhance the productivity of the looms. Another primary motion of a loom is the beating or beat-up process, which is defined as the pushing or forwarding of the newly inserted weft yarns to a certain point of the already woven fabric, known as the fell of the cloth. This process is accomplished by the reciprocating motion of the reed [19, 20].

The let-off motion of the warp yarns from the warping/weaving beam and the take-up motion of the already woven cloth onto the cloth beam are known as the secondary motions of the loom. The warp yarns from the warping beam are released

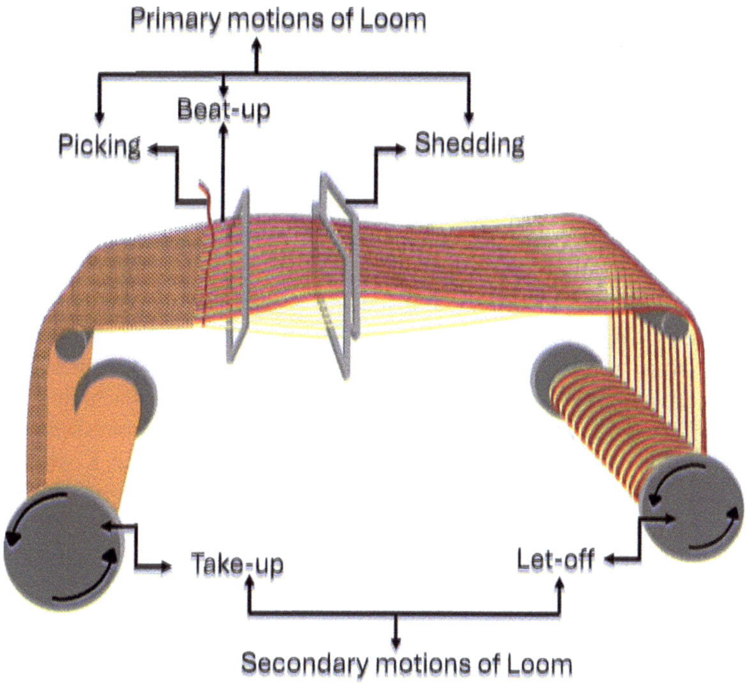

Fig. 2 Primary and secondary motions of a loom [14]

at a constant rate and tension to the weaving area and it is known as the let-off motion [11, 14]. On the contrary, the woven cloth is carried to the cloth roller under a constant rate and tension with required pick spacing, which is termed the take-up motion. Without the primary and secondary motions, the weaving process is not possible [7–9].

There are some other motions such as warp stop motions, weft stop motions, and warp protector motions known as tertiary motions [11, 21]. Without these motions of a loom, the weaving is possible, however, these are important to ensure a higher production rate and lower defects in the fabric. For example, if there is no warp stop motion then maybe a few warp ends will be missed into the fabric, and there will be an empty line present on the fabric in the vertical direction. Similarly, without the weft stop motion, there will be some missed picked areas, and a vacant area will be visible on the fabric. Therefore, it can be concluded that it is a must to have all the basic principles or motions of a loom to produce export-quality woven fabrics and achieve customer satisfaction [10, 21].

3 Sustainability Issues of Weaving

3.1 Energy Consumption

Minimising the energy consumption is the primary concern in weaving industries, encompassing both electrical and thermal energy. Energy consumption is calculated by measuring the amount of energy required to manufacture a unit (weight or length) of the fabric, generally expressed like kilowatt hour per kilogram (kWh/kg), Megajoule per kilogram (MJ/kg), and specific energy consumption (SEC) etc. In weaving, electrical energy is utilised in every step of the manufacturing processes, from preparatory process to finishing, primarily for the operation of machinery [22]. Consumption of energy varies depending on the type of the machineries. Moreover, electricity is employed to the compressors that provide compressed air for the processes including winding, warping, sizing, and operating looms. Koç and Çinçik stated the tentative energy requisite (in specific energy consumption) for weaving preparatory machines shown in Table 1 [23].

Furthermore, thermal energy is necessary in sizing process to produce steam from boilers, primarily use the fossil fuels [24]. In addition to manufacturing activities, energy is employed for other tasks, for instances, temperature control, humidification and material handling (Table 2). Figure 3 depicts the percentage share of the energy consumption of a weaving industry [22].

Table 1 Compare between various commercial weaving machines [22]

Weaving machines	Reed width (m)	Speed (rpm)	Weft insertion rate (m/min)	Energy consumption (KWh/m^2)
Shuttle loom	Upto 1.9	150–200	350–450	0.8
Projectile	1.9–6.55	300–450	1200–1500	0.26
Rapier (flexible)	1.2–4.0	350–550	1000–1200	0.52
Rapier (rigid)	1.4–3.6	350–700	1000–1200	0.52
Air jet	1.4–4.3	400–1150	1200–2500	0.65
Water jet	Upto 3.0	800–1000	1500–2000	0.35

Table 2 Energy consumption in weaving preparatory processes [22]

Process	Winding (autoconers)	Direct warping	Indirect warping	Sizing
Electrical energy consumption (kWh/kg)	0.1–0.4	0.1–0.2	0.2–0.3	0.05–0.08
Thermal energy consumption (MJ/kg)	None	None	None	5

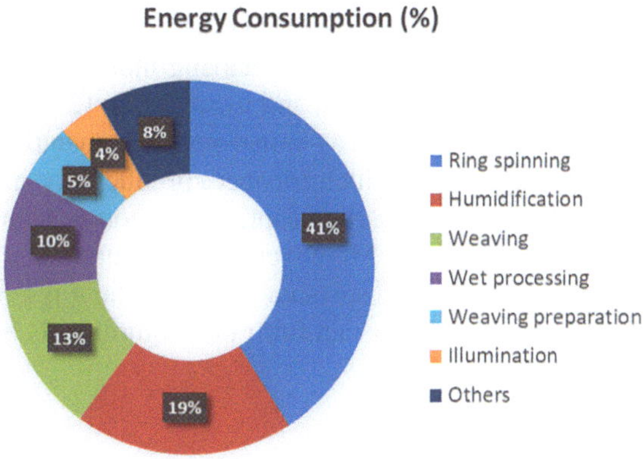

Fig. 3 Share percentage of the energy consumption of a composite weaving industry [22]

3.2 *Carbon Footprint*

In this era of industrial revolution 4.0, carbon footprint is a crucial concern in context of sustainability, where modern machineries are mostly responsible for carbon footprint. In weaving, it requires a series of modern machineries from preparation to product manufacturing. Development and adaptation of energy-efficient looms and machinery minimize energy consumption and consequently can contribute to lowering the carbon footprint in weaving. The technologies need to design considering some factors including minimisation of the waste generation, conserve energy and carbon neutrality etc. Kazan et al. studied the life cycle assessment of cotton woven shirts with four manufacturing techniques, and they analysed the common environmental impact categories [25]. The carbon footprint results for the different sceneries are presented in Table 3.

In another study, the carbon footprint for Denim manufacturing is investigated where one pair of jeans (weighing 570 g) with a medium stonewash is considered as the functional unit [26]. The total carbon footprint is reported to CO_2 53.5 kg equivalent for the whole denim garments manufacturing process, whereas for the weaving process, it is 14.4 kg eq which is 27% of the total.

Initiatives like carbon offsetting, the use of renewable energy, and sustainable logistics can be crucial to achieving carbon neutrality. In addition, sustainable and

Table 3 Carbon footprint for cotton woven shirts (for the unit of 1000 pcs) [25]

Manufacturing techniques	Total (kg CO_2 eq.)	Fabric production (kg CO_2 eq.)	Fabric production (%)
Conventional cotton cultivation	4113	1127	27.40
Organic cotton cultivation	1221	705	57.73
Using natural dyes	3652	666	18.23
Recycled cotton lint	2169	462	21.30

eco-friendly materials, for instances, organic fibres, recycled and biobased fibres can reduce the carbon footprint. Weaving industries generally produce wastes like left over yarn, hard sized yarn and waste fabric [27]. However, implementing measures to reduce greenhouse gas emissions from manufacturing operations, such as optimizing the process, which enhances productivity, improves waste management practices, and invests in carbon capture and storage technologies, can decrease the effect of carbon footprint. Moreover, manufacturers can showcase their commitment to carbon neutrality and sustainability by obtaining certifications like the Carbon Trust Certification and they can fulfil the requirements of international standards such as ISO 14001.

3.3 Water Footprint

Minimisation of the water consumption is another point of interest for the sustainable practice in weaving industries. Despite water being used only in the sizing process, there are some substantial consequences as most of the sizing chemicals contain synthetic materials, and these need to be washed out during the finishing process. As a result, the water with pollutants is transported either to the effluent treatment plant (ETP), increasing the cost or to the freshwater sources and affecting the environment. For this, minimising or eliminating the de-sizing process has the potential to be highly advantageous for the sustainable development of the weaving industries is essential. For sizing, biocompatible eco-friendly chemicals, for instances, starch-based and reactive sizing agents should be employed for sustainable practice. Moreover, the development of the sizing process with less water consumption like plasma, foam, electrostatic, solvent sizing etc., while maintaining the same performance can mitigate the environmental effect.

3.4 Process and Parameter Optimisation

With the development of sustainable materials, it requires the optimisation of the processes and parameters in every stage of weaving. In winding stage, tension setting needs to be optimised. If recycled yarn is utilised in weaving, the yarn tension and other parameters should be adjusted as the properties of the recycled yarn inherently weaker compared to the yarn with virgin fibres. In case of warping, introducing energy efficient process including versatile and interactive movable creel, rapid pattern changes mechanism are necessary. For sizing, different processes and advanced machines are adopted for sustainable practices, for instance, besides starch, biodegradable and reactive sizing agents, one shot sizing agents are used. Additionally, Advanced techniques, including plasma, foam, and solvent sizing, reduce water, energy, and waste. Moreover, industry 4.0 technologies optimize parameters, improving efficiency. Innovations like hot melt sizing and integrated machines save resources, reduce costs, and minimize waste, supporting environmentally friendly and cost-effective processes. Utilising automatic machines for drawing-in, denting, knotting and quick style change can reduce the overall set-up time. Advancements in weaving machinery emphasise user-friendly interfaces, energy efficiency, and sustainability. Innovative technologies reduce raw material waste, while high-speed machines process recycled yarns effectively. Some modern looms incorporate real-time data to enhance machine performance and maintenance. These innovations enhance production efficiency, lower expenses, and facilitate environmental objectives in contemporary weaving techniques. In summary, optimisation of the process and machine parameters underscore the potential in fostering a more sustainable and economically viable weaving industry. Moreover, special weaving technology can ensure near net zero waste in terms of raw materials by overcoming key technological challenges.

3.5 Social Sustainability Issue

Social sustainability issues are growing consumer demand for ethical products. Social sustainability practices, including fair wages, worker safety, and community engagement, align environmental efforts with ethical standards, fostering a more sustainable and responsible weaving industry. Woven fabric manufacturers and brands should prioritise the mentioned social sustainable issues. They should consider the health issues of the workers and their working conditions. For instance, dust and noise pollution are substantial problems on the weaving floor. The sound of the weaving floor is usually above the normal label of 90 dB, and the workers need to stay there for long hours (8 h daily) continuously. Therefore, necessary measures should be taken to mitigate this pollution, including providing sound resistive devices etc. Human toxicity potential (HTP Inf.) is an indicator which implies the use and production of harmful substances and their toxic effects on human body. In a study, it was found that the significant sources of HTP in the cotton shirt manufacturing process are in the fabric manufacturing step, which is 70% of the total process [25].

4 Advancement of Weaving Technologies in the Context of Sustainability

The textile industry is undergoing a significant transition towards sustainability, primarily motivated by increasing demand and the pressing necessity to tackle environmental issues. In response to growing environmental concerns and consumer demand, producers of textile machinery are actively working on creating technologies that focus on sustainability. These technologies aim to minimize production waste, conserve energy and water usage, and produce yarns and textiles using recycled and biobased fibres [28]. Weaving industry is adopting new technologies to contribute to sustainable development and eco-friendly practices with the integration of man, machines, and materials [5].

4.1 Advancement in Weaving Preparatory Processes

Weaving technology generally follows a sequence of manufacturing steps to convert yarn into fabric. In the sample preparation process, traditional weavers used production machines for developing prototypes that create significant waste in raw materials as well as loss of production. Advanced prototyping machine can reduce this loss as well as ensured diversified production of small and medium orders. The key features of the MULTIMATIC 32 Compact include its efficient production capabilities, versatile and interactive movable creel, rapid pattern changes, and minimal setup times. Productivity is enhanced by eliminating the pause for lease and sizing separation. LINK-MATIC reduces material waste and requires one operator. Karl Mayer's warp preparation systems and other machinery are cloud-connected for quick data access and remote support. Knot Master and Warp master plus by Groz-Beckert KG make drawing-in process faster and reduce set-up time. The main goal of sizing is to apply size material to the warp sheet. Selection of the sizing material is very important in terms of environmentally friendly and recyclability approach [29]. Normally starch-based sizing agents have been used for both natural and manmade fibres. Besides this, biodegradable and reactive sizing agents are also used to overcome some limitation on the starch-based sizing agent. Energy saving approaches in sizing process have been taken with the help of advanced technologies like plasma, foam, electrostatic, solvent sizing. Those sizing processes are characterized by less water consumption, lower drying energy consumption, less waste generation and lower processing cost. Also, reactive sizing agent reduced de-sizing process that leading to environment friendly method as well as savings on energy, water, time, and chemicals. One shot sizing chemical development is a time saving approach as it is not required to add native starch, softener, lubricant, anti-foaming agent etc. Besides sizing agent, sizing machines also develop to control the desired quality of sized yarn. The recent advancements in sizing machine technology have been centred around the fourth industrial revolution, also known as Industry 4.0. These developments involve integrating automation, artificial intelligence, and user-friendly interfaces. For example, machines now include features like pneumatic creel loading to control various parameters such as tension, size box level, squeeze roller pressure, stretch control, moisture content of sized yarn, beam pressure, beam tension, and speed of draw roll. Hot melt sizing, pre-wet sizing, high pressure sizing also reduced necessity of sizing cooking and dumping, reduced amount of size chemical and ensured lower energy consumption. Moreover, design of special size box can significantly reduce the amount of ingredients needed, resulting in a 10% decrease in energy consumption during de-sizing [30]. This not only lowers costs but also minimizes the amount of waste generated. Moreover, Integrating the sizing/slashing machine with other operational machines, like dyeing and resin treating, can result in substantial savings in terms of floor space, energy consumption, time, and cost.

4.2 Sustainable Advancement in Fabric Production Stage

Innovation in weaving machines is continuously addressing to make the machine more user friendly, efficient in terms of energy, cost, and increased rate of production. The interface of A2 air jet machine is designed to be user-friendly and high-resolution, allowing for efficient monitoring and quick changeover of woven fabric parameters. Itema has developed the mechatronic device iSAVER® to reduce raw material waste in weaving. This device has been successfully implemented in denim weaving and has now been extended to other fabric types, contributing to sustainability efforts. Three machines at the Itema booth were equipped with the iSAVER® device. This technology eliminates the need for additional selvage yarns and cutters, resulting in a reduction of fringe on the left-hand side of the fabric. The iSAVER® can be used with a maximum of six different filling yarns. The OmniPlus-i Connect-4-P-190 weaving machine efficiently produces bottom weight fabric using 100% recycled warp and weft yarns. The fabric had a fibre content of 67% polyester and 33% cotton, and it was woven at an impressive speed of 1,500 ppm (2,850 m/min RFI). Yarns made from recycled fibres have lower performance and wearability. However, Picanol has shown dedication to the development of high-speed machines capable of handling yarns with short fibres, in support of sustainability efforts. The factory management support system FACT developed by Toyota has been upgraded to FACT-plus. The system gathers real-time data from weaving preparation, weaving machines, and the inspection room, and suggests actions to machines and operators [31]. The system monitors various sensors, including speed, pressure, temperature, and compressor pressure data. It then provides operators with instructions for machine maintenance and their next assignments.

Innovations in sustainable woven textiles concentrate on minimising energy, chemical, and water consumption, preventing waste production, and fostering healthy working environments. Many environmental, social, and economic effects of weaving operations render the advancement of the woven industry essential for global sustainable development initiatives. Innovation in weaving must be founded on the principles of sustainability, as well as environmental, social, and economic opportunities. Additional significant domains include the advancement of bio-derived fabrics and the application of eco-friendly sizing representatives [32]. It is essential to rapidly adopt new technologies and to cultivate collaboration throughout the value chain and across many disciplines [22]. The Table 4 exhibits overall improvement upon advances on sustainable weaving techniques.

Table 4 Analysis on the advantages of sustainable improvement [16]

Features	Online production	Contribution on environmental sustainability
• Minimal setup time	• User-friendly machine interface	• Waste reduction practices
• Eliminating the stop for leasing and sizing separation	• Precise control on process parameters by robots	• Environmental compliance
• Data cloud and remote access for maintaining accuracy and time	• Shortens changeover of fabric parameters	• Size ingredients management and substitutions
• Less waste of raw material (yarn)	• Tune Navigation delivers the best machine setting on loom	• Eco-design and production process
• Auto reverse function on double end detections	• Compatibility of fabric with machines from recycled and bio-based fibres	• Reduction of noise pollution
• Savings on size ingredients by special size box	• Short-liquor sizing process	• Solid waste reduction
• Using renewable and hydropower energy source	• Atmospheric air-plasma for yarn wettability	• Recycling and reusing technology
• Energy efficient	• Lower processing cost	• Resource saving technology
• Elimination on the need of secondary selvedge yarn	• Carbon dioxide-based sizing process reduced water consumption	• Use of natural energy source as well as biodegradable raw materials
• Higher production	• Ensuring higher quality in fabric	• Lower emission of greenhouse gas

5 Near-Net Shape Weaving—A Step Towards Zero Waste Weaving

5.1 Concept of Near Net Shape Weaving

Near-net-shape weaving is an advanced textile manufacturing technique where fabrics are woven as close as possible to their final shape and dimensions, reducing the need for additional cutting, trimming, or sewing. This method minimizes material waste and post-production processes, making it an efficient and sustainable approach, particularly useful for industries requiring precise, shaped textiles, such as aerospace, automotive, medical, and increasingly, fashion.

Zero-waste weaving refers to a sustainable approach in textile manufacturing that aims to eliminate or minimize waste generated during the weaving process. Traditional textile production can result in significant waste, particularly during the cutting and shaping stages when fabrics are made into garments or other products. Zero-waste weaving addresses these inefficiencies through innovative design and production techniques.

(a) (b) (c)

Fig. 4 Development of Pi-shaped composites using 2D to 3D weaving concept: **a** Flat 3D structure in 2D form, **b** unfolded flanges to give 3D shape to the flat structure and **c** final Pi-composite with top and front view [34]

5.2 *Key Technologies*

5.2.1 2D to 3D Weaving

2D to 3D weaving is a customised textile manufacturing technique used to create multidimensional woven structures by extending traditional two-dimensional weaving into three dimensions [33]. This approach is increasingly applied in industries such as aerospace, automotive, medical, and fashion for its ability to produce complex, near-net-shape fabrics with enhanced structural integrity and performance.

The 2D-to-3D weaving methodology involves transforming traditional two-dimensional woven fabrics into complex, three-dimensional shapes by modifying weaving techniques and loom configurations. This approach uses multilayer weaving, spacer fabrics, angle interlocks, and post-weaving forming techniques to create volumetric structures on standard 2D looms, bypassing the need for specialized 3D weaving equipment. For instance, multilayer weaving enables the production of layered fabrics that are interconnected through binding points, adding thickness and structure to the material. Spacer fabrics incorporate pile yarns between layers, allowing for adjustable thickness and cushioning, which is valuable in applications like automotive upholstery and footwear. By carefully programming weaving patterns and adding selective interlocking, the fabric can achieve shape memory and rigidity, providing load-bearing or flexible 3D shapes without cutting and sewing (Fig. 4).

5.2.2 Fully Integrated 3D Weaving

Fully integrated 3D weaving techniques are developed to manufacture 3D woven structures with more design versatility, minimise the material waste and to reduce the

(a) (b)

Fig. 5 Fully integrated 3D loom at Robotics and Textiles Composites Group @ University of Manchester and **a** Stationery multi shed and multi-weft section and **b** 'T' structure in near-net-shape weaving [39]

per unit production time [33]. Fukuta et al. [35] developed an apparatus for manufacturing 3D woven preforms where warp, weft and binders are mutually in orthogonal/perpendicular position to each other to manufacture rectangular or square block 3D shaped cross-sections. Mohamed and Zhang [36] have developed a novel weaving machine with multiple weft insertions simultaneously. This machine produces a unit of non-crimp multi-layered fabric at every weaving cycle. This method removed the involvement of building-up of the fibre layers as the warps are arranged in several layers, and wefts are inserted in multiple layers simultaneously. As a result, a thick, non-crimp multi-layered preform is developed in each cycle of weaving. Potluri et al. [37] have developed another novel technique that allows flexibility in architectures in 3D woven preforms. This novel machine can manufacture all the common 3D weaves architectures such as orthogonal, angle interlock in both through-the-thickness and layer to the layer binding pattern. This machine is also able to produce preforms of various thicknesses as well as near-net-shape preforms 'I', 'T', 'Π', 'H' (Fig. 5) [38].

5.3 *Current Trends on Near-Net Shape Weaving Practices*

5.3.1 Fashion Industry: Seamless and Sustainable Garment Production

Near-net-shape weaving is gaining traction in fashion, particularly in the creation of seamless garments. By weaving fabrics in shapes that closely resemble the final garment, the fashion industry can reduce waste and eliminate much of the labour involved in cutting and sewing.

Fully fashion woven garment has attracted the attention of academia and industry due to their ability to minimise the material waste and labour cost. To date, a few

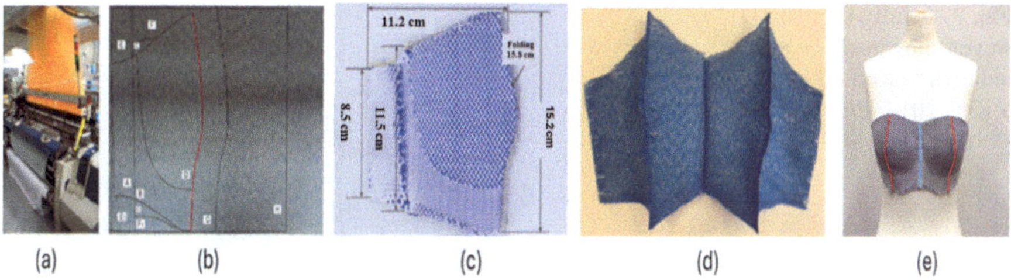

Fig. 6 Development of woven 'Bra' using the novel 3D-to-2D-to-3D: © 2024 The Authors **a** Picanol Optimax rapier loom with Bonas Jacquard; **b** 2D form with segments; **c** The flattened 2D form of bra; **d** Four-layer weaving sample using rapier loom and **e** The final 3D woven bra front block [44]

practice-led research initiatives have reported on the emergence of the integration of weaving and garment manufacturing for fully fashioned woven garment production. This process also remove the need of the cut and sew technique which is liable for up to 40% of material waste during woven garment production [40].

Mcquillan et al. [41] used a digital tool—CLO3D—for pattern construction to reduce the material waste and remove the spreading process. Piper and Townsend [42] developed a concept woven garment which is made on a hand dobby loom and can be worn directly after cutting from the loom. Lefferts [43] developed a novel woven garment on loom where they used heat treatment on synthetic yarns to avoid the stitching after cutting. Shi et al. [44] developed more sustainable technique combining the state-of-the-art 3D weaving and technical weave software which remove the need for any additional sewing or joining. They used an innovative concept named as 3D-to-2D-to-3D process to manufacture a woven 'Bra' where the entire woven garment is developed in two phases: 3D-to-2D geometric process and 2D-to-3D manufacturing process. Phase 1 involves the flattening the 3D scanned geometry to a 2D flat geometric models with shape reference and other weaving parameters. On the next stage, 2D geometries are woven using a Jacquard loom and transmitted back to 3D woven apparel forming the shape targeted for. This novel approach seamlessly weaves 3D woven garments in a single weaving cycle eliminating the need for any further assembly processes (Fig. 6).

5.3.2 Footwear Industry: Custom-Fit and High-Performance Shoes

Footwear manufacturers have integrated near-net-shape weaving to create shoes that conform to the foot's shape with minimal material waste. This approach allows for customization, lightweight design, and efficient production. The virtual Textiles Research Group at Rhode Island School of Design, USA in collaboration with TEAM, USA and Cornell University, USA has developed a series of volumetric 3D woven structures and finally developed a fully formed shoe using the multi-layered 3D weaving technique and custom built 'Weave craft' software [6]. They

Fig. 7 A fully formed 3D woven shoe made on an industrial Jacquard loom (© VTRG 2019. *Photo* Emily Holtzman) [6]

used different types of yarns and their sizes to enhance the aesthetic property of that shoe (Fig. 7). Pinski et al. reported a practice-led research, where they developed a zero-waste and stitch free woven sandal using craft-based 3D weaving method [45].

5.3.3 Lightweight and Complex Structural Composites

The scope of near-net 3D woven structures in composite industrial applications are ever-increasing, thanks to their excellent structural properties. The first 3D woven composite is reported to have been developed by AVCO in the 1970s to replace the metal alloys in aircraft brakes. In 1983 a carbon–carbon sandwich composite was used in some parts of the scramjet engine of a hypersonic missile [46]. Orthogonal 3D woven composites were used in the LEAP (Leading Edge Aviation Propulsion) engine in the A320neo and Boeing 737-Max as fan blades and engine casings (Fig. 8a) to improve the damage tolerance [47]. The General Electric (GE) Company developed airframe components and fan casings for the next generation aero engine. They used a combination of multiaxial orthogonal 3D weave with 2D weaves [48]. The structure offered a high weight reduction (about 450 kg). Orthogonal structures were also used in the engine fan blades in Boeing 787 and Airbus A350 [49]. Mohamed and Wetzel [50] reported the use of orthogonal weaves in wind turbine rotor blades. They have integrated $\prod$ joints to I-beams to connect the spar cap and the shear web to improve the damage property by resisting delamination. Sharp et al. also reported the use of orthogonal weaves in rotor blades [51]. Orthogonal structures are also used in thermal applications. Hemrick reported the use of carbon fibre orthogonal 3D woven structures in the heat exchangers for vehicle radiators [52]. NASA has recently used 3D orthogonal woven composites for thermal protection and pyro-shock resistance in the Orion test vehicle (Fig. 8b) [53].

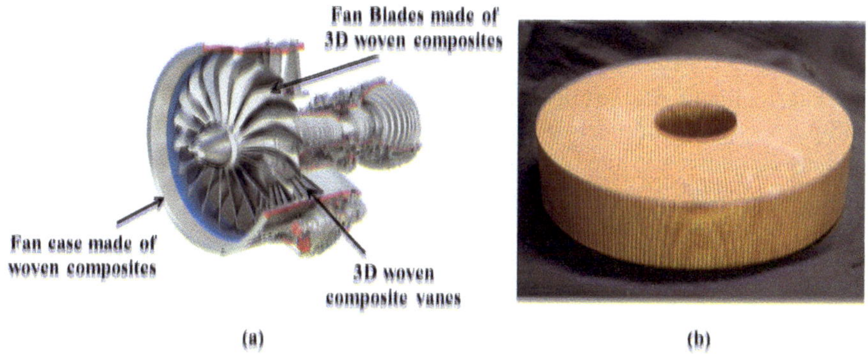

Fig. 8 **a** LEAP engine parts equipped with 3D woven composites [47] and **b** Quartz fibre 3D woven composite used by NASA for Orion test vehicle [53]

5.3.4 Medical and Other High-Performance Textiles

Near-net-shape weaving enables to produce fabrics that are pre-shaped for specific protective applications such as ballistic vests. This process ensures that the textile fibres are optimally aligned for strength, providing a lightweight yet durable structure [54]. 3D weaving is also used in the creation of customized implants and prosthetics, including vascular grafts and tissue scaffolds. It allows textiles to be woven to the specific anatomical shapes required, which is crucial for compatibility and functionality within the human body [55].

5.4 Key Challenges of Near-Net Shape Weaving

Although near-net shape weaving offers a more sustainable textile practice to design and develop woven textiles for versatile uses minimising the material waste and post-processing, it faces several challenges. One key issue is the complexity of designing precise weave patterns that align with the intended three-dimensional geometry, requiring advanced computational tools and expertise. Achieving consistent tension and fabric stability during weaving is also critical, as any deviation can compromise the integrity and accuracy of the final product. Additionally, the integration of different material types or functional properties within a single weave can be difficult to manage without introducing defects. Scalability is another concern, as near net shape weaving often requires highly specialized looms and setups, which may not be feasible for mass production. Finally, addressing waste reduction while maintaining structural and aesthetic quality remains a delicate balance in this innovative approach.

6 Conclusion

This chapter highlights the critical role of weaving technology in advancing sustainability across the environmental, economic, and social dimensions. From the intricate motions of weaving to the adoption of energy-efficient and waste-reducing practices, the industry's journey reflects its adaptability and commitment to sustainable growth. Innovations such as near-net-shape weaving and 3D weaving exemplify how technology can minimize resource use while enhancing product quality and versatility. However, achieving full sustainability in weaving is not without challenges. Complexities in integrating diverse materials, maintaining fabric stability, and scaling advanced techniques underline the need for interdisciplinary collaboration and continued innovation. A major step towards sustainability has been achieved with the incorporation of SDG 12 into weaving technology. Sustainable consumption and production may be further advanced by the textile industry through the adoption of resource-efficient techniques and waste reduction strategies. This will lead to a future that is both economically and environmentally sustainable. Addressing these hurdles requires concerted efforts across the textile value chain, from raw material production to post-consumer waste management. Ultimately, the chapter underscores the imperative for weaving technology to align with global sustainability goals. By embracing advancements in materials, processes, and social responsibility, the weaving industry can not only meet contemporary demands but also set benchmarks for future sustainable practices.

References

1. Y. Kyosev, F. Boussu (eds.) *Advanced Weaving Technology* (Springer International Publishing, Cham, 2022). https://doi.org/10.1007/978-3-030-91515-5
2. C. Popescu, M.D. Stanescu, Eco-friendly processing of wool and sustainable valorization of this natural bioresource. Sustainability **16**(11) (2024). https://doi.org/10.3390/su16114661
3. B.H. Firlus, R. Martin, J.O. Rypestøl, Weaving sustainability: asset modification and green path development in Norway's outdoor textile industry. Nor. Geogr. Tidsskr. - Nor. J. Geogr. **77**(5), 269–280 (2023). https://doi.org/10.1080/00291951.2023.2271494
4. A.E. Asmah, V. Okpattah, H. Dominic, *Improving the Efficiency of the Traditional Loom*, vol. 2, no. 4 (2015)
5. C. Var, S. Palamutcu, Sustainable approaches in textile-sizing process, in *Sustainable Textiles: Production, Processing, Manufacturing & Chemistry* (Springer Nature Switzerland, Cham, 2024), pp. 55–74. https://doi.org/10.1007/978-3-031-51362-6_3
6. C. Harvey et al., Weaving objects: spatial design and functionality of 3D-woven textiles. Leonardo **52**(4), 381–388 (2019). https://doi.org/10.1162/leon_a_01780
7. S. Adanur, *Handbook of Weaving* (CRC Press, 2020). https://doi.org/10.1201/9780429135828
8. M.J. Yun, S.I. Cha, S.H. Seo, H.S. Kim, D.Y. Lee, Insertion of dye-sensitized solar cells in textiles using a conventional weaving process. Sci. Rep. **5**(1), 11022 (2015). https://doi.org/10.1038/srep11022
9. S.T.A. Hamdani, Introduction to weaving, in *Structural Textile Design* (CRC Press, Boca Raton, FL, 2017), pp. 31–46. https://doi.org/10.1201/9781315390406-4

10. F. Boussu, C. Dufour, F. Veyet, M. Lefebvre, Weaving processes for composites manufacture, in *Advances in Composites Manufacturing and Process Design* (Elsevier, 2015), pp. 55–78. https://doi.org/10.1016/b978-1-78242-307-2.00003-8
11. R. Marks, A. Robinson, *Principles of Weaving* (Textile Institute Manchester, UK, 1976)
12. R. Marsden, *Cotton Weaving: Its Development, Principles, and Practice* (G. Bell & Sons)
13. K.L. Gandhi, The fundamentals of weaving technology, in *Woven Textiles* (Elsevier, 2020), pp. 167–270. https://doi.org/10.1016/b978-0-08-102497-3.00005-2
14. P.R. Lord, M.H. Mohamed, *Weaving: Conversion of Yarn to Fabric* (Woodhead Publishing, 1982)
15. S.K. Neogi, *Role of Yarn Tension in Weaving* (WPI Publishing, 2016). https://doi.org/10.1201/b19896
16. M.S. Hoque, L.M. Degenstein, P.I. Dolez, Innovations in woven textiles, in *Sustainable Innovations in the Textile Industry* (Elsevier, 2024), pp. 125–163. https://doi.org/10.1016/b978-0-323-90392-9.00019-7
17. M. Decrette, J.-Y. Drean, Shedding principles and mechanisms, in *Advanced Weaving Technology* (Springer International Publishing, Cham, 2022), pp. 115–166. https://doi.org/10.1007/978-3-030-91515-5_3
18. U.S. Modi, B. Modi, Design and analysis of picking cam for high speed shuttle loom, in *National Conference on Machines and Mechanisms (iNaCoMM2013)* (IIT, Roorkee, India, 2013)
19. H.İ Çelik, M. Topalbekiroğlu, Kinematic analysis and synthesis of the beat-up mechanism for handmade carpet looms. J. Text. Inst. **101**(10), 882–889 (2010). https://doi.org/10.1080/00405000902934570
20. A. Stemheim, P. Grosberg, Weaving with constant beat-up force. J. Text. Inst. **82**(3), 317–323 (1991). https://doi.org/10.1080/00405009108659215
21. R.B. Kabir, Exploration of the advancement in warp & weft stop motion: primitive to electronic system. Eng. Technol. **4**(1), 1–8 (2017)
22. A.K. Pattanayak, 3-Sustainability in fabric manufacturing, in *Sustainable Technologies for Fashion and Textiles*, ed. by R. Nayak (Woodhead Publishing, 2020), pp. 57–72. https://doi.org/10.1016/B978-0-08-102867-4.00003-7
23. E. Koç, E. Çinçik, Analysis of energy consumption in woven fabric productions. Fibres Text. East. Eur. (2010) [Online]. https://api.semanticscholar.org/CorpusID:208805558
24. Md.T. Rahaman, A.D. Pranta, Md.R. Repon, Md.S. Ahmed, T. Islam, Green production and consumption of textiles and apparel: importance, fabrication, challenges and future prospects. J. Open Innov. Technol. Mark. Complex. **10**(2), 100280 (2024). https://doi.org/10.1016/j.joitmc.2024.100280
25. H. Kazan, D. Akgul, A. Kerc, Life cycle assessment of cotton woven shirts and alternative manufacturing techniques. Clean Technol. Environ. Policy **22**(4), 849–864 (2020). https://doi.org/10.1007/s10098-020-01826-x
26. A.P. Periyasamy, G. Duraisamy, Carbon footprint on denim manufacturing, in *Handbook of Ecomaterials*, ed. by L.M.T. Martínez, O.V. Kharissova, B.I. Kharisov (Springer International Publishing, Cham, 2019), pp. 1581–1598. https://doi.org/10.1007/978-3-319-68255-6_112
27. M.Md. Khairul Akter, U.N. Haq, Md.M. Islam, M.A. Uddin, Textile-apparel manufacturing and material waste management in the circular economy: a conceptual model to achieve sustainable development goal (SDG) 12 for Bangladesh. Clean. Environ. Syst. **4**, 100070 (2022). https://doi.org/10.1016/j.cesys.2022.100070
28. S.S. Muthu, Evaluation of sustainability in textile industry, in *Sustainability in the Textile Industry*, ed. by S.S. Muthu (Springer Singapore, Singapore, 2017), pp. 9–15. https://doi.org/10.1007/978-981-10-2639-3_2
29. Z. Kamble, B.K. Behera, Upcycling textile wastes: challenges and innovations. Text. Prog. **53**(2), 65–122 (2021). https://doi.org/10.1080/00405167.2021.1986965
30. S.M.F. Kabir, S. Haque, A mini review on the innovations in sizing of cotton. J. Nat. Fibers **19**(13), 6993–7007 (2022). https://doi.org/10.1080/15440478.2021.1941486

31. R. Davis, Innovation in weaving lights the road to industry 4.0. [Online]. https://www.textileworld.com/textile-world/features/2023/09/innovation-in-weaving-lights-the-road-to-industry-4-0/. Accessed 18 Nov 2024
32. R. Paul, T. Gries, Sustainable innovations in textiles, in *Sustainable Innovations in the Textile Industry* (Elsevier, 2024), pp. 1–9. https://doi.org/10.1016/b978-0-323-90392-9.00013-6
33. K. Sowrov, *Processing and Manufacturing Mechanics of 3D Woven Composites Influenced by Z-Binder Architecture* (UK, 2022)
34. H. Oeij, *Novel Infusion Technique for Optimized 3D Woven Pi Shaped Textile Composites* (UK, 2018)
35. M.S.K. Fukuta, R. Miyashita, J. Sekiguti, Y. Nagatsuka, S. Tsuburaya, E. Aoki, Three-dimensional fabric, and method and loom construction for the production thereof. US3834424A (1973)
36. Z. Mohamed, M.H. Zhang, Method of forming variable cross-sectional shaped three-dimensional fabrics. US5085252A (1990)
37. P. Potluri, D. Jetavat, S. Sharma, Method and apparatus for weaving a three-dimensional fabric. US9598798B2 (2013)
38. M. Matveev, V. Koncherry, S. S. Roy, P. Potluri, A. Long, Novel textile preforming for optimised fibre architectures. https://doi.org/10.1088/1757-899X/406/1/012050
39. P. Potluri et al., Novel manufacturing techniques for optimized 3D multiaxial orthogonal preform, in *22nd International Conference on Composite Materials (ICCM22)* (2019)
40. M. Carvalho, H. Carvalho, L.F. Silva, F. Ferreira, Sewing-room problems and solutions, in *Garment Manufacturing Technology* (Elsevier, 2015), pp. 317–336. https://doi.org/10.1016/b978-1-78242-232-7.00012-6
41. H. McQuillan, Digital 3D design as a tool for augmenting zero-waste fashion design practice. Int. J. Fash. Des. Technol. Educ. **13**(1), 89–100 (2020). https://doi.org/10.1080/17543266.2020.1737248
42. A. Piper, K. Townsend, Crafting the composite garment: the role of hand weaving in digital creation. J. Text. Des. Res. Pract. **3**(1–2), 3–26 (2015). https://doi.org/10.1080/20511787.2015.1127037
43. J. Lefferts, Gestalt process. MA Dissertation, R. Coll. Arts, 2016
44. Y. Shi, L.W. Taylor, A. Kulessa, V. Cheung, A.S.M. Sayem, Re-engineer apparel manufacturing processes with 3D weaving technology for efficient single-step garment production. iScience **27**(8), 110315 (2024). https://doi.org/10.1016/j.isci.2024.110315
45. J. Pinski, M. Evans, F. Kane, Designing through craft practice: a woven textile approach for footwear. Craft. Res. **10**(1), 41–67 (2019). https://doi.org/10.1386/crre.10.1.41_1
46. H. Carlson, *Preliminary Scramjet Design for Hypersonic Airbreathing Missile Application*
47. J. Jewell, R. Kennedy, A. Menard, Full-scale LEAP fan blade-out rig test yields outstanding results; advanced LEAP fan endurance test complete, in *CFM Power Flight* (2011)
48. G. Marsh, Aero engines lose weight thanks to composites. Reinf. Plast. **56**(6), 32–35 (2012). https://doi.org/10.1016/S0034-3617(12)70146-7
49. T. Huang, Y. Wang, G. Wang, Review of the mechanical properties of a 3D woven composite and its applications. Polym.-Plast. Technol. Eng. **57**(8), 740–756 (2017). https://doi.org/10.1080/03602559.2017.1344857
50. M.H. Mohamed, K.K. Wetzel, 3D woven carbon/glass hybrid spar cap for wind turbine rotor blade. J. Sol. Energy Eng. **128**(4), 562–573 (2006). https://doi.org/10.1115/1.2349543
51. K. Sharp, A. Bogdanovich, R. Boyle, J. Brown, D. Mungalov, Wind blade joints based on non-crimp 3D orthogonal woven Pi shaped preforms. Compos. Part Appl. Sci. Manuf. **49**, 9–17 (2013). https://doi.org/10.1016/J.COMPOSITESA.2013.01.012
52. J.G. Hemrick, E. Lara-Curzio, E.R. Loveland, K.W. Sharp, R. Schartow, Woven graphite fiber structures for use in ultra-light weight heat exchangers. Carbon **49**(14), 4820–4829 (2011). https://doi.org/10.1016/J.CARBON.2011.06.094
53. E. Vitug, First 3D woven composite for NASA thermal protection systems (2016)
54. M.K. Bannister, Development and application of advanced textile composites. Proc. Inst. Mech. Eng. J. Mater. Appl. **218**(3), 253–260 (2004). https://doi.org/10.1177/146442070421800310

55. L. Limmer, G. Weissenbach, D. Brown, R. McIlhagger, E. Wallace, The potential of 3-D woven composites exemplified in a composite component for a lower-leg prosthesis. Compos. Part Appl. Sci. Manuf. **27**(4), 271–277 (1996). https://doi.org/10.1016/1359-835x(95)00040-9

Chapter 6
Advanced Knitting Technology for Sustainable Fashion

Dewan Murshed Ahmed, Afsana Mobin, Nadvi Mamun Pritha, Arif Istiak Ador, Safi Ahmed, and Md. Raju Ahmed

Abstract The knitting industry has already emerged as a game-changer for sustainable fashion. This chapter illustrates the advancements made in the field of the knitting sector such as 3D and 4D knitting, seamless knitting, and computerized fully automated knitting that are making significant contributions towards building an eco-friendly sustainable future for the fashion industry. Alongside ensuring sustainability, these advancements are opening new doors to possibilities for the production of smart technical textile products with high-performance functional benefits. Nonetheless, these advancements are associated with significant challenges that are being pointed, discussed and resolved accordingly with a view to fully exploiting these technologies for the intended purpose. By continuing research, collaboration and investments, these advanced knitting technologies can be turned into a more effective solution to a sustainable fashion future while proving the fact that sustainability and innovation can go hand in hand.

Keywords Sustainability · Fashion · 3D knitting · 4D knitting · Whole garment · Automation

1 Introduction

Sustainability is one of the major aspects that need to be ensured by any industry worldwide. Sustainability in environmental and social contexts have drawn serious attention in recent years from textile manufacturers, fashion industries, customers, and scholars [1]. Consumers are increasingly aware of the environmental impact

D. M. Ahmed (✉) · A. Mobin · A. I. Ador · S. Ahmed
Department of Fabric Engineering, Bangladesh University of Textiles, Dhaka, Bangladesh
e-mail: murshed@fe.butex.edu.bd

D. M. Ahmed · Md. R. Ahmed
Materials Engineering, University of Manchester, Manchester, UK

N. M. Pritha
BGMEA University of Fashion and Technology, Dhaka, Bangladesh

A. S. Muhammad Sayem and P. Goswami (eds.), *SDG 12 and Global Fashion Textiles Production*, SDGs and Textiles, https://doi.org/10.1007/978-981-96-9051-0_6

of fashion, demanding transparency and sustainability from brands. Social media platforms like Instagram and TikTok have amplified this movement, with influencers and activists highlighting the importance of sustainable fashion. To meet this demand, brands are looking for traceability solutions, such as blockchain, to allow consumers to track the environmental and social footprint of their garments from production to disposal [2].

Over the past 10 years, the textile market has become much more varied, entering sectors including the military, medical care, sports, fashion, and automotive. This change has had an impact on the knitting industry, which today serves a variety of purposes, mirroring the development of the larger market and the need for creative textile solutions [3]. This industry has been thriving for contributing to sustainable fashion by introducing various innovative features in the manufacturing processes that have greatly enhanced the accuracy as well as the speed of manufacturing with minimal human intervention [4–6]. The advanced knitting technologies do not just represent a technical evolution, they symbolize the potential for a more sustainable, creative, and ethical fashion industry.

This chapter is basically focused on discussing the advanced technologies adopted in both circular and flat knitting industries that promote sustainable fashion. It highlights how these new technologies reduce waste, improve resource efficiency, and enable on-demand production, which eventually minimizes overproduction and reduces the environmental impact of unsold inventory as compared to the conventional methods of knitting production. This chapter also provides an overview of how the integration as well as the potential application of eco-friendly technologies in the knitting industry can support the transition to a more sustainable fashion industry. Lastly, the chapter sheds light on the challenges associated with the technological advances made in knitting production and also the prospective solutions to resolve those challenges alongside discussing the future research opportunities in knitting for ensuring a greener and better fashion outcome.

2 Overview of Knitting Technology

Knitting is a technology that has evolved from a hand-operated completely manual method. Now in the market, there are many fully computerized modern circular and flat-bed weft knitting machines and modernized high-speed warp knitting machines. The fashion industry is taking full advantage of the modern world of flat-bed knitting machines due to their notable technological as well as software advancements [7].

Once a largely static and domestic activity that was only practiced by older women, knitting has expanded to include people from all walks of life in both urban and rural areas [8]. The rise in recreational hand knitting and crocheting, especially among younger city dwellers, has driven these crafts into lucrative marketplaces driven by demand for creative and hobby handicrafts. Today, hand knitting offers therapeutic and medical advantages in addition to being a creative outlet and a means of social and international connection. Its versatility, portability, and ease of use (just yarn and needles are needed) make it possible to knit almost anywhere. Hand-knits are distinctively individual expressions since they are colorful, adjustable, and can be even three-dimensional [9]. When it comes to machine knitting, tubular single jersey, rib, interlock, and fleece fabrics can be produced by using circular knitting machines for certain applications [10–12]. Another largely popular technique for knitted fabric production is flat knitting. Flat knitting is distinguished from circular knitting by its increased process flexibility and product range [13].

3 Advances in Machine Knitting Technology

Knitting is mainly divided into weft knitting and warp knitting on the basis of the method of fabric formation. Weft knitting is employed in the production of fabrics like single jersey, rib, interlock, fleece, terry, etc. whereas warp knitting is used to produce a range of technical textile products and apparel wear fabrics such as lingerie, net, mesh, etc. The knitting machines come in either circular-bed or flat-bed arrangements for the knitting needles. The machines may vary in diameter and needle bed arrangements which make them feasible for the production of fabrics of varying styles and structures. These technologies, by integrating biodegradable and recycled fibers, reducing energy and water consumption, minimizing waste, and ensuring high efficiency contribute to eco-friendly textile production as well as sustainable fashion [14].

As the fashion industry seeks to reduce its environmental footprint, innovative knitting technologies are playing a key role in promoting sustainability. Advances such as 3D knitting, seamless knitting, and digital knitting minimize waste, improve material efficiency, and streamline production. This section explores these cutting-edge technologies and their impact on creating more sustainable fashion, from reducing carbon emissions to optimizing resource use [14].

3.1 Innovations in Circular Knitting for Fashion Applications

The advancements in circular knitting technology are significantly altering the landscape of fabric production, facilitating a transition towards more sustainable mass production practices. The circular knitting machine manufacturers are continuously developing their technologies and making available machines that are specifically engineered for the efficient production of fabrics on a larger scale. Latest developments encompass circular weft-knitted machines that have been developed specifically to manufacture fabric for cut-and-sew shoe applications. Modern circular knitting machines come with computerized control systems featuring quick style change and stitch transfer system [14, 15]. These innovations contribute to a reduction in material waste while simultaneously enhancing efficiency, versatility and precision. The Italian manufacturer Santoni continues to lead in the field of seamless knitting technology, especially in the production of intimate apparel, beachwear, and fine gauge sportswear, all of which necessitate a reduction in seams. It is noteworthy that Santoni has ventured into the production of warp seamless apparel, thereby creating new opportunities for seamless sportswear applications and advancing the frontiers of creative and feasible textile manufacturing [14].

3.2 Precision and Customization Capabilities of Computerized Flat Knitting

Computerized flat knitting machines utilize servomotors, electromagnets, and needle selection mechanisms to perform knitting processes. With the help of sophisticated computer-driven mechanisms, flat knitting machines can control individual needles to produce intricate textures, patterns, and three-dimensional structures with minimal human intervention. The capacity to digitally program designs significantly enhances customization. The design data undergoes preparation on a computer, followed by conversion into designated formats and is subsequently transferred to the machine through storage devices such as USB drives [16]. This data undergoes processing and is subsequently transmitted to facilitate the control of knitting operations. This high degree of digital control makes it possible to manipulate yarns precisely, which makes it easier to produce bespoke clothing or fabrics with different stitch patterns, colors, and structures while preserving production efficiency and reducing waste to an unprecedented amount which is less likely to happen in circular knitting. The latest flat knitting machines come with multiple gauge system, split cam-carriage system and electro-optical tension control system that makes it possible to knit almost fault-free more versatile patterns [15]. Nonetheless, compared to V-bed machines, circular knitting machines have a higher production speed.

3.3 Advances in Warp Knitting for Ensuring Sustainability

Specialized warp knitting techniques such as crochet knitting, circular warp knitting, and multiaxial knitting broaden the application of warp knitting in the production of unique fabric structures having enhanced functionality. Crochet machines, for example, produce lace, ribbons, and technical textiles using electronically controlled guide bars. Circular warp knitting enables seamless and tubular fabric production, making it ideal for hosiery and medical textiles. Multiaxial warp knitting integrates reinforcement yarns in various directions, enhancing fabric strength and stability for technical applications such as composites and protective textiles. By incorporating computerized control and automation, specialized warp knitting continues to evolve, providing innovative solutions for both fashion and industrial applications [15].

3.4 3D (Three Dimensional) and 4D (Four Dimensional) Knitting

Cutting and sewing procedures are responsible for approximately 50% of waste production in traditional knitwear cut-and-sew designs [17]. In order to create affordable and sustainable apparel, the invention of 3D and 4D knitting have been essential in cutting waste, transportation-related emissions, and post-knitting processes [18–21]. The cut and sew steps are removed from these knitting technologies and only the necessary amount of yarn is needed to knit. End bits and cut wastes (tiny portions of cloth outside patterns) are among the many wastes produced in the cutting room during the cut and sew process which are removed in these technologies [21]. Moreover, by facilitating on-demand production and minimizing overproduction and surplus inventories, these two technologies enhance sustainability [22–24]. They streamline manufacturing by doing away with several production phases and reducing the need for manual labor in sewing and assembly, leading to more efficient use of human resource and waste reduction from unsold stock [21]. This efficiency makes them more environmentally friendly options than traditional clothing manufacturing since they cut energy usage and the overall carbon footprint [23, 24]. Without sacrificing functionality, these initiatives encourage sustainable fashion [25].

3D knitting can be done by utilizing either flat bed or circular knitting machines. Initially, the target surface is interpreted into a digital pattern and subsequently converted into the knitting structure. This pattern is predominantly generated in flat bed machines through the activation or deactivation of needles via an electronic needle selector. The process also involves controlling sinker height, managing back bed racking movement, modifying stitch cams, employing partial knitting technology [26, 27]. Appropriate programming of computer systems can efficiently convert yarn into a three-dimensional structure in a short time. Using loose and tight patterns, darts are incorporated into the design process to produce 3D volume from a 2D fabric. These textiles can be utilized in automobile clothing, medical equipment the production of composites, filtration systems and composite manufacturing [28, 29]. Additionally, 3D knitting enhances sustainability in footwear production by optimizing efficiency and reducing waste. Optimization of 3D-knitted footwear uppers can enhance sustainability by reducing production time by 45% (from 21:42 to 11:53 min) and minimizing waste while increasing material consumption by 12–15%, making it ecologically and economically viable [30].

The 4D warp knitting technology utilizes a double needle bar Raschel frame. This technology brings together an advanced needle bar setup with a piezo-jacquard mechanism, allowing for the creation of intricate, reversible patterns and spacer fabrics. This unique approach not only adds interesting textures and functional benefits but also opens up new possibilities for design and performance in warp-knitted [15].

3.5 Seamless Knitting

'Complete garment knitting' or 'seamless knitting' represents a specific category of 3D knitting, characterized by the production of an entire garment as a singular entity, thereby eliminating the need for conventional cutting and sewing processes. Seamless garments are created by broadening or narrowing a piece of cloth using loop transferring method to increase or decrease the number of wales. The process of seamless garment knitting involves the utilization of multiple feeders to produce a fully formed garment, significantly eliminating the need for cutting and sewing operations thus making it less dependent on human labor [29, 31]. The manufacturing process of seamless garments requires 30–40% less time compared to the traditional lab intensive cut and sew process [32]. One-piece clothing manufacturing also improves the garment's durability along with reducing fabric wastage which eventually leads to less environmental impact [33]. Furthermore, seamless clothing has fewer weak spots and is more resilient, which extends its lifespan and lowers the need for replacements while advancing sustainability.

Digital Stitch Control System (DSCA) implemented on the Shima Seiki (Japan) machine accurately determines the yarn required for each stitch, thereby minimizing yarn wastage. Additional benefits of seamless knitting include mass customisation, just-in-time production, improved trim-ability for finished edge lines, and more consistent product quality. Compared to conventional V-bed flat knitting machines, Santoni® circular knitting machines use less energy and make clothing with fewer seams [29, 34, 35]. These devices have quicker turnaround times, which increases their sustainability and efficiency [36]. Therefore, because of its advantages for sustainability and speed of production, seamless clothing, especially undergarments like bras and briefs, are anticipated to experience substantial market expansion [37, 38].

3.6 Digital Knitting and Design Automation

Digital knitting has revolutionized production by improving precision and sustainability. With CAD and advanced knitting software, complex patterns can be produced efficiently and with minimal waste [39]. Digital knitting enables seamless garment production, which reduces material waste by up to 30% compared to traditional methods [40]. This aligns with sustainable fashion goals and helps reduce the environmental impact of garment manufacturing [41].

The creation of knitting structures and patterns is the main emphasis of computer graphics in the knitting business, namely CAD/CAM systems. CAD means Computer Aided Design and CAM means Computer Aided Manufacturing [42], that is a project assisted by a computer. Fully-fashion (FF) knitting is made possible by the latter, in which machines produce finished goods or fabric cuts without chopping raw materials. This minimizes waste and streamlines operations. By incorporating shaping directly into the knitting process, FF knitting, which was inspired by hand knitting, improves efficiency and makes production more resource-efficient and sustainable [48, 49]. CAD promotes sustainability by simplifying design procedures, reducing material waste, minimizing prototypes, optimizing fabric usage, and increasing production efficiency. By enhancing communication, expediting product improvements, and doing away with manual labor, it makes textile production more efficient, flexible, and sensitive to shifting consumer needs through automation, seamless CAD-CAM integration [43–45].

4 Sustainability Aspects of Advanced Knitting Technologies

4.1 Eco-design

Eco-design approaches aim for minimizing negative environmental impacts of any product right from the beginning to end of its life [46]. This applies to knitting machines, or knitting technology and software, as well as knitted items [47]. The degree of overlap and congruence between these two spheres predicts the relative success or failure of ecological design. Functional and technical textiles are rapidly evolving as a result of the combination of eco-design and technology in textiles, especially in knitting. To rethink eco-design concepts, this fusion foster collaboration between technologists, fashion designers, and industrial designers. Knitting is a prime example of this innovation due to its versatility in design [48]. A variety of materials, such as natural fibers and recycled yarns, are used to create fully fashioned knitwear. This method lowers energy use and gets rid of cutting waste [49, 50]. Applying eco-design principles in knitting combined with proper selection of materials, i.e., recycled yarns, improves fabric performance and endorses sustainability.

4.2 On-Demand Production and Customization

Prior research on mass customization in the fashion sector has primarily examined how consumers feel about the modification process and why they choose to purchase customized goods [51–53]. Intelligent manufacturing is now the primary focus of the textile industry's current strategic transformation, thanks to the use and advancement of intelligent scheduling theory and robotics. Modern knitting machines that have computerized systems, such Shima Seiki's WHOLEGARMENT, enable highly flexible designs and rapid production adjustments. This feature facilitates on-demand manufacturing, which reduces unsold inventory and overproduction of necessary goods by producing goods in response to customer demand rather than in excess [54]. In order to assure effective yarn utilization, minimize material waste, and enable manufacturers to maximize production, advanced knitting technologies include computerized stitch patterning and accurate yarn tension control [55, 56]. This improves resource efficiency and reduces the possibility of overproduction.

By dynamically matching output with real sales data, advanced knitting machines with IoT (Internet of Things) and real-time analytics allow for real-time production monitoring and demand-based modifications, eliminating overproduction. IoT is transforming the textile industry by advancing efficiency, quality, and sustainability. It makes possible a data-centric strategy that maximizes a number of textile production process elements stated that quality control, inventory management, real-time process monitoring, and machine predictive maintenance are examples of IoT uses in the textile industry [57]. BMSvision, a Belgian company known for its production monitoring systems, developed a solution named Knit Master, which helps with monitoring and controlling knitting mill equipment in addition to managing yarn supplies [58]. The "Inteos Networker" is a web-based knitting machine monitoring tool offered by Memminger-IRO. Using controller sensors for yarn feeding, needle breakage, and machine speed, this can keep an eye on output and performance [59]. To improve performance evaluation, the cutting-edge online knitting machine monitoring system combines machine learning and the Internet of Things. This solution greatly helps knitting mills by enabling real-time productivity tracking and preventive maintenance, which results in increased operational efficiency, lower costs, and consistently high-quality output [59]. High scalability is provided by modern knitting systems, which effectively handle both small and big production runs. Because of this flexibility, producers can produce targeted orders in batches, lowering the risk of overproduction and excess inventory [60, 61]. Just-in-time (JIT) production is made possible by advanced knitting technologies, which reduce stocking and do away with the need for vast stocks.

4.3 Increased Productivity

Due to the rise of technical textiles and a move away from excessive consumerism, the knitting industry is changing and placing an increasing emphasis on sustainability and cross-industry collaboration. This trend is being driven by industries such as automotive, aerospace, sports, and medicine. Multipurpose, seamless knitwear from both circular and V-bed knitting machines is being promoted by increased communication among designers, yarn developers, machine manufacturers, and users [33, 62, 63]. This technology's primary benefits include reduced waste, increased productivity, time and money savings, comfort, and ideal fit [66]. Increasing productivity, using new technologies like seamless knitting, or combining techniques like hybrid spin-knit and sock toe-closing can all help to speed up the production of knitted goods. Seamless technology reduces the need for the conventional production process and can directly produce almost finished goods and garments without side seams [63–65]. The production process is optimized by doing away with the fabric cutting and sewing step, which results in seamless production that is significantly faster than traditional ones [66]. Consequently, when compared to traditional knitting, this technology can cut production costs by up to 40% [33].

The rapid response carriage return mechanism, also known as the R2CARRIAGE System, offers lighter and thinner carriages combined with advanced software that facilitates faster knitting speeds, which has improved the productivity of Shima Seiki knitting machines [67]. Mayer & Cie (Germany) believe that new machines with upgraded bearded needles may knit more quickly than those with latch needles [11, 68]. The bearded needle's reinvention emphasizes the importance of restoring ancient concepts and substantially contributes to the knitting industry's sustainability. Most problems in high-fashion production were resolved when a new integrated design process for computerized seamless V-bed knitting systems was created. Additionally, it shortens the time to market, lowers design expenses, and increases the range of creative options available for designing high-end knitwear [67, 69]. The incorporation of robotics into flat knitting and hybrid spinning-knitting technology, which lowers expenses and resources, are important innovations. Unconventional materials like foils and rigid yarns are being worked with by machines that have been modified [33].

4.4 Energy and Resource Efficiency

Traditional knitting methods, despite their longstanding presence in the textile industry, raise several sustainability concerns. Environmental issues such as excessive material waste, high energy consumption, and pollution have been historically associated with conventional knitting techniques. Additionally, many workers involved in traditional knitting processes face poor working conditions and low wages, posing a significant challenge to social sustainability. However, advancements in knitting technologies are beginning to address these issues, offering a path toward more environmentally and socially responsible manufacturing processes. For instance, technologies like 3D knitting and seamless knitting have significantly reduced material waste, improved energy efficiency, and even opened avenues for on-demand production, which lowers the need for excessive inventories and thus reduces pressure on factory workers [69].

The integration of circular economy principles in knitting technologies is revolutionizing sustainability. Notably, companies like Adidas are incorporating recycled polyester and establishing closed-loop systems for product end-of-life recycling, further enhancing the sustainability of the entire production cycle. Waterless dyeing and the reduction of chemical treatments are also pushing the boundaries of eco-friendly practices in the knitting industry [70, 71].

5 Case Studies of Sustainable Knitting Applications Adopted by Major Brands

Recent years have witnessed significant advancements in the footwear and fashion industries, particularly in terms of advanced knitting techniques. In the footwear industry, major brands such as Nike and Adidas have led this transformation by introducing knit-based footwear in their manufacturing processes. Among these processes, Nike's Flyknit technology and Adidas' Primeknit technology have revolutionized the footwear industry in terms of performance, comfortability, sustainability, and mass-manufacturing. Nike first introduced Flyknit technology in 2012. Unlike conventional methods where shoes are made by attaching multiple pieces of materials glued or stitched together, Nike developed a technique that uses one single continuous stitching to produce a shoe (see Fig. 1), which makes it lightweight and provides less in-shoe movement and more breathability [72]. To mimic the human foot shape, the shoe upper is manufactured by a complex knitting structure, which then undergoes a thermoforming process. Using a single knitting process to produce a custom-shaped shoe upper results in 60% less waste compared to conventional cutting-sewing methods [73].

Adidas has introduced Primeknit technology, which, like Nike's Flyknit, uses seamless knitting technology to produce shoe uppers. It helps to reduce waste in cutting and sewing and also promotes sustainability by adding recycled polyester to the production system [75].

Given the high price of their products, luxury fashion brands often overlook the importance of sustainable production processes. However, a fashion brand named Stella McCartney set an industry standard for sustainable production. In 2019, they collaborated with Adidas to combine fashion with function [76]. They leveraged seamless knitting to produce sports garments that use 'Lycra Fitsense+', which offers lightweight, breathable, and wicking properties. Over 70% of the fabric in Adidas x Stella McCartney clothing in The Sports Edit is recycled from plastic. 85% of Adidas footwear styles in The Sports Edit in 2021 are PRIMEBLUE (performance fabric

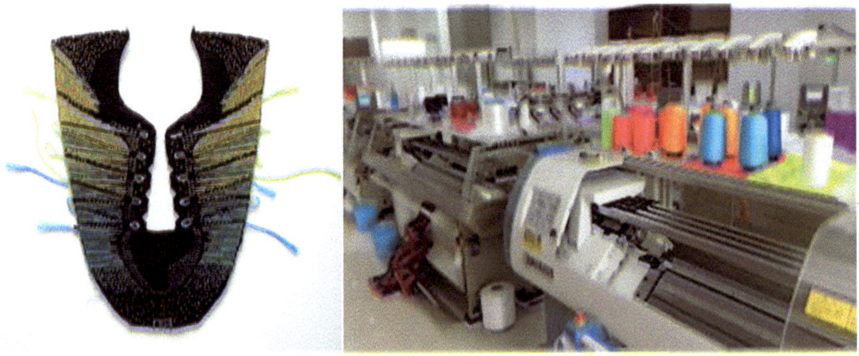

Fig. 1 4D Knit Flyknit™ shoe construction [74]

with Parley Ocean Plastic) or PRIMEGREEN (performance fabric with zero virgin plastic, incorporating a high proportion of sustainable content). Alongside Parley Ocean Plastic, the footwear selection utilizes uppers from the same source in the selection, and the goal is to use only recycled polyester in products by 2024. The 'Made to be Remade' process is a closed-loop, so the products don't end up in the ocean, and also don't end up in the ground. Instead, the raw materials are remade into high-performing sporting wear. Products created through the use of the process are starting to turn up in the selection, such as the Infinite Hoodie, and there are many, many more to come. The research into biodegradable fabric is in its infancy but serves to illustrate the desire to push the limits of what's possible in terms of sustainability to the extreme [78].

Both Nike Flyknit and Adidas Primeknit have set new industry standards for sustainable knitting applications. The key sustainability benefits of these technologies include:

- Significant waste reduction: Seamless knitting eliminates excess fabric scraps.
- Enhanced efficiency: Fewer materials and reduced production steps lead to lower energy consumption.
- Eco-friendly materials: Recycled polyester and other sustainable fibers minimize reliance on virgin materials.
- Lightweight and durable designs: Ensuring both performance and longevity, reducing product turnover.

The success of the Flyknit and Primeknit lines has also seen the broader use of seamless knitting technology in the fashion world, in apparel, accessories, and sportswear. Stella McCartney, Uniqlo, and Lululemon are among the fashion houses and apparel manufacturers who are starting to integrate cutting-edge knitting techniques into product lines to produce high-performing, zero-waste clothing. The uses showcase how eco-friendly knitting is spreading from the world of footwear to mainstream fashion, providing a scalable and eco-friendly solution to the old textile-making ways [77].

While current Flyknits and Primeknits are predominantly manmade fibers, including nylon and polyester, the future also offers the promise of biodegradable fibers from plants and bioengineered proteins, including bamboo, hemp, and bioengineered fibers. Researchers are also developing fully biodegradable knit-based footwear, further advancing circular economy models in the footwear industry. Other emerging developments, including 3D knits and bioengineered fibers in conjunction with organic dye, could shift the world of eco-friendly fashion to 100% biodegradable knitted products [74].

6 Challenges Encountered in Advanced Knitting

The innovations as well as the additions made to the knitting technologies are accompanied by a number of challenges that should be addressed and resolved accordingly with a view to making sure that the benefits of these additions outnumber the drawbacks so that these can be fully exploited to ensure a sustainable production outcome.

6.1 Technical and Economic Challenges

Modern knitting machines incorporate advanced functionalities such as automated pattern generation, integrated three-dimensional knitting capabilities, precise control over stitch density, tension regulation, and yarn-feeding mechanisms. As shown in Fig. 2, the integration of a sophisticated servo motor system, along with sensors and a computerized control interface, significantly improves the operational capabilities of the machine. However, these advanced techniques come with a high initial investment and ongoing operational costs, such as regular servicing, software upgrades, and machine downtime management [35]. A significant number of manufacturers appear unprepared to confront this economic challenge.

Advanced knitting technologies incorporate computerized interfaces, automated controls, and continuous tracking systems, necessitating employees who possess proficiency in both programming and machine operation. Furthermore, it is essential for technicians to possess the necessary expertise to address problems in real

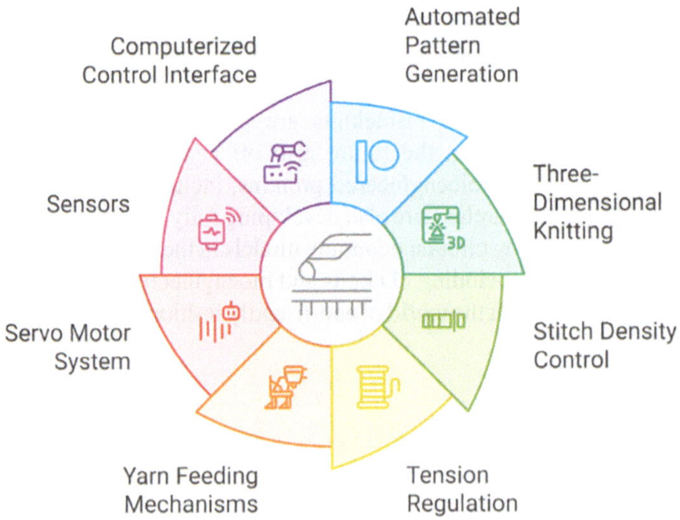

Fig. 2 Features of modern knitting machines

time. As these technologies develop further, there is a growing need for people with interdisciplinary expertise in mechanical systems, software engineering, and textiles; therefore, ongoing training and upskilling are essential to their deployment.

6.2 *Industry Adoption Barriers*

The optimization of traditional knitting methods and processes has evolved over time, focusing on enhancing cost effectiveness and productivity. Workers of all ages, however, do not react identically in embracing new technology. The adoption of new technology is more prevalent among younger workers than among senior workers [78]. In order to effectively implement new technology, decision-makers must adopt an entrepreneurial mindset rather than a managerial one [79]. The implementation of any new technology requires sustainable modifications in the processing unit, such as rearranging production lines, upgrading equipment, and transforming operational protocols.

According to the profitability model, an industry usually implements a significant delay before adopting new technology to mitigate the danger of embracing unproductive innovations. The model of technological uncertainty posits that for an industry to achieve success, it must balance the opportunity cost of adopting new technology against the potential costs incurred from selecting an inappropriate technology [80]. So, it is essential for manufacturers to strategically allocate their resources and time towards the upgrading of their workforce, ensuring that this initiative occurs with minimal disruption to current production activities. The transition from small-scale operations to full-scale industrial production of advanced knitting technology necessitates significant customization within the production unit. The introduction of new technology presents challenges in sustaining high-volume production environments, where the maintenance of consistency, speed, and cost control emerges as critical factors.

7 Conclusion

The fashion industry is at a turning point, and the need for change has never been more important. With its significant impact on the environment- through waste, pollution, and resource consumption the industry must find new, innovative ways to create fashion without compromising health and environment. The good news is that the advanced knitting technologies, like 3D knitting, seamless knitting, and computerized flat knitting, are leading the charge towards a more sustainable future. These technologies help reduce fabric waste, use resources more efficiently, and enable on-demand production, addressing many of the industry's most pressing challenges, including overproduction and unsold inventory.

The potential for knitting technologies to reshape fashion manufacturing is enormous, not only from an environmental perspective but also as a business opportunity that can meet the evolving expectations of today's mindful consumer. Looking ahead, the continued adoption of these advanced knitting techniques will play a vital role in making fashion more sustainable. By embracing these technologies, the industry has a chance to create a future where fashion is both beautiful and responsible. It's about reimagining how clothes are made, consumed, and disposed of, making fashion a positive force for change, while also ensuring that the planet's resources are respected for generations to come.

References

1. J. Kim, C.R. Taylor, K.H. Kim, K.H. Lee, Measures of perceived sustainability. J. Glob. Scholars Market. Sci. **25**, 182–193 (2015). https://doi.org/10.1080/21639159.2015.1015473
2. TEM-20-0621.R1
3. E.J. Power, Advanced knitting technologies for high-performance apparel, in *High-Performance Apparel* (Elsevier, 2018), pp. 113–127
4. GmbH KMT STOLL knitting machines KARL MAYER
5. WHOLEGARMENT Patents. https://www.shimaseiki.com/wholegarment/patents/. Accessed 17 Nov 2024
6. D.J. Spencer, *Knitting Technology: A Comprehensive Handbook and Practical Guide* (Woodhead, Technomic, 2001)
7. ER J (2008) Developments in apparel knitting technology. Adv. Apparel Prod. **178**
8. S.J. Vachhani, (Re)creating objects from the past—affect, tactility and everyday creativity. Manag. Organ. Hist. **8**, 91–104 (2013). https://doi.org/10.1080/17449359.2013.750052
9. A. Pavko-Čuden, Multiple faces of contemporary hand knitting, in *IOP Conference Series: Materials Science and Engineering*, vol. 254 (2017), p. 192014. https://doi.org/10.1088/1757-899X/254/19/192014
10. Mayer & Cie, *Machine Study for Weft Knitting*
11. ITMA, Mayer & Cie focuses on sport, regeneration and new customer experiences—Mayer & Cie, Rundstrickmaschinen und Flechtmaschinen aus Albstadt (2019)
12. Double Jersey/Interlock/Rib Circular Knitting Machine-REL-TEX
13. H. Hu, Z. Wang, S. Liu, Development of auxetic fabrics using flat knitting technology. Text. Res. J. **81**, 1493–1502 (2011). https://doi.org/10.1177/0040517511404594
14. J. Fernando, G.-N. Agustín (eds.), *C-C SDGs and Textiles Textile Industry and the SDGs Exploring Synergies for a Better Future*
15. K. Fayez, A. Mobin, D.M. Ahmed, *Advanced Technology in Fabric Manufacturing* (2023), pp. 65–95
16. C. Zhang, X. Wu, C. Zhang, Control data preparation of computerize flat knitting machine, in *Advanced Materials Research* (2013), pp. 422–427
17. S.I. Tanvir, T. Mahmood, Solid waste for knit fabric: quantification and ratio analysis. J. Environ. Earth Sci. **4**, 68–80 (2014)
18. S. Maiti, S. Maity, P. Pandit et al., Sustainability analysis for knitting process and products, in *Advanced Knitting Technology* (Elsevier, 2022), pp. 657–671
19. E.J. Power, Sustainable developments in knitting. Int. J. Bus. Glob. **9**, 1–11 (2012). https://doi.org/10.1504/IJBG.2012.047519
20. A. Morandi, C. Tonelli, The impact of 3D technology on sustainability in fashion luxury product design: evidences from an exploratory analysis
21. *How Is 3D Knitting Technology More Sustainable in Fashion Business? Online Clothing Study* (2024)

22. Seemsay|*Find 3D Clothing Models and Fabric & Trim Supplier Here*
23. A. Henderson, *The Future of Garment Production: 3D Knitting Fashion Takes Action* (2023)
24. E.I. Stefan, M. Blaga, M. Penciuc, *Digital Solutions for Education in Sustainable Knitting* (Bucharest, RO, 2021), pp. 233–240
25. C. Tyler, A. Gorea, Eco-conscious design process: natural dyed knitwear with innovative 3D motifs, in *Making Waves Toward a Sustainable and Equitable Future* (Iowa State University Digital Press, 2025)
26. C. Peiner, H. Löcken, L. Reinsch et al., 3D knitted preforms using large circular weft knitting machines. Appl. Compos. Mater. **29**, 273–288 (2022). https://doi.org/10.1007/s10443-021-09956-1
27. H. Hong, A.A. Filho, R. Fangueiro, M.D. De Araujo, The development of 3D shaped knitted fabrics for technical purposes on a flat knitting machine (1994)
28. B.K. Dejene, M.A. Abtew, A.A. Birlie, M.E. Woldeab, Three-dimensional (3D) knitted spacer textile materials for advanced healthcare solutions: a comprehensive review. J. Ind. Text. **54** (2024). https://doi.org/10.1177/15280837241290169
29. W. Choi, N.B. Powell, Three dimensional seamless garment knitting on V-bed flat knitting machines. **4** (2005)
30. J.J. Hinojo Pérez, J. Valdés-Martínez, B. Juan-Fernandez et al., Innovative and efficient three-dimensional design strategies for optimizing the industrial production of knitted footwear. Processes **12**, 2698 (2024). https://doi.org/10.3390/pr12122698
31. L. Lu, G. Jiang, G. Wu, The knitting methods for seamless garments based on four-needle bed computerized flat machine. Text. Res. J. **92**, 479–497 (2022). https://doi.org/10.1177/00405175211035139
32. J. Sarkar, N.M. Rifat, Md. Sakib-Uz-Zaman et al., *Advanced Technology in Apparel Manufacturing* (2023), pp. 177–231
33. S. Maity, K. Singha, P. Pandit, Production of seamless knitted apparels, in *Advanced Knitting Technology* (Elsevier, 2022), pp. 203–218
34. S.G. Hayes, P. Venkatraman, *Materials and Technology for Sportswear and Performance Apparel* (CRC Press, Boca Raton, FL, 2016)
35. K. Sayer, J. Wilson, S. Challis, Seamless knitwear—the design skills gap. Des. J. **9**, 39–51 (2006). https://doi.org/10.2752/146069206789377113
36. A. Gorea, F. Baytar, E.A. Sanders, Challenges and design opportunities in prototyping seamless knitted apparel: a case study. Int. J. Fash. Des. Technol. Educ. **14**, 127–138 (2021)
37. D. Atalie, Application of digital CAD in WHOLE-GARMENT knitting manufacturing: a review. Int. J. Sci. Technol. (STECH) AFRREV **9** (2020)
38. F. Lau, W. Yu, Seamless knitting of intimate apparel, in *Advances in Women's Intimate Apparel Technology* (Elsevier, 2016), pp. 55–68
39. K. Buss, *Big Book of Knitting* (Sterling Publishing Company, Inc., 2001)
40. J. Hiatt, *The Principles of Knitting* (Simon and Schuster, 2012)
41. Mayer & Cie, Rundstrickmaschinen und Flechtmaschinen aus Albstadt—Schwäbische Ingenieurskunst seit (1905)
42. A. Dwivedi, A. Dwivedi, Role of computer and automation in design and manufacturing for mechanical and textile industries: CAD/CAM. Int. J. Innov. Technol. Explor. Eng. (IJITEE) **3**, 8 (2013)
43. E. Zaharieva-Stoyanova, *Project of CAD Module for Knitting Pattern Design* (2009), pp. 22–24
44. A. Kaspar, L. Makatura, W. Matusik, Knitting skeletons: a computer-aided design tool for shaping and patterning of knitted garments, in *Proceedings of the 32nd Annual ACM Symposium on User Interface Software and Technology* (ACM, New Orleans, LA, USA, 2019), pp. 53–65
45. N. Sun, B. Cao, X. Mu, The transformative role of computer-aided design technology in innovation in the apparel industry. Appl. Math. Nonlinear Sci. **9**, 20242230 (2024). https://doi.org/10.2478/amns-2024-2230
46. S. Walker, S. Ryn, S. Cowan, Ecological design. Des. Issues **13**, 87 (2007). https://doi.org/10.2307/1511945
47. P. Ghezzo, *Ecodesign in the Textile Sector*

48. A. Pavko Čuden, Knitting towards sustainability, circular economy and Industry 4.0. Appl. Res. **2**, e202200087 (2023). https://doi.org/10.1002/appl.202200087
49. N. Venkatasamy, *A Sustainable Design Practice in Fully Fashioned Knitwear* (2023)
50. Why is recycled polyester considered a sustainable textile? OEcotextiles (2009)
51. P.V. Ulrich, L. Jo Anderson-Connell, W. Wu, Consumer co-design of apparel for mass customization. J. Fash. Market. Manag. Int. J. **7**, 398–412 (2003). https://doi.org/10.1108/13612020310496985
52. C. Huffman, B.E. Kahn, Variety for sale: mass customization or mass confusion? J. Retail. **74**, 491–513 (1998). https://doi.org/10.1016/S0022-4359(99)80105-5
53. K. Sülük, S. Coskun, Environmental risk analysis: an energy transmission line installation activity, in *Academic Research and Reviews in Engineering Sciences*, ed. by B. Kılıç (Platanus Publishing®, Ankara, 2023), pp. 169–179. ISBN: 978-625-6517-61-5
54. WHOLEGARMENT SHIMA SEIKI Computerized Flat Knitting Machines, Design Systems, CAD/CAM Systems
55. K. Fletcher, *Sustainable Fashion and Textiles: Design Journeys* (2012), pp. 1–237. https://doi.org/10.4324/9781849772778
56. SHIMA SEIKI Computerized Flat Knitting Machines, Design System/Software, CAD/CAM Systems
57. M. Bhanushali, Application of IOT in textile industry, *Textile Magazine, Textile News, Apparel News, Fashion News* (2020)
58. Company TBR, *Use of IoT and 3D Knitting in the Knitted Fabrics Industry—Latest Global Market Insights* (2019)
59. S. Elkateb, A. Métwalli, A. Shendy, *An Innovative Online Monitoring System in Knitting Industry* (2023)
60. V. Bhardwaj, A. Fairhurst, Fast fashion: response to changes in the fashion industry. Int Rev Retail Distrib Consum Res **20**, 165–173 (2010). https://doi.org/10.1080/09593960903498300
61. A. De La Haye, *The Handbook of Fashion Studies: International Perspectives*, 1st edn. (Bloomsbury, London; New York, 2013)
62. A. Pavko Čuden, Recent developments in knitting technology, in *Advanced Knitting Technology* (Elsevier, 2022), pp. 13–66
63. C. Viorica, Seamless technology on circular knitting machines
64. S. Yang, *Computerized Seamless V-bed Knitting Systems*
65. N. Nawaz, R. Nayak, Seamless garments, in *Garment Manufacturing Technology* (Elsevier, 2015), pp. 373–383
66. K. Foster, C. Istook, *3D Knitting Technology: A Decision-Making Model* (Ames, 2017)
67. H. Davis, Shima Seiki to exhibit at Techtextil. Knit. Trade J. (2019)
68. E. Team, Mayer & Cie takes positive stock of ITMA 2019. Knitting Views (2019)
69. M. Gamage, N. Amarasinghe, M. Kgh et al., Occupational health and safety in the textile industry. J. Res. Technol. Eng. **4**, 127–140 (2023). https://doi.org/10.13140/RG.2.2.15821.61921
70. M. O'Connell, *The Quadruple Helix in Fashion Paradigms: Bridging Innovation, Sustainability, and Societal Impact for a Regenerative Future* (Seneca Polytechnic). https://doi.org/10.5281/zenodo.14820251
71. T. Rahaman, S.H. Khan, Green merchandising of textiles and apparel in a circular economy: recent trends, framework, challenges and future prospects towards sustainability. J. Open Innov. Technol. Mark. Complex. **11**, 100457 (2025). https://doi.org/10.1016/J.JOITMC.2024.100457
72. C. Fink, *Nike: Sustainability and Innovation through Flyknit Technology* (Research Scholar, NYU Stern Center for Sustainable Business, New York, NY, USA, 2016)
73. Nike Launches Next Generation of Flyknit for Footwear—NIKE, Inc. https://about.nike.com/en/newsroom/releases/next-generation-flyknit-footwear-official-images. Accessed 22 Oct 2024
74. 4D Knitting Flyknit^{TM} Shoe Construction: How Shoes Are Made. https://shoemakersacademy.com/4d-knitting-flyknit-shoe-construction/?srsltid=AfmBOopkQjlTWjHVmHT2-DYFry7pxNEoUhwPVCtPHsYFViO-K-VxRyZT. Accessed 24 Feb 2025

75. Glow in the Dark 4D Running Shoe: The Ultra4D. https://news.adidas.com/running/glow-in-the-dark-4d-running-shoe--the-ultra4d/s/7a73fc3e-acc9-457d-b79a-81c7baa5de0d. Accessed 22 Oct 2024
76. Change Creators: The New Adidas by Stella McCartney Campaign | Stella McCartney US. https://www.stellamccartney.com/us/en/stellas-world/change-creators-the-new-adidas-by-stella-mccartney-campaign.html. Accessed 23 Oct 2024
77. Adidas x Stella McCartney Sustainability | The Sports Edit EU. https://eu.thesportsedit.com/pages/adidas-stella-mccartney-sustainability. Accessed 24 Feb 2025
78. M.G. Morris, *Age Differences in Technology Adoption Decisions: Implications for a Changing Work Force* (2000)
79. Y.-C. Gagnon, J.-M. Toulouse, *The Behavior of Business Managers When Adopting New Technologies* (1996)
80. U. Doraszelski, Innovations, improvements, and the optimal adoption of new technologies. J. Econ. Dyn. Control **28**, 1461–1480 (2004). https://doi.org/10.1016/S0165-1889(03)00112-X

Chapter 7
Green Colouration Processes Targeting Sustainability

Eshita Datta, Md. Razib Sheikh, Mohammad Mahmudur Rahman Khan, and Sadia Mehrin

Abstract Conventional textile colouration processes are often harmful for the environment. Mainly because of the use of excess water, synthetic dyes, and toxic chemicals. These affect human health and aquatic life badly. Moreover, traditional dyeing processes consume high energy which contributes to carbon footprint of textile production. Increasing awareness among consumers and stricter environmental standards imposed by the regulatory bodies demand the use of sustainable eco-friendly colouration processes. Methods including digital printing, spray dyeing, low liquor ratio dyeing, supercritical fluid dyeing, biological colouration are investigated, developed, and used to implement more sustainable practices in textile industries. These methods show promising performance in reducing water and energy consumption, lowering the use of toxic chemicals, and minimizing negative impact on the environment.

Keywords Green colouration · Digital printing · Spray dyeing · Low liquor ratio dyeing · Supercritical fluid dyeing · Biological colouration

1 Introduction

Various methods of colouration process have been in use since ancient times to make textiles attractive [1]. However, traditional methods of textile colouration, particularly the widespread use of synthetic dyes, are often harmful to the environment [2]. These methods usually consume a lot of water, which is wasteful and contributes to lowering underground water levels in many areas [3]. Additionally, the use of toxic chemicals in textile dyeing pollutes water, harming aquatic life and human health [4]. The production and disposal of synthetic dyes generate significant amounts of waste, including hazardous by-products that are difficult to manage. This waste contributes to soil and air pollution, further exacerbating environmental degradation

E. Datta (✉) · Md. Razib Sheikh · M. Mahmudur Rahman Khan · S. Mehrin
Department of Textile Engineering, Bangladesh University of Business and Technology (BUBT), Dhaka, Bangladesh
e-mail: eshita@bubt.edu.bd

A. S. Muhammad Sayem and P. Goswami (eds.), *SDG 12 and Global Fashion Textiles Production*, SDGs and Textiles, https://doi.org/10.1007/978-981-96-9051-0_7

[5]. Furthermore, the energy-intensive nature of traditional dyeing processes adds to the carbon footprint of textile production, contributing to climate change [6]. These problems with traditional dyeing methods make it difficult to maintain a sustainable environment.

As awareness of environmental issues grows, the textile industry is under increasing pressure to adopt more sustainable practices. Consumers are becoming more conscious of the environmental impact of their purchases, demanding eco-friendly products [7, 8]. Regulatory bodies are also imposing stricter environmental standards on textile manufacturers [9]. In response, the industry is exploring innovative solutions to reduce its ecological footprint while still delivering the vibrant, high-quality colours that consumers expect [10].

This chapter highlights innovative, eco-friendly methods that aim to reduce the ecological footprint of textile production. These methods, including digital printing, spray dyeing, low liquor ratio dyeing, supercritical fluid dyeing, and biological colouration, represent a shift towards more sustainable practices in the industry. By understanding and implementing these sustainable colouration processes, we can contribute to a cleaner and greener future.

2 Textile Colouration Processes and Their Environmental Impact

Textile industry is one of the largest industries which includes spinning, weaving or knitting, textile colouration, apparel manufacturing etc. The colouration process of textiles has a significant impact on the environment (Fig. 1). In textile colouration process huge amount of fresh water is essential. The textile industry's dyeing and finishing processes consume significant resources, using up to 150 L of water per kilogram of fabric and accounting for an average of 2.58 kWh/kg of energy. It also relies on approximately 8,000 synthetic chemicals for dyeing processes [11]. The industry contributes to environmental pollution with an average carbon footprint of 1.64 kg CO_2 per kilogram of fabric [12]. Dyeing and finishing procedures are significant sources of water pollution and greenhouse gas emissions, underscoring the necessity of sustainable practices during these phases [13], contributing between 56.3% and 85.6% of the effects from chemical consumption. Pretreatment comes in second, contributing 13.8–28.3% [14]. It is stated that the World Bank estimates textile dyeing and finishing treatment produce 17–20% of total global effluent to the environment [15, 16]. Additionally, dyeing process of textiles have generated more concern as many hazardous chemicals and synthetic dyes are used in this process. These toxic elements causes skin and respiratory irritation on human health [17]. For 50,000 m fabric 1–2 million litre water is required for regular production and this produces enormous amount of wastewater which is harmful for the human health as well as whole eco system [18]. Textile dyeing industries uses different types of chromophore containing dyes and pigments like azo, xanthene, anthraquinone etc.

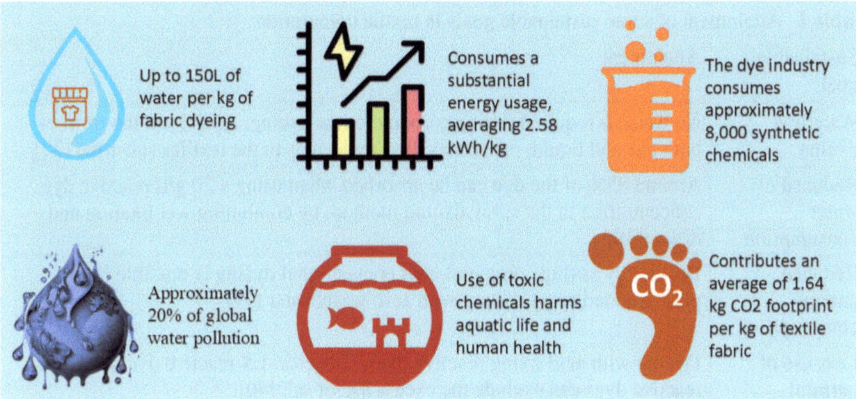

Fig. 1 Impact of conventional textile colouration process [11, 12]

which has complex aromatic structures. These colourants are cancer-causing and have adverse impression on the aquatic network in a long run [19–21]. Besides these colourants, various dyeing auxiliaries, colour fixing agents, fabric softeners etc. are used which affect the water parameters. These nonbiodegradable chemicals increase turbidity of water which hinders the photosynthesis process.

The European Environment Agency reported that 9 m^3 water and 391 kg raw materials are required for per person on average which causes 270 kg carbon footprint. They also reported that textile sector is the third largest source of water degradation in 2020. It is also mentioned that for making a single cotton T-shirt, 2700 litters of clean water are required. 700,000 microplastic fibres can be discharged by single washing of polyester clothes [22]. In a research work, carbon, water and energy footprint analysis of dyed cotton fabric in China is conducted and it is found that 49.1% carbon footprint is reduced by waterless dyeing processes [14]. In a work the carbon footprints of some disperse dyes are calculated, such as Disperse Blue 79, Disperse Red 167 etc. [23]. For global climate heating several aspects of textile colouration like energy consumption, use of chemicals, treatment time, material and liquor ratio are responsible [12]. Hence dyeing with less water, application of ultrasonic energy, super critical fluid, employment of natural colourants can be the alternative which is comparatively less harmful and reduce the environmental contamination [24–26].

3 Eco-Friendly Colouration Processes

In the early 1990s, people started to take the harmful effects of textile manufacturing seriously [27, 28]. As awareness grew about the environmental impact of water pollution, toxic chemicals, and high energy use in traditional dyeing, the textile industry began adopting eco-friendly dyeing methods to promote cleaner production. Eco-friendly colouration processes refer to the minimization of environmental

Table 1 Attainment of some sustainable goals in textile colouration

Sustainable goal	Attainment
Waterless dyeing	No water is required during carbon dioxide dyeing. As CO_2 in the form of both gas and liquid, penetrates dyes molecules to the textiles substrate [37]
Reduced of water consumption	Around 85% of the dye can be absorbed when using a 20 g/L reactive dye concentration in the spray dyeing method, by combining wet fixation and drying[38]
Reduced energy consumption	Big energy saving compared with conventional dyeing is possible by plasma-aided spray dyeing with zero wastewater [39]
Less use of harmful chemicals	Dyeing with acid fixing reactive dyes, Cibacron-LS-reactive dyes, cationic reactive dyes can exclude the excess use of salt [40]
Minimum wastes	One-step one-bath dyeing of acrylic, wool, nylon, silk, acrylic/wool (50:50) with cationic reactive dyes ensure minimum waste water footprint to the environment [41]
Faster colouration	Digital printing offers faster colouration of textile surface compared to traditional dyeing process [42]
Material and energy saving	Mixture of minimum water with pressurized air in the coloration of denim processing saves 75% water, 47% energy, 80% salt, and 50% chemical consumptions. [43]

impact through reduction or exclusion the uses of excess water, elevated temperature, energy, toxic chemicals in textile colouration processes. [29–31]. Thus, these techniques reduce water pollution as well as decrease the need for harmful chemicals, ensuring the stable and balanced aquatic ecosystem and human health [32]. The examples of eco-friendly colouration processes for textiles are dyeing with natural or biodegradable colourants, waterless dyeing, digital printing, spray dyeing, biological colouration, enzyme-based colour fixation, UV and LED curing are to fulfil the aim of eco-friendly colouration [30, 33]. Foam dyeing is shown as better than conventional pad dyeing system regarding production cost and sustainability [34]. These green techniques fulfil the growing demand for environmental sustainability by lowering the impact on the environment [35, 36]. Table 1 states some goals regarding green colouration and their attainment.

3.1 *Digital Printing*

Digital textile printing is currently the most advanced printing technique [44]. It involves creating a digital image using graphic design software, which is then directly printed onto various substrates such as textile fabrics, paper, plastics, and other materials. This modern method, also known as ink-jet printing, offers precision and versatility across a wide range of applications [45]. Digital textile printing has revolutionized the fabric and fashion industry by offering a flexible, efficient, and creative approach to printing on fabrics with huge number of colours [46, 47]. Although the first patent for a digital printing system was obtained by Lord Kelvin in 1867, the technology was applied to textiles in the early 1970s. In the 1980s, a digital laser version of the electrophotography transfer system for garments, especially T-shirts and other accessories, was invented [48]. This modern method of printing uses digital technology to apply designs directly onto textiles, providing numerous advantages over traditional printing methods like screen printing.

3.1.1 Sublimation Digital Printing

Sublimation printing is a type of digital printing where the design is first printed on sublimation paper and then transferred onto polyester fabrics using high heat and pressure [49]. This process is typically done by printing the design on paper, placing it on the fabric, and feeding it through a heat press for short period of time, where the high temperature and pressure transfer the ink from the paper to the fabric [50, 51].

A design developer creates a colourful design based on the buyer's requirements using graphic software like Adobe, with the CMYK colour model as the base [52]. The design is then printed as a mirrored image on specially coated sublimation paper. The print is produced as the print head moves back and forth across the paper, layering the ink over multiple passes [53]. In the heat transfer machine, the paper and fabric are heated and pressed between two large rollers. This process turns the solid dye into gas without going through a liquid state [54]. The heat and pressure make the dye penetrate deep into the fabric, forming a permanent colour bond. When the heat is removed, the dye solidifies, completing the process. Because the colours are embedded within the fabric, the images will not fade or crack, even after many washes [54, 55]. Sublimation printing is the popular method in the textile coloration industries, known for its vibrant colours, long-lasting results, and precise details [45].

3.1.2 Direct Digital Printing

Direct digital printing of fabric is a new and innovative process that involves printing a design, a pattern or an image directly from the computer onto the fabric by way of large format digital printing machine and inkjet printer. This type of digital printing involves the direct application of colour onto flexible materials, such as textile fabrics, to create various designs [56]. Direct-to-Fabric (DTF) printing uses ink-jet technology and ink cartridges which is almost identical to colour printing on paper. A number of print heads pass back and forth across the flexible textile fabric injecting microdroplets of colours onto it [47, 48, 57]. Specific colours are applied only to the designated areas as required [58]. No screen preparation or grooved cylindrical roller preparation is needed in this type of printing, which saves the cost and time associated with preparing screens or grooved rollers. Moreover, unlike screen or rotary printing, colour waste can be minimized, fulfilling the sustainable goals of textile colouration. With this method it is possible to print most detailed artwork and design at reduced amount of materials and processing time [59, 60]. By this process water and energy consumption can be reduced and thus 1 kg carbon dioxide emission per kg fabric can be reduced compared to the traditional dyeing [47, 61–63]. This technology is also suitable for producing high-quality prints on a wide range of fabrics, including cotton, silk, polyester, nylon and blends [47, 50, 64]. As a result, this technique has gained significant popularity.

Before the direct type of digital printing process, it is necessary to pretreat the fabric with liquid solutions [46]. Generally, cotton and viscose fabrics are inkjet printed with reactive dye [65]. During this method, thickener, alkali, and urea are used to prepare the fabric to accept the dye and ink and better absorb the colour [57]. The profile is chosen according to the fabric and RIP (Raster Image Processing) the design to get it ready for printing [46, 66]. With the help of graphic design software, digital copy of printing design is created and the digital copy is taken in a suitable format which is compatible to the printer [46]. The pretreated fabric is then fed into the direct digital printer for printing. The fabric is placed on a sticky printer belt. Digital textile printing uses inkjet technology and ink cartridges, where the machine sprays ink or dye with tiny droplets as directed by the design and software [58]. A series of print heads pass back and forth across the fabric, spraying dyestuff onto it [45]. A printer driver, using a digital version of the design as a reference, controls which colour is printed and when the printed fabric is dried by the heater attached to the digital printer. The printed fabric undergoes steaming, washing, drying, and ironing [67]. This is the most sensitive process. Each fabric requires a different length and quantity of steam depending on several variables, including fabric type and fabric length. Digital textile printing is highly flexible, environmentally friendly, and offers high-quality printing [68]. Depending on the type of fabric, various inks are used. Pigment ink is used for cotton fabrics, acid dyes work primarily with nylon and silk, reactive dyes provide the best results on cotton and other viscose materials, and disperse dyes are almost exclusively used with polyester. Direct digital printing is widely used in various industries, including fashion and apparel, home textiles, promotional products, and specialty fabrics. It offers versatility, vibrant designs, and durability, making it ideal for custom clothing, home decor, branded items, and technical textiles [56].

3.2 Spray Dyeing

Spray dyeing technology applies capillary forces and natural inherent absorbency of textile fibres to get deeper penetration [69, 70]. The spray dyeing machine comprises a closed chamber where the dye dispersion is allocated through spray valves. The revolutionary part of this dyeing machine is the spray cassettes which consists of spray nozzles through which dyes, chemicals and auxiliaries are applied in vapor form in a controlled, precise way [70]. Nozzles sprinkle the accurate quantity of dye particles and required finishing auxiliaries directly onto the surface of the fabric. Therefore, in spray dyeing technology, the liquor and the chemical consumption is significantly lower than the conventional dyeing processes. Moreover, the process is highly efficient because of having the facility of digital controlling system [69].

Spray dyeing method is applicable in jet-type overflow dyeing machine which is generally used in exhaust dyeing process with a reduced liquor ratio to lessen the total water, chemical and utility consumption. Spray dyeing system includes a dye fixation unit downstream of the dye spray station and minimum one rinse station downstream of the dye fixation unit. A spreader facilitates the opening of the fabric to keep the fabric straight and flat [71]. Spray dyeing technology includes continuous dyeing method that comprises of spontaneous movement of the fabric according to machine direction, removal of any folds or crease by the help of spreader, preparation of the dye chemical solution; spraying the dye solution on the fabric surface, exposing the dyed fabric surface to atmospheric steam, conducting dye fixation treatment, removal of the unfixed dye from the dyed fibre and finally drying the dyed substrate[38, 71].

Spray dyeing system is considered as the fruitful one for the processing of the textile fabrics due to the easy installation, utilization of sustainable chemical solution as well as retaining actual fabric structure and properties due to less chemical reaction. Spray process does not require the urea in the liquor as like conventional pad-batch or pad-dry-bake dye operation. Moreover, salt which is the unavoidable chemical for the pad-steam or pad-dry-chemical process, can easily be eliminated in this sustainable process. Besides, combined action of wet fixation and drying treatment for a high dye fixation rate in a short time is applied in this system that saves dye fixation time and reduce chemical, water and energy consumption compared to that of the pad-steam-dry system [38].

The spray dyeing method is a continuous hydraulic atomizing system which sprays required chemical solution in the fabric through an atomizer. Because of using less water during processing, consumption of chemical is low. As a result, total usage of utility for the dyeing production unit is lessened. Moreover, amount of wastewater discharge as well as wastewater treatment process also gets reduced. However, inherent property of treated dyed substrate does not get affected due to having no chemical or physical interaction [72].

Various companies are trying to develop the spray dyeing technology. Among them, DyeMax, Alchemie and Imogo are the remarkable ones. These companies are coming forward to mitigate the negative impact of the environment and promote the sustainability for future generation. But these emerging technologies has not been yet developed to the extent level properly because of the lack of interest of the business community as these processes involves limitations such as high installation cost to setup new machines and equipment, low capacity batch process, limitations of some textile fibres, economic viability, arranging different training sessions for professionals and manpower etc. [72, 73]. However, to improve the quality of the dyed fabrics, dyestuffs, chemical, auxiliaries as well as to develop the dyeing technology is highly needed to facilitate the sustainable methods. Planning for the theoretical framework and supporting arrangements should be developed for less-water or non-aqueous dyeing technologies [74].

3.3 Super Critical Fluid Dyeing

Textile dyeing industry consumes huge amount of water for washing, rinsing, pretreatment, dyeing and finishing process and large scale of water contamination also produced by this industry which includes various types of unused colourants, surface active agents etc. [75, 76]. So, there is always a demand of developing new technology to colour textile material without using water as a media. Super critical carbon dioxide or super critical fluid dyeing has unlocked the new entrance to waterless dyeing [76, 77]. It can be a remarkable innovation for textile manufacturing which increases productivity, reduce energy consumption, decreases effluent and also saves water. It is harmless, economical, nonexplosive and for application it requires some critical conditions as seen in Fig. 2, like critical temperature (TC) 31.1 °C and critical pressure (PC) 7.38 MPa [78–80].

This method was first introduced in Germany, 1988 by Professor Schollmeyer and his research group as a sustainable medium for textile dyeing, which was an alternative of conventional textile dyeing with lots of advantages [82]. Several groups like Josef Jasper (Germany), Ciba-Geigy (Basle, Switzerland), Praxair (USA), Kherson Cotton Pape (USA), DTNW (Krefeld, Germany), Batelle National Institute (USA), Weatherly (USA), Obem (Germany), Separex (France), Chematur Ltd (U.K.), of United States and Europe are exploring SC-CO_2 processing in textile finishing industry [83, 84]. For several years effective pilot plant technologies have been exhibited at ITMA 1995 (Milan, Italy), Otemas 1997 (Osaka, Japan) [85].

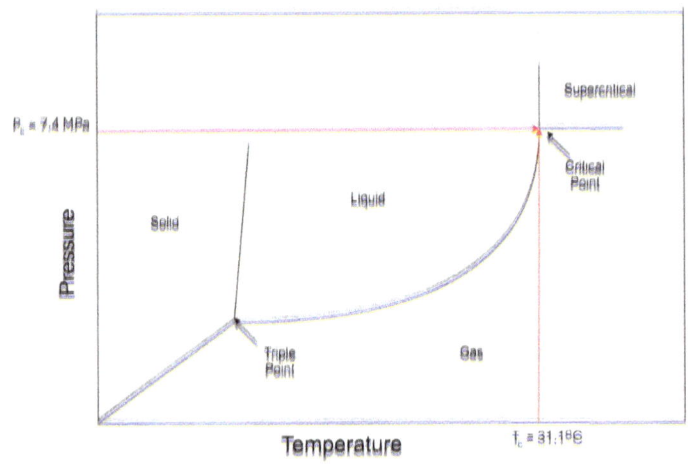

Fig. 2 The phase diagram of CO_2 [81]

Super Critical Fluids (SCF) are the substances which remain neither at vapor state nor at liquid state above their critical temperature and pressure. CO_2 is the most widely used SCF as it has lower critical condition which is convenient than other SCFs. Moreover, it has lots of benefits like economically feasible, non-irritating, non-flammable, chemically inert etc. [81, 86].

The viability and commercial efficiency of super critical carbon dioxide in dyeing of polyester fibres has been studied in a pilot plant. It is reported that the categories of dyes influence the diffusion of disperse dyes. Several scientists explored the solubility models of various dye and their incorporation with the fibres which will help to improve the SC-CO_2 dyeing process [75, 87–89]. Generally natural fibres like cotton, viscose, wool and silk has difficulty to dye by this method as carbon dioxide is incapable to break hydrogen bond present in these type of fibres which hinders the diffusion of dyestuff, but this problem can be overcome by using modified CO_2 soluble disperse dyes [90, 91]. In current years, many scientists continue their research on the application of SCF dyeing on natural fibres also. Sawada et al. (2002) stated that water soluble dyes can be easily solubilized in super critical carbon dioxide by reverse miceller system. In this work it is found that wool and silk fibre can be acceptably dyed by this method [92].

A novel pilot plant was designed by a group of researchers for polyester fabric rope dyeing in SC-CO_2 method and they reported that about 20% energy can be minimized by this method than conventional dyeing process. In this work Disperse scarlet S-BWFL dye was used and they proved that the designed plant can produce commercially acceptable fabric in this method [82]. Dyeing of polyester/cotton blended fabric in one bath by SCF dyeing process was also compared with thermosol dyeing process. It was found that, the colour strength and fastness properties were excellent for SCF dyeing [93]. Another pilot experiment was designed to assess the K/S value and colourfastness of dyed polypropylene fabric with azo based disperse dyes by super critical fluid dyeing method [94]. Zheng et al. [95] reported that wool fibre can be dyed by some disperse dye with special structures in super critical fluid dyeing method [95]. Guzel and Akgerman (2000) also stated that dyed wool fibres with mordant dyes by SCF process showed excellent fastness properties [96]. Some researchers proposed curcumin-based dyes to dye cotton, wool and silk fibre by SCF process [97]. SCF dyeing technology is a great alternative to conventional water based dyeing technology. Table 2 shows various approaches to apply SCF dyeing technology in textile dyeing and their impacts on colour yield and colourfastness properties. More investigations are required on natural fibre and solubility of dyes in super critical carbon dioxide for increasing commercial acceptance.

Table 2 Various approaches to apply SCF dyeing technology and their impacts

Dyes used in SCF dyeing	Material used	Optimum condition	Findings
Hydrazonopropanenitrile dyes (Disperse)	100% polyester woven fabric	The optimum dyeing temperature and pressure are 120 °C and 15 MPa, and the optimum dye concentration ranges from 2 to 6% on the weight of fabric	No wastewater is produced compared to the conventional dyeing, The colour strength and fastness properties, antibacterial efficiency against G +ve and G −ve bacteria were excellent [98]
C.I. Disperse orange 30	Polyester fabric	Optimum condition of supercritical carbon dioxide at 95 °C and 300 bar	The requirement for the diffusion of disperse dyes within the fibre is at temperature below the required temperature for dyeing in conventional method, and it does not require any washing or drying [99]
C.I. Disperse blue 14 and CI disperse orange 25	Polyester fibres	Dye uptake increased at 125 °C and 230 bar	Fibers having larger crystallites showed greater dye uptake and better sublimation fastness [100]
Disperse orange 30	PET fibres	Optimum condition of experiment was 373 K and 20 MPa	SCO_2 method is suitable for polyester dyeing [101]
Several thiadiazolyl azo dyes	Polyester fibres	Better results showed at 80 °C and 3500 psi	Excellent colourfastness and dye uptake were obtained in SCF dyeing [102]

(continued)

Table 2 (continued)

Dyes used in SCF dyeing	Material used	Optimum condition	Findings
Kayalon polyester orange R-SF200, Kayaset red 802 and Kayaset blue N	Polyester and polypropylene composite fibres	Dyeing was done at varying temperatures ranges from 60 °C to 120° C, at 250 bar for 60 min	Pressure impacts the dye uptake [103]
Disperse blue 79	Polylactide fibres	The pressure and temperature were kept 20 MPa and 100–120 °C	Not only good wash fastness was achieved but also it showed that breaking strength of PLA fibre decreases during conventional dyeing than SCF dyeing [104]
Disperse–reactive yellow dye and CI disperse yellow 3	Nylon 66 fabric	The pressure and temperature were kept 400 bar and 200 °C	Disperse-reactive dyestuff shows better colour yield and colourfastness [105]
C.I. Disperse red 60, C.I. Disperse yellow 114, and C.I. Disperse blue 79	Scoured meta-aramid fabrics	The dyed samples were treated for 10–90 min keeping dyeing temperatures and pressures, ranging from 80 to 160 °C, 18 to 34 MPa	Dyed meta-aramid gives good colour fastness and colour depth with carrier [106]
C.I. Disperse blue 3, C.I. Disperse red 13, and C.I. Disperse blue 56, C.I. Natural yellow 27, C.I. Natural yellow 27	Bleached and unfinished cotton fabric treated with PEG 400 and Benzamide	Fabric is treated at 98°C for 30 min under a 200 ± 250 bar pressure	Dyed fabric gives good result of dye uptake, wet fastness, and light fastness [107]
APAN [1-(4aminophenylazo)-2-naphthol] and DY82 (CI disperse yellow 82)	Modified cotton fabric with benzoyl chloride and sodium benzoylthioglycollate	100 °C and 300 bar pressure is used	Colour yields were high [108]

(continued)

Table 2 (continued)

Dyes used in SCF dyeing	Material used	Optimum condition	Findings
Fluorotriazine reactive dyes and chlorotriazine dyes	Mercerized cotton fabric pre-soaked in a protic solvent prior dyeing and cosolvents was applied during dyeing	Use SC CO_2 at 300 bars at different temperatures	Fluorotriazine dyes were more reactive with cotton and methanol was found to be the best cosolvent for dyeing cotton in supercritical carbon dioxide [109]
Disperse red 54, disperse blue 183, disperse red 153, disperse blue 148, disperse red 92, and disperse yellow 82 are used	Pretreated wool fibre	Optimum dyeing condition for wool fibres in supercritical carbon dioxide is 110 °C and pressure 25 MPa	Higher K/S values and fixation rates were attained with Disperse Red 153 and Disperse Blue 148, but the breaking strength and tensile strength of the dyed wool fibres were reduced slightly after SC CO_2 dyeing [95]
CI reactive red 2, CI reactive blue 74, CI acid orange 7, CI acid blue 9, CI acid red 52 and CI acid green 25	Unmodified wool, silk, and cotton fibre	–	Wool and silk fabrics can successfully be dyed by reverse miceller system, but cotton fabric needs pretreatment to improve dyeability [92]
Reactive disperse blue and reactive disperse red	Cotton fabric is used and Tetraethylene glycol dimethylether (TEGDME) and N-methyl-2-pyrrolidinone (NMP) are used as the solvents for pretreatment	Optimum dyeing temperature and pressure are about 393 K and 20 MPa	Colour strength and colourfastness of dyed fabric is acceptable [110]

(continued)

Table 2 (continued)

Dyes used in SCF dyeing	Material used	Optimum condition	Findings
Bi-acyl fluoride reactive disperse dye	Cotton fabric without using any co-solvent for pretreatment	Dyeing temperature was kept from 60 to 120 °C and time between 1 and 3 h	Dyed fabric possesses good colourfastness [111]
Azo based reactive disperse dye	Cotton and wool fibres	90°C and 140 bar were kept with 60 min dyeing time	Cotton and wool fibres both shows good colour strength and fixation properties [112]
CI disperse red 74	Ramie fibre treated with benzoyl Chloride	Optimum dyeing condition was 403.15 K and 20 MPa	Colour yield and wash fastness were excellent but new method of pretreatment is required for ramie fibre [113]
CI basic blue 47	Acrylic fibres	–	Dye uptake was high than conventional dyeing [114]
Disperse azo dyes	Nylon 6	Dyeing was done at 120°c 15 MPa for 60 min	Shows excellent colour strength, fastness properties and antimicrobial properties [115]

3.4 Foam Application Technology

Foam technology is a type of low wet pick-up method in textile finishing industry. Various finishing agents, dyes etc. can be applied by foam technology using very less amount of water. Compared to conventional dyeing process this technology has some benefits such as reducing dyeing time, increasing production capacity, saving energy consumption, producing less effluent. This technique is not just efficient but also sustainable than traditional dyeing. Foam technology was first introduced in industry in 1979, and the first effective trial was by Dawson in 1981. Figures 3 and 4 show the process sequence and process arrangements respectively.

Usually, dispersion foam is suitable for textile application. For application it is required to maintain steady and uniform bubble size, and level spreading of foam on textile substrate.

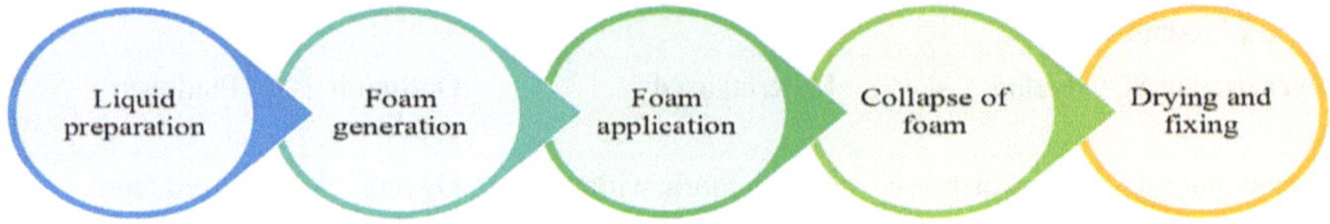

Fig. 3 Process sequence of foam application on textile substrate [116]

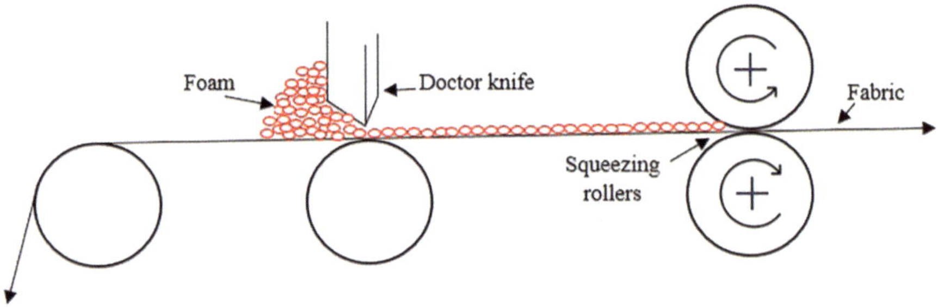

Fig. 4 A schematic diagram of foam application process [117]

Moreover, foam must destruct rapidly to diffuse dye into textile substrate[81, 116]. In a research work it is reported that the dyed fabric by foam technology shows better colourfastness to washing and cracking than conventional one [118]. Foam dyeing can be an effective method for polyamide filament which is reported in a work. Filaments can achieve better colour fastness with higher colour yield [119]. Many attempts were undertaken to compare the various aspects like water and energy consumption, amount of chemicals required, K/S value. colourfastness properties for cotton fabric using various reactive dyes and pigments by foam dyeing method and conventional dyeing method. It is also reported that foam dyeing method can reduce 80% cost of energy required for drying as well as 30–90% water consumption leading to the 50–60% reduction in the treatment of effluent [116].

3.5 Low Liquor Ratio Dyeing

The batch dyeing process, also known as exhaust dyeing process requires high liquor ratio, energy and time. The dye liquor contains various types of chemicals and their amount of use, depends on the substrate and dye liquor ratio. Water,energy and chemical consumption can be reduced by reducing M:L ratio [120, 121]. Researchers are developing several water conservation processing for textile colouration. Low liquor ratio dyeing is one of the promising technologies beside SCF dyeing for water saving. In exhaust dyeing, material and liquor ratio has huge impact on water and chemical consumption during dyeing. By this technique not only the dye uptake

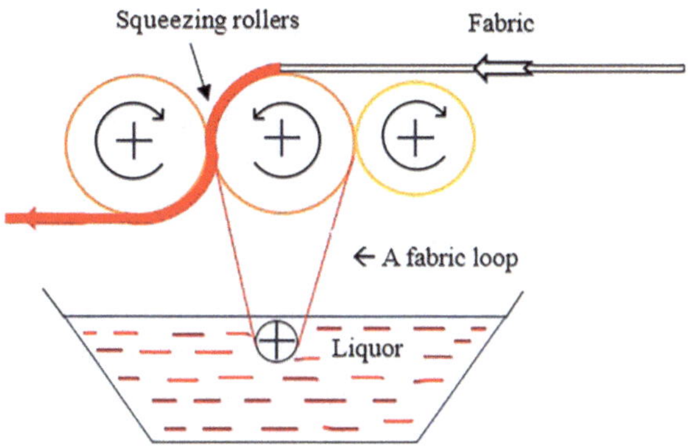

Fig. 5 Schematic diagram of loop transfer process [117]

can be increased but also it acts as the alternative of overflow rinsing. Liquor ratio also has the impact on product quality, energy, costing and environment [78, 122]. Though there is greater concentration of dye particles than conventional colouration process, but it does not show any adversative consequence on quality of dyed material and environment [123]. When more water is used then the material requires more energy consumption to dry the fabric and this large amount of evaporated water causes migration of finishing chemical during drying. Various types of method for low wet pick up is reported like loop transfer system (seen in Fig. 5), nip padding system, foam application etc. [117]. In current years several papers reported on the investigation on special machines which not only reduce the energy consumption but also increase production. Ultra-low liquor dyeing machine is one of them in which dyeing mechanism includes special pump and pulsating current control system. It shows that this dyeing machine saves 33.3% dyeing times with 1:3 bath ratio [124].

The effect of low liquor ratio on wool fibre dyeing and its feasibility to apply in large scale production is also studied. It is stated that M:L ration less than 1:10 may produce uneven shade due to agglomeration of dyestuffs, but some auxiliaries can minimize this problem and increase solubility of dyes in bath [120]. Another research investigated that how material and liquor ration in dyeing influence the various colourfastness properties. It showed that colour strength is decreased with increased liquor ratio [125]. In a research work, the microwave irradiation, at lower liquor ratio is also used to colour cotton fabric by batch dyeing process and it is compared with the conventional dyeing process. It is observed that this technology decrease the dyeing time and increase the values of exhaustion percentage, fixation percentage, and colour yield [121]. Another research reported that nano bubble technology, low liquor ratio of 1:1, may overcome some problem of uneven dyeing, fastness properties etc.; all the cotton fabric dyeing with this technology showed even dyeing performance with zero discharge of liquid [126].

3.6 Biological Colouration

In biological colouration, microbe-based dyes have been demonstrated to be scalable and cost-effective, addressing a key limitation of many sustainable solutions. The textile and fashion industry uses approximately 25% of the world's chemicals and contributes to 5–10% of global carbon emissions, making microbe-based dyes an essential alternative. The traditional dyeing process consumes around 5 trillion liters of water annually and often leads to water pollution, especially in developing countries. Microbial dyes offer a solution with reduced water usage. The microbial dyeing process reduces the carbon footprint of dyeing cotton by at least 31% and saves up to 49% of water compared to conventional methods [127]. Some microorganisms like bacteria, fungi, algae etc. can yield pigments by their natural metabolic reactions. These pigments are termed as microbial pigments [128–130]. Carotenoids, produced by bacteria such as Rhodotorula and Serratia marcescens, are pigments that range from yellow to red and are renowned for their antioxidant properties [131]. Melanins, produced by fungi and some bacteria, are dark brown or black pigments with high UV resistance, making them particularly useful for textiles exposed to sunlight [130]. Quinones, generated by fungi like Penicillium and Monascus, can produce a spectrum of colours, including reds and oranges [132]. Polyketides, complex pigments produced by fungi and some bacteria, offer a wide variety of colours and shades, adding to their versatility in textile applications [133]. Microbial dyes present a promising alternative to synthetic dyes by offering vibrant colours essential for textiles while ensuring a sustainable and eco-friendly production process [134]. These dyes are renewable and biodegradable, making them a much more environmentally friendly option [135]. Additionally, the production of microbial dyes typically requires fewer resources, such as water and energy, compared to traditional dye manufacturing [136]. Their non-toxic nature further enhances their appeal, as they are safe for both the environment and human health [137].

In a regulated environment providing optimal conditions–temperature, pH, nutrients, time–microorganisms are cultivated for their growth [138]. These microbes digest the nutrients in their surroundings as they expand. As a byproduct of metabolism, secondary metabolites are obtained which contain colouring substances, pigments [139]. For instance, the bacteria Chromobacterium violaceum creates the violet pigment violacein, and Serratia marcescens produces the red pigment prodigiosin [140]. The microbes are removed from the culture media once they have produced an adequate quantity of colours. Then the pigments are taken out of the microbial cells. Solvents or other extraction methods can be used for this. To concentrate the colourant and get rid of contaminants, the pigments are frequently purified. After the pigments are extracted, they are turned into dyes that can be used to colour textiles [141]. These dyes are applied using methods similar to traditional dyeing but may need some adjustments to help the natural pigments fix better to the fabric. To improve the bond between the dyes and the fibres, substances called fixatives or mordants are often used [142]. Some microbial pigments naturally bond well with certain fibres, so they do not need as many chemical fixatives [143]. The lasting quality of the dye, like how well it resists fading from washing or sunlight, depends on both the type of pigment and the dyeing method [140].

4 Conclusion

Traditional textile coloration methods pose significant environmental challenges, highlighting the urgent need for sustainable alternatives. Challenges like excessive water consumption, chemical pollution, and high energy demands have long contributed to environmental degradation and public health risks. However, increasing consumer awareness, stricter regulations, and advancements in eco-friendly technologies offer an opportunity to revolutionize the textile industry.

Eco-friendly coloration processes described in this chapter, such as digital printing, waterless dyeing, and the use of natural colorants, align directly with the United Nations Sustainable Development Goal 12 (SDG 12) which promotes sustainable production practices. SDG 12 emphasizes the importance of reducing resource consumption and minimizing waste, particularly hazardous chemicals, to ensure sustainable production practices. By adopting these innovative techniques, the textile industry can significantly reduce its ecological footprint, conserve natural resources, and mitigate harmful environmental impacts.

Among these methods, direct digital printing of fabric reduces color waste, energy, CO_2 footprint, and water consumption. Spray dyeing technology minimizes wastewater treatment processes, while supercritical fluid dyeing is a waterless alternative. Other promising methods include low liquor ratio dyeing and foam technology, both of which require minimal water. The use of natural microorganisms in biological colorations further reduces the need for toxic chemicals. These green coloration processes are making a positive impact in creating a sustainable environment by significantly lowering pollution, conserving water, and minimizing the use of harmful substances.

The shift to sustainable practices not only addresses ecological concerns but also contributes to creating more resilient and ethically aligned supply chains. It ensures that the vibrant, high-quality textiles consumers demand is produced responsibly, without compromising the health of the planet or future generations. Through continued research, investment in green technologies, and collaborative efforts across stakeholders, the textile industry can lead the way in achieving SDG 12 and fostering a cleaner, more sustainable future.

References

1. M. Yusuf, M. Shabbir, F. Mohammad, Natural colorants: historical, processing and sustainable prospects. Natl. Prod. Bioprospecting **7**(1), 123–145 (2017). https://doi.org/10.1007/s13659-017-0119-9

2. T. Islam, T. Islam, T. Islam, R. Repon, Synthetic dyes for textile colouration: process, factors and environmental impact. Text. Leather Rev 327–373 (2022)
3. F. Uddin, Environmental hazard in textile dyeing wastewater from local textile industry. Cellulose **28**(17), 10715–10739 (2021). https://doi.org/10.1007/s10570-021-04228-4
4. S. Dutta et al., Contamination of textile dyes in aquatic environment: adverse impacts on aquatic ecosystem and human health, and its management using bioremediation. J. Environ. Manag. **353**, 120103 (2024). https://doi.org/10.1016/j.jenvman.2024.120103
5. A. Garg, L. Chopra, Dye waste: a significant environmental hazard. Mater. Today Proc. **48**, 1310–1315 (2022). https://doi.org/10.1016/j.matpr.2021.09.003
6. K. Farhana, K. Kadirgama, A.S.F. Mahamude, M.T. Mica, Energy consumption, environmental impact, and implementation of renewable energy resources in global textile industries: an overview towards circularity and sustainability. Mater. Circ. Econ. **4**(1), 15 (2022). https://doi.org/10.1007/s42824-022-00059-1
7. L. Rothenberg, D. Matthews, Consumer decision making when purchasing eco-friendly apparel. Int. J. Retail Distrib. Manag. **45**(4), 404–418 (2017). https://doi.org/10.1108/IJRDM-06-2016-0099
8. N. Oelze, Sustainable supply chain management implementation–enablers and barriers in the textile industry (2017). https://doi.org/10.3390/su9081435
9. L. Siliņa, I. Dāboliņa, E. Lapkovska, Sustainable textile industry–wishful thinking or the new norm: a review. J. Eng. Fibers Fabr. **19**, 15589250231220360 (2024). https://doi.org/10.1177/15589250231220359
10. A. Iannuzzi, *Greener Products: The Making and Marketing of Sustainable Brands*, 3rd edn. (CRC Press, 2024). https://doi.org/10.1201/b11276
11. A.A. Mamun et al., An assessment of energy and groundwater consumption of textile dyeing mills in Bangladesh and minimization of environmental impacts via long-term key performance indicators (KPI) baseline. Textiles **2**(4), 511–523 (2022). https://doi.org/10.3390/textiles2040029
12. D. Zhu, Y. Bao, X. Ding, X. Wu, Unveil the carbon footprint of textiles dyed with different reactive dyestuff recipes from an industrial manufacturing perspective. Int. J. Life Cycle Assess. **29**(9), 1598–1620 (2024). https://doi.org/10.1007/s11367-024-02323-9
13. M. Kim et al., Reduction of greenhouse gas emissions by optimizing the textile dyeing process using digital twin technology. Fash. Text. **11**(1), 17–17 (2024). https://doi.org/10.1186/s40691-024-00384-w
14. C. Li et al., Carbon–water–energy footprint impacts of dyed cotton fabric production in China. J. Clean. Prod. **467**, 142898–142898 (2024). https://doi.org/10.1016/j.jclepro.2024.142898
15. R. Kant, Textile dyeing industry an environmental hazard. NS **04**(01), 22–26 (2012). https://doi.org/10.4236/ns.2012.41004
16. T. Toprak, P. Anis, Textile industry's environmental effects and approaching cleaner production and sustainability: an overview. JTEFT **2**(4) (2017). https://doi.org/10.15406/jteft.2017.02.00066
17. A.A.U. Jorhat, M. Konwar, Textile Industry and Its Environmental Impacts: A Review. Ind. J. Pure App. Biosci. **8**(3), 134–139 (2020). https://doi.org/10.18782/2582-2845.7848. (Department of Textiles and Apparel Designing, College of Community Science)
18. M.T. Islam, T. Islam, T. Islam, M.R. Repon, Synthetic dyes for textile colouration: process, factors and environmental impact. Text Leather Rev **5**, 327–373 (2022). https://doi.org/10.31881/TLR.2022.27. (Department of Textile Engineering, Mawlana Bhashani Science and Technology University, Tangail, Bangladesh et al.)
19. M. Berradi et al., Textile finishing dyes and their impact on aquatic environs. Heliyon **5**(11), e02711 (2019). https://doi.org/10.1016/j.heliyon.2019.e02711
20. M. Shabbir (ed.), *Textiles and Clothing: Environmental Concerns and Solutions* (John Wiley & Sons Inc., Hoboken, New Jersey, 2019)
21. S. Sudarshan et al., Impact of textile dyes on human health and bioremediation of textile industry effluent using microorganisms: current status and future prospects. J. Appl. Microbiol. **134**(2), lxac064 (2023). https://doi.org/10.1093/jambio/lxac064

22. The impact of textile production and waste on the environment (infographics). European Parliament. https://www.europarl.europa.eu/topics/en/article/20201208STO93327/the-impact-of-textile-production-and-waste-on-the-environment-infographics. Accessed 28 Dec 2024
23. X. Li, L. Zhu, X. Ding, X. Wu, L. Wang, Climate change and the textile industry: the carbon footprint of dyes. AATCC J. Res. **11**(2), 109–123 (2024). https://doi.org/10.1177/24723444231212954
24. T.A. Khattab, M.S. Abdelrahman, M. Rehan, Textile dyeing industry: environmental impacts and remediation. Environ. Sci. Pollut. Res. **27**(4), 3803–3818 (2020). https://doi.org/10.1007/s11356-019-07137-z
25. F.M. Drumond Chequer, G.A.R. De Oliveira, E.R. Anastacio Ferraz, J. Carvalho, M.V. Boldrin Zanoni, D.P. De Oliveir, Textile dyes: dyeing process and environmental impact, in *Eco-Friendly Textile Dyeing and Finishing*, ed. by M. Gunay (InTech, 2013). https://doi.org/10.5772/53659
26. D.M. Lewis, Coloration in the next century. Rev. Prog. Color. Relat. Top. **29**(1), 23–28 (1999). https://doi.org/10.1111/j.1478-4408.1999.tb00124.x
27. G.C. de Oliveira Neto, J.M. Ferreira Correia, P.C. Silva, A.G. de Oliveira Sanches, W.C. Lucato, Cleaner production in the textile industry and its relationship to sustainable development goals. J. Clean. Prod. **228**, 1514–1525 (2019). https://doi.org/10.1016/j.jclepro.2019.04.334
28. A. Desore, S.A. Narula, An overview on corporate response towards sustainability issues in textile industry. Environ. Dev. Sustain. **20**(4), 1439–1459 (2018). https://doi.org/10.1007/s10668-017-9949-1
29. S. Saxena, A.S.M. Raja, Natural dyes: sources, chemistry, application and sustainability issues BT-roadmap to sustainable textiles and clothing: eco-friendly raw materials, technologies, and processing methods, ed. by S.S. Muthu (Springer Singapore, Singapore, 2014), pp. 37–80. https://doi.org/10.1007/978-981-287-065-0_2
30. B. Jena, B.P. Das, A. Khandual, S. Sahu, L. Behera, Ecofriendly processing of textiles. Mater. Today Proc. **2**(4), 1776–1791 (2015). https://doi.org/10.1016/j.matpr.2015.07.020
31. J.M. Jabar, T.E. Adedayo, Y.A. Odusote, Green, eco-friendly and sustainable alternative in dyeing cotton fabric using aqueous extract Mucuna slonaei F dye: effects of metal salts pre-mordanting on color strength and fastness properties. Curr. Res. Green Sustain. Chem. **4**, 100151 (2021). https://doi.org/10.1016/j.crgsc.2021.100151
32. M.R. Islam, M.G. Mostafa, Textile dyeing effluents and environment concerns-a review. J. Environ. Sci. Natl. Resour. **11**(1–2) SE-Articles, 131–144 (2019). https://doi.org/10.3329/jesnr.v11i1-2.43380
33. S. Saxena, A.S.M. Raja, A. Arputharaj, Challenges in sustainable wet processing of textiles, ed. by S.S. Muthu (Springer Singapore, Singapore, 2017), pp. 43–79. https://doi.org/10.1007/978-981-10-2185-5_2
34. M. Mohsin, S. Sardar, Multi-criteria decision analysis for textile pad-dyeing and foam-dyeing based on cost, performance, productivity and sustainability. Cellulose **26**(6), 4143–4157 (2019). https://doi.org/10.1007/s10570-019-02337-9
35. M.L. Parisi, E. Fatarella, D. Spinelli, R. Pogni, R. Basosi, Environmental impact assessment of an eco-efficient production for coloured textiles. J. Clean. Prod. **108**, 514–524 (2015). https://doi.org/10.1016/j.jclepro.2015.06.032
36. M.M. Islam, P. Perry, S. Gill, Mapping environmentally sustainable practices in textiles, apparel and fashion industries: a systematic literature review. J. Fash. Mark. Manag. Int. J. **25**(2), 331–353 (2021). https://doi.org/10.1108/JFMM-07-2020-0130
37. K.M. Babu, Overview on significant approaches to waterless dyeing technology, in *Emerging Technologies for Textile Coloration* (CRC Press, 2022), pp. 171–188
38. L. Lin et al., Combination of wet fixation and drying treatments to improve dye fixation onto spray-dyed cotton fabric. Sci. Rep. **11**(1), Art. no. 1 (2021). https://doi.org/10.1038/s41598-021-94885-z

39. A.N.M.A. Haque, M. Naebe, Zero-water discharge and rapid natural dyeing of wool by plasma-assisted spray-dyeing. J. Clean. Prod. **402**, 136807 (2023). https://doi.org/10.1016/j.jclepro.2023.136807
40. A. Khatri, M.H. Peerzada, M. Mohsin, M. White, A review on developments in dyeing cotton fabrics with reactive dyes for reducing effluent pollution. J. Clean. Prod. **87**, 50–57 (2015). https://doi.org/10.1016/j.jclepro.2014.09.017
41. H. Xiao, T. Zhao, C.-H. Li, M.-Y. Li, Eco-friendly approaches for dyeing multiple type of fabrics with cationic reactive dyes. J. Clean. Prod. **165**, 1499–1507 (2017). https://doi.org/10.1016/j.jclepro.2017.07.174
42. J. Moltchanova, *Digital Textile Printing*. (Metropolia Ammattikorkeakoulu, 2011)
43. A. Wahab, T. Hussain, Eco-Friendly garment processing using aerosol technology. Water Resour. Ind. **23**, 100127–100127 (2020). https://doi.org/10.1016/j.wri.2020.100127
44. J. Lavanya, N. Kishore, Different textile printing techniques-hand block printing, screen printing, and digital printing. Webology **19**(2), 787–802 (2022)
45. H.M. Yesuf, A.K. Jhatial, P.K. Gianchandani, A. Siddique, A.A. Simair, 2-Digital printing mechanisms, in *The Textile Institute Book Series*, ed. by H. Wang, H.B.T.-D.T.P. Memon (Woodhead Publishing, 2023), pp. 21–40. https://doi.org/10.1016/B978-0-443-15414-0.00005-4
46. D.J. Tyler, Textile digital printing technologies. Text. Prog. **37**(4), 1–65 (2005). https://doi.org/10.1533/tepr.2005.0004
47. D. Javoršek, A. Javoršek, Colour management in digital textile printing. Color. Technol. **127**(4), 235–239 (2011). https://doi.org/10.1111/j.1478-4408.2011.00304.x
48. H. Ujiie, 19-Design and workflow in digital inkjet printing, in *Woodhead Publishing Series in Textiles*, ed. by H.B.T.-D.P.T. Ujiie (Woodhead Publishing, 2006), pp. 337–354. https://doi.org/10.1533/9781845691585.4.337
49. S. Stojanović, J. Geršak, D. Trajković, N. Ćirković, Influence of sublimation transfer printing on alterations in the structural and physical properties of knitted fabrics. Color. Technol. **137**(2), 108–122 (2021). https://doi.org/10.1111/cote.12508
50. B. Sarkodie, B. Tawiah, C. Agbo, J. Wizi, Status and development of transfer printing in textiles—a review. AATCC J. Res. **5**(2), 1–18 (2018)
51. M.C. Silva et al., Digital sublimation printing on knitted polyamide 6.6 fabric treated with non-thermal plasma. Polymers **13**(12) (2021). https://doi.org/10.3390/polym13121969
52. B. Gooby, The development of methodologies for color printing in digital inkjet textile printing and the application of color knowledge in the ways of making project. J. Text. Des. Res. Pract. **8**(3), 358–383 (2020). https://doi.org/10.1080/20511787.2020.1827802
53. C. Williams, Ink-jet printers go beyond paper. Phys. World **19**(1), 24 (2006). https://doi.org/10.1088/2058-7058/19/1/32
54. A.M. Adel, N.M. Ahmed, M.A. Diab, F.N. El-Shall, N. El-Shinnawy, Exploration on ability of printable modified papers for the application in heat sublimation transfer printing of polyester fabric. Sci. Rep. **13**(1), 6536 (2023). https://doi.org/10.1038/s41598-023-33546-9
55. Z. Özomay, M. Özomay, The effect of temperature and time variables on printing quality in sublimation transfer printing on nylon and polyester fabric TT-Naylon ve Polyester Kumaş Üzerine Sublimasyon Transfer Baskıda Sıcaklık ve Süre Değişkeninin Baskı Kalitesine Etkisi. Avrupa Bilim ve Teknoloji Dergisi **23**, 882–891 (2021). https://doi.org/10.31590/ejosat.889147
56. D.M. Kumelachew, B.T. Wagaye, B.F. Adamu, 13-Digital textile printing innovations and the future, in *The Textile Institute Book Series*, ed. by H. Wang, H.B.T.-D.T.P. Memon (Woodhead Publishing, 2023), pp. 241–259. https://doi.org/10.1016/B978-0-443-15414-0.00006-6
57. M.M. Ragab, H. Othman, A. Hassabo, An overview of printing textile techniques. Egypt. J. Chem. **65**(8), 749–761 (2022). https://doi.org/10.21608/ejchem.2022.131477.5793
58. C.W. Kan, C.W.M. Yuen, Digital Ink-jet Printing on Textiles. Res. J. Text. Appar. **16**(2), 1–24 (2012). https://doi.org/10.1108/RJTA-16-02-2012-B001
59. A. Russell, *The Fundamentals of Printed Textile Design* (A&C Black, 2011)

60. M.A. Shah, D.-G. Lee, B.-Y. Lee, S. Hur, Classifications and applications of inkjet printing technology: a review. IEEE Access **9**, 140079–140102 (2021). https://doi.org/10.1109/ACCESS.2021.3119219
61. B. Zamani, The environmental impact of screen printing, digital printing, and fabric dyeing. Carbonfact. https://www.carbonfact.com/blog/research/the-environmental-impact-of-screen-printing-digital-printing-and-fabric-dyeing#:~:text=The%20comparisons%20indicate%20that%20digital,compared%20to%20traditional%20dyeing%20methods. Accessed 30 Dec 2024
62. H.M. El-Hennawi, A.A. Shahin, M. Rekaby, A.A. Ragheb, Ink jet printing of bio-treated linen, polyester fabrics and their blend. Carbohyd. Polym. **118**, 235–241 (2015). https://doi.org/10.1016/j.carbpol.2014.10.067
63. S.M.A. Hoque et al., Environmental sustainability analysis of rotary-screen printing and digital textile printing. AATCC J. Res. **11**(6), 464–473 (2024). https://doi.org/10.1177/24723444241275996
64. E. Pei, J. Shen, J. Watling, Direct 3D printing of polymers onto textiles: experimental studies and applications. Rapid Prototyp. J. **21**(5), 556–571 (2015). https://doi.org/10.1108/RPJ-09-2014-0126
65. Y. Yang, V. Naarani, Improvement of the lightfastness of reactive inkjet printed cotton. Dyes Pigm. **74**(1), 154–160 (2007). https://doi.org/10.1016/j.dyepig.2006.01.030
66. M. El-Kashouti, S. Elhadad, K. Abdel-Zaher, Printing technology on textile fibers: review. J. Text. Coloration Polym. Sci. **16**(2), 129–138 (2019). https://doi.org/10.21608/jtcps.2019.15989.1027
67. K.W. Lo, Digital printing in unconventional knitwear design. PolyU Electronic Theses (2018)
68. B. Tawiah, E.K. Howard, B.K. Asinyo, The chemistry of inkjet inks for digital textile printing-review. BEST **4**, 61–78 (2016)
69. J.V.D. Bergen, G. Parker, Textile processing guide: pretreatment, colouration & finishing. Fashion for Good (Jul 2023)
70. A. Samanta, Sustainable dyeing technologies for polyester. Int. J. Sci. Res. Eng. Manag. **07**(08) (2023)
71. M.D. Abbott, R.A. Miller, R.E. May, System and method for spray dyeing fabrics. US 8,814,953 B1, Aug. 26, 2014
72. R. Mulder, M.N. Morshed, S. Seipel, U. Norén, E. Niit, V. Nierstrasz, Study on hydraulic spray atomizing system as a new resource-efficient dyeing-finishing method for wool fabric. Sci. Rep. **12**(1), 21814 (2022). https://doi.org/10.1038/s41598-022-26172-4
73. L. Lara, I. Cabral, J. Cunha, Ecological approaches to textile dyeing: a review. Sustainability **14**(14), 8353 (2022). https://doi.org/10.3390/su14148353
74. W. Wei, J. Bolin, M. Zhiping, Review of new dyeing technologies for reactive dyes and disperse dyes. J. Text. Res. **44**(5) (2023)
75. E. Ramsey, Q. Sun, Z. Zhang, C. Zhang, W. Gou, Mini-Review: Green sustainable processes using supercritical fluid carbon dioxide. J. Environ. Sci. **21**(6), 720–726 (2009). https://doi.org/10.1016/S1001-0742(08)62330-X
76. M.-W. Park, H.-K. Bae, Dye distribution in supercritical dyeing with carbon dioxide. J. Supercrit. Fluids **22**(1), 65–73 (2002). https://doi.org/10.1016/S0896-8446(01)00102-4
77. L. Miah, N. Ferdous, Textiles material dyeing with supercritical carbon dioxide (CO2) without using water
78. T. Hussain, A. Wahab, A critical review of the current water conservation practices in textile wet processing. J. Clean. Prod. **198**, 806–819 (2018). https://doi.org/10.1016/j.jclepro.2018.07.051
79. S. Sicardi, L. Manna, M. Banchero, Diffusion of disperse dyes in PET films during impregnation with a supercritical fluid. J Supercrit. Fluids **17**(2), 187–194 (2000). https://doi.org/10.1016/S0896-8446(99)00055-8
80. L. Zheng, H. Zheng, B. Du, J. Wei, S. Gao, J. Zhang, Dyeing procedures of polyester fiber in supercritical carbon dioxide using a special dyeing frame. J. Eng. Fibers Fabr. **10**(4), 155892501501000 (2015). https://doi.org/10.1177/155892501501000414

81. M.T. Abate, M.G. Tadesse, Airflow, foam, and supercritical carbon dioxide dyeing technologies, in *Innovative and Emerging Technologies for Textile Dyeing and Finishing*, 1st ed., ed. by L.J. Rather, A. Haji, M. Shabbir (Wiley, 2021), pp. 137–164. https://doi.org/10.1002/9781119710288.ch5
82. J.-J. Long, H.-M. Xu, C.-L. Cui, X.-C. Wei, F. Chen, A.-K. Cheng, A novel plant for fabric rope dyeing in supercritical carbon dioxide and its cleaner production. J. Clean. Prod. **65**, 574–582 (2014). https://doi.org/10.1016/j.jclepro.2013.08.008
83. "CO_2 Dyeing," DyeCoo. https://dyecoo.com/co2-dyeing/. Accessed 30 Dec 2024
84. "CO_2 Dyeing," Wrap. https://ckh.wrap.org.uk/colouration/CO2Dyeing?breadcrumb=Explore+by%3A+Process+Step. Accessed 30 Dec 2024
85. G.A. Montero, C.B. Smith, W.A. Hendrix, D.L. Butcher, Supercritical fluid technology in textile processing: an overview. Ind. Eng. Chem. Res. **39**(12), 4806–4812 (2000). https://doi.org/10.1021/ie0002475
86. M. Van Der Kraan, M. Vanesa Fernandez Cid, G.F. Woerlee, W.J.T. Veugelers, G.-J. Witkamp, Equilibrium study on the disperse dyeing of polyester textile in supercritical carbon dioxide. Text. Res. J. **77**(8), 550–558 (2007). https://doi.org/10.1177/0040517507077483
87. V.F. Cabral, W.L.F. Santos, E.C. Muniz, A.F. Rubira, L. Cardozo-Filho, Correlation of dye solubility in supercritical carbon dioxide. J. Supercrit. Fluids **40**(2), 163–169 (2007). https://doi.org/10.1016/j.supflu.2006.05.004
88. D. Tuma, B. Wagner, G.M. Schneider, Comparative solubility investigations of anthraquinone disperse dyes in near- and supercritical fluids. Fluid Phase Equilib. **182**(1–2), 133–143 (2001). https://doi.org/10.1016/S0378-3812(01)00387-9
89. A. Ferri, M. Banchero, L. Manna, S. Sicardi, Dye uptake and partition ratio of disperse dyes between a PET yarn and supercritical carbon dioxide. J. Supercrit. Fluids **37**(1), 107–114 (2006). https://doi.org/10.1016/j.supflu.2005.07.001
90. G. Brunner, Applications of supercritical fluids. Annu. Rev. Chem. Biomol. Eng. **1**(1), 321–342 (2010). https://doi.org/10.1146/annurev-chembioeng-073009-101311
91. M. Perrut, Supercritical fluid applications: industrial developments and economic issues. Ind. Eng. Chem. Res. **39**(12), 4531–4535 (2000). https://doi.org/10.1021/ie000211c
92. K. Sawada, T. Takagi, J.H. Jun, M. Ueda, D.M. Lewis, Dyeing natural fibres in supercritical carbon dioxide using a nonionic surfactant reverse micellar system. Color. Technol. **118**(5), 233–237 (2002). https://doi.org/10.1111/j.1478-4408.2002.tb00105.x
93. S. Maeda, K. Kunitou, T. Hihara, K. Mishima, One-bath dyeing of polyester/cotton blends with reactive disperse dyes in supercritical carbon dioxide. Text. Res. J. **74**(11), 989–994 (2004). https://doi.org/10.1177/004051750407401109
94. T. Abou Elmaaty et al., Optimization of an eco-friendly dyeing process in both laboratory scale and pilot scale supercritical carbon dioxide unit for polypropylene fabrics with special new disperse dyes. J CO2 Util. **33**, 365–371 (2019). https://doi.org/10.1016/j.jcou.2019.06.018
95. H. Zheng, Y. Xu, J. Zhang, X. Xiong, J. Yan, L. Zheng, An ecofriendly dyeing of wool with supercritical carbon dioxide fluid. J. Clean. Prod. **143**, 269–277 (2017). https://doi.org/10.1016/j.jclepro.2016.12.115
96. B. Guzel, A. Akgerman, Mordant dyeing of wool by supercritical processing. J. Supercrit. Fluids **18**(3), 247–252 (2000). https://doi.org/10.1016/S0896-8446(00)00065-6
97. M. Liu, H. Zhao, J. Wu, X. Xiong, L. Zheng, Eco-friendly curcumin-based dyes for supercritical carbon dioxide natural fabric dyeing. J. Clean. Prod. **197**, 1262–1267 (2018). https://doi.org/10.1016/j.jclepro.2018.06.202
98. T. Abou Elmaaty, J. Ma, F. El-Taweel, E. Abd El-Aziz, S. Okubayashi, Facile bifunctional dyeing of polyester under supercritical carbon dioxide medium with new antibacterial hydrazono propanenitrile dyes. Ind. Eng. Chem. Res. **53**(40), 15566–15570 (2014). https://doi.org/10.1021/ie502088r
99. A.S. Özcan, A. Özcan, Adsorption behavior of a disperse dye on polyester in supercritical carbon dioxide. J. Supercrit. Fluids **35**(2), 133–139 (2005). https://doi.org/10.1016/j.supflu.2004.12.006

100. Y. Kawahara, T. Kikutani, K. Sugiura, S. Ogawa, Dyeing behaviour of poly(ethylene terephthalate) fibres in supercritical carbon dioxide. Color. Technol. **117**(5), 266–269 (2001). https://doi.org/10.1111/j.1478-4408.2001.tb00073.x
101. L. Cardozo-Filho et al., Dyeing of polyethylene terephthalate fibers with a disperse dye in supercritical carbon dioxide. Text. Res. J. **84**(12), 1279–1287 (2014). https://doi.org/10.1177/0040517513515317
102. M.R.D. Giorgi, E. Cadoni, D. Maricca, A. Piras, Dyeing polyester ®bres with disperse dyes in supercritical CO2. Dyes Pigment. (2000)
103. T. Hori, A. Kongdee, Dyeing of PET/co-PP composite fibers using supercritical carbon dioxide. Dyes Pigment. **105**, 163–166 (2014). https://doi.org/10.1016/j.dyepig.2014.01.023
104. H. Wen, J.J. Dai, Dyeing of polylactide fibers in supercritical carbon dioxide. J Appl. Polym. Sci. **105**(4), 1903–1907 (2007). https://doi.org/10.1002/app.26234
105. S.K. Liao, Y.C. Ho, P.S. Chang, Dyeing of nylon 66 with a disperse-reactive dye using supercritical carbon dioxide as the transport medium. Color. Technol. **116**(12), 403–407 (2000). https://doi.org/10.1111/j.1478-4408.2000.tb00018.x
106. H. Zheng, L. Zheng, Dyeing of meta-aramid fibers with disperse dyes in supercritical carbon dioxide. Fibers Polym. **15**(8), 1627–1634 (2014). https://doi.org/10.1007/s12221-014-1627-4
107. P.L. Beltrame et al., Dyeing of cotton in supercritical carbon dioxide. Dyes Pigment. **39**(4), 335–340 (1998). https://doi.org/10.1016/S0143-7208(98)00016-3
108. A.S. Özcan, A.A. Clifford, K.D. Bartle, P.J. Broadbent, D.M. Lewis, Dyeing of modified cotton fibres with disperse dyes from supercritical carbon dioxide. J. Soc. Dyers Colour. **114**(5–6), 169–173 (1998). https://doi.org/10.1111/j.1478-4408.1998.tb01977.x
109. M.V. Fernandez Cid, J. Van Spronsen, M. Van Der Kraan, W.J.T. Veugelers, G.F. Woerlee, G.J. Witkamp, A significant approach to dye cotton in supercritical carbon dioxide with fluorotriazine reactive dyes. J. Supercrit. Fluids **40**(3), 477–484 (2007). https://doi.org/10.1016/j.supflu.2006.07.011
110. S. Maeda, S. Hongyou, K. Kunitou, K. Mishima, Dyeing cellulose fibers with reactive disperse dyes in supercritical carbon dioxide. Text. Res. J. **72**(3), 240–244 (2002). https://doi.org/10.1177/004051750207200310
111. D. Yang, X. Kong, D. Gao, H. Cui, T. Huang, J. Lin, Dyeing of cotton fabric with reactive disperse dye contain acyl fluoride group in supercritical carbon dioxide. Dyes Pigment. **139**, 566–574 (2017). https://doi.org/10.1016/j.dyepig.2016.12.050
112. X. Luo et al., Novel sustainable synthesis of dyes for clean dyeing of wool and cotton fibres in supercritical carbon dioxide. J. Clean. Prod. **199**, 1–10 (2018). https://doi.org/10.1016/j.jclepro.2018.07.158
113. Z.-T. Liu et al., Supercritical CO_2 dyeing of ramie fiber with disperse dye. Ind. Eng. Chem. Res. **45**(26), 8932–8938 (2006). https://doi.org/10.1021/ie061093h
114. J.H. Jun, K. Sawada, T. Takagi, G.B. Kim, C.H. Park, M. Ueda, Effects of pressure and temperature on dyeing acrylic fibres with basic dyes in supercritical carbon dioxide. Color. Technol. **121**(1), 25–28 (2005). https://doi.org/10.1111/j.1478-4408.2005.tb00244.x
115. T. Elmaaty, E. El-Aziz, J. Ma, F. El-Taweel, S. Okubayashi, Eco-friendly disperse dyeing and functional finishing of nylon 6 using supercritical carbon dioxide. Fibers **3**(3), 309–322 (2015). https://doi.org/10.3390/fib3030309
116. A. Siddique et al., Foam dyeing: a review of its potential to revolutionize sustainable textile processing (25 Jul 2024). https://doi.org/10.20944/preprints202407.2011.v1
117. W.D. Schindler, P.J. Hauser, *Chemical finishing of textiles*. (CRC, Woodhead, Boca Raton, Cambridge, England, 2004)
118. H. Yu, Y. Wang, Y. Zhong, Z. Mao, S. Tan, Foam properties and application in dyeing cotton fabrics with reactive dyes. Color. Technol. **130**(4), 266–272 (2014). https://doi.org/10.1111/cote.12088
119. Q. Wang et al., Application of foam dyeing technology on ultra-fine polyamide filament fabrics with acid dye. Text. Res. J. **89**(23–24), 4808–4816 (2019). https://doi.org/10.1177/0040517519839377

120. M.M. Hassan, M. Bhagvandas, Sustainable low liquor ratio dyeing of wool with acid dyes: effect of auxiliaries on agglomeration of dye molecules in a dyebath and dyeing uniformity. J. Clean. Prod. **152**, 464–473 (2017). https://doi.org/10.1016/j.jclepro.2017.03.139
121. M. Irfan, H. Zhang, U. Syed, A. Hou, Low liquor dyeing of cotton fabric with reactive dye by an eco-friendly technique. J. Clean. Prod. **197**, 1480–1487 (2018). https://doi.org/10.1016/j.jclepro.2018.06.300
122. K.A. Rade, V.A. Pharande, D.R. Saini, *Effect of Mass to Liquor Ratio on Dyeing Process* (2015)
123. C. Chakraborty, A.B.M. Foisal, *A Study on Dyeing of Cotton Knit Fabric with Reactive dye in variable Liquor Ratio*, vol. 1, no. 2 (2021)
124. S.I. Editor, A novel high temperature and high pressure yarn dyeing machine with ultralow liquor ratio. Int. J. Simul. Syst. Sci. Technol. (2016). https://doi.org/10.5013/IJSSST.a.17.23.03
125. R. Saha, A.S.M. Naser, Effects of liquor ratio on cotton knit fabric with reactive dye. Southeast Univ. J. Text. Eng. **1**(1), Art. no. 1 (2021)
126. M. Mohsin, S. Sardar, M. Hassan, N. Akhtar, A. Hassan, M. Sufyan, Novel, sustainable and water efficient nano bubble dyeing of cotton fabric. Cellulose **27**(10), 6055–6064 (2020). https://doi.org/10.1007/s10570-020-03187-6
127. I. Technologies, A revolutionary dyeing process to help the textile industry reduce its environmental impact Results in Brief Natural dyeing process delivers environmental benefits.
128. H. Agarwal et al., Bacterial pigments and their multifaceted roles in contemporary biotechnology and pharmacological applications (2023). https://doi.org/10.3390/microorganisms11030614
129. L. Dufossé, Chapter 19-Current and potential natural pigments from microorganisms (bacteria, yeasts, fungi, and microalgae), in *Woodhead Publishing Series in Food Science, Technology and Nutrition*, 2nd edn. ed. by E. Schweiggert (Woodhead Publishing, 2024), pp. 419–436. https://doi.org/10.1016/B978-0-323-99608-2.00001-X. (R.B.T.-H. on N. P. in F. and B.)
130. L.J. Rather, S.S. Mir, S.A. Ganie, Shahid-ul-Islam, Q Li, Research progress, challenges, and perspectives in microbial pigment production for industrial applications-a review. Dyes Pigment. **210**, 110989 (2023). https://doi.org/10.1016/j.dyepig.2022.110989
131. K. Joshi, P. Kumar, R. Kataria, Microbial carotenoid production and their potential applications as antioxidants: a current update. Process Biochem. **128**, 190–205 (2023). https://doi.org/10.1016/j.procbio.2023.02.020
132. W. Elkhateeb, M.O. Elnahas, G. Daba, *Wide Range Applications of Fungal Pigments in Textile Dyeing BT-Fungi and Fungal Products in Human Welfare and Biotechnology*, ed. by T. Satyanarayana, S.K. Deshmukh (Springer Nature Singapore, Singapore, 2023), pp. 289–304. https://doi.org/10.1007/978-981-19-8853-0_10
133. M. Afroz Toma, Md.H. Rahman, Md.S. Rahman, M. Arif, K.H.M.N.H. Nazir, L. Dufossé, Fungal pigments: carotenoids, riboflavin, and polyketides with diverse applications (2023). https://doi.org/10.3390/jof9040454
134. A.D. Pranta, Md.T. Rahaman, Extraction of eco-friendly natural dyes and biomordants for textile coloration: a critical review. Nano Struct. Nano Objects **39**, 101243 (2024). https://doi.org/10.1016/j.nanoso.2024.101243
135. K. Singh, P. Kumar, N.V. Singh, Natural dyes: an emerging ecofriendly solution for textile industries. Poll Res **39**(2), S87–S94 (2020)
136. R. Fried, I. Oprea, K. Fleck, F. Rudroff, Biogenic colourants in the textile industry–a promising and sustainable alternative to synthetic dyes. Green Chem. **24**(1), 13–35 (2022). https://doi.org/10.1039/D1GC02968A
137. E. Di Salvo et al., Natural pigments production and their application in food, health and other industries (2023). https://doi.org/10.3390/nu15081923
138. J. M. Gonzalez, B. Aranda, Microbial growth under limiting conditions-future perspectives (2023). https://doi.org/10.3390/microorganisms11071641

139. M.P. Narsing Rao, M. Xiao, W.-J. Li, Fungal and bacterial pigments: secondary metabolites with wide applications. Front. Microbiol. **8**, 1113 (2017)
140. V.T. Orlandi, E. Martegani, C. Giaroni, A. Baj, F. Bolognese, Bacterial pigments: a colorful palette reservoir for biotechnological applications. Biotechnol. Appl. Biochem. **69**(3), 981–1001 (2022). https://doi.org/10.1002/bab.2170
141. C.K. Venil, Z.A. Zakaria, W.A. Ahmad, Bacterial pigments and their applications. Process Biochem. **48**(7), 1065–1079 (2013). https://doi.org/10.1016/j.procbio.2013.06.006
142. C.K. Venil, P. Velmurugan, L. Dufossé, P. Renuka Devi, A. Veera Ravi, Fungal pigments: potential coloring compounds for wide ranging applications in textile dyeing (2020). https://doi.org/10.3390/jof6020068.
143. B. Balachandran, P.C. Sabumon, *Scope of Natural Dyes and Biomordants in Textile Industry for Cleaner Production BT-Novel Sustainable Process Alternatives for the Textiles and Fashion Industry*, ed. by S.S. Muthu (Springer Nature Switzerland, Cham, 2023), pp. 73–106. https://doi.org/10.1007/978-3-031-35451-9_4

Chapter 8
Industrial-Scale Plasma Processing for Sustainable Textiles

Mehbuba Manir Nova, Md. Najmul Islam, Syeda Maliha Fairuz, Emtiaz Ahmed Pappu, Abdullah Sayam, Gaffar Hossain, and Abu Sadat Muhammad Sayem

Abstract Conventional methods of chemical processing of textiles rely on water-intensive processes and hazardous chemicals, leading to ecological damage through toxic wastewater, high energy consumption, and significant carbon emissions. Plasma technology offers a transformative solution as a dry, energy-efficient, and chemically neutral alternative. Industrial plasma treatment enables precise modification of textile surface properties at the molecular level, enhancing dyeability, water resistance, antimicrobial properties, and adhesion without affecting the structural integrity of the textiles. This technology utilizes ionized gas to achieve functionalities that traditional methods struggle to provide sustainably.

Keywords Plasma · Textile · Surface modification · Sustainability · Industrial application

1 Introduction

Traditional textile treatments, which rely heavily on water-intensive processes and hazardous chemicals, have long been associated with significant ecological degradation, including toxic wastewater discharge, excessive energy consumption, and

M. M. Nova
Department of Wet Process Engineering, Bangladesh University of Textiles, Tejgaon, Dhaka, Bangladesh

Md. N. Islam · S. M. Fairuz · E. A. Pappu · A. Sayam
Department of Textile Engineering, Ahsanullah University of Science and Technology, Tejgaon, Dhaka, Bangladesh

G. Hossain
V-Trion GmbH Textile Research, Lustenau, Austria

A. S. M. Sayem (✉)
Manchester Fashion Institute Manchester, Metropolitan University Manchester, Manchester, UK
e-mail: asm.sayem@mmu.ac.uk

A. S. Muhammad Sayem and P. Goswami (eds.), *SDG 12 and Global Fashion Textiles Production*, SDGs and Textiles, https://doi.org/10.1007/978-981-96-9051-0_8

carbon emissions. Against this backdrop, plasma technology emerges as a transformative solution, offering a dry, energy-efficient, and chemically neutral alternative to modify textile surfaces at the molecular level. Plasma treatment of textiles improves surface properties, offering enhanced wettability, better adhesion, and the ability to introduce anti-microbial features. This surface modification method is highly efficient, boosting the productivity and performance of textiles [1].

2 Plasma Technology

Plasma, often called the fourth state of matter, distinct from solid, liquid, and gas, due to its unique electrical conductivity and responsiveness to magnetic fields. When a gas is exposed to appropriate level of energy, whether in the form of heat, light, or electricity, it transforms into plasma, consisting of a mix of electrons, ions, free radicals, and metastable atoms or molecules.

Typically, plasma is generated using electrical energy. The high-speed electrons in the electric field collide with gas molecules, knocking electrons from the outer shells of the atoms. This process produces a mixture of reactive particles like ions, free radicals, and electrons that can interact with various materials. Plasmas are classified as hot or cold based on their temperature ranges. Plasma can be generated using various power supplies, such as (a) Low frequency (LF): 1–500 kHz, (b) Radiofrequency (RF): 2–60 MHz, and (c) Microwave (MW): 915 MHz–2.45 GHz [2, 3].

2.1 Low-Pressure Plasma

Low-pressure systems typically operate in the RF or MW band with pressures ranging from 1–100 Pa (0.01–0.10 mbar) [3]. LTPs are excellent for surface modification of heat-sensitive polymeric and textile materials. An LTP can be formed by applying an electric field to two electrodes separated by a gas or by generating an RF resonant current in a coil. This can be done in a closed vessel with either decreased or ambient pressure [2]. Reactive species found in LTPs include electrons, ions, free radicals, metastables, and photons in the short-wave ultraviolet area. These organisms can cause physical and chemical interactions on substrate surfaces. Plasma only affects the substrate's outermost thin layer, creating changes of a few nanometers. LTP treatments can enhance the surface of textile fibres without compromising bulk properties. Plasma treatments have a wide range of uses, including surface cleaning and coating, as well as improving moisture absorption, printability, dyeability, and fungal adhesion promotion. Plasma treatments produce modifications based on gas type, pressure, frequency, power, treatment period, and substrate nature [2].

2.2 *Atmospheric Pressure Plasma*

Atmospheric pressure plasma (APP) presents a promising solution for the textile sector. It offers several advantages over low-pressure plasma systems, such as operation at atmospheric pressure, continuous material processing, and compatibility with existing textile processing equipment [4]. Following the discovery that stable plasma could be generated at ambient temperature and pressure, various cold plasma technologies were developed—including corona discharge, dielectric barrier discharge (DBD), and plasma jets. Notably, plasma reactors can operate without the need for costly vacuum systems [5].

(a) Corona discharge

Corona is the simplest and most ancient plasma method for remodelling polymer surfaces. Corona discharge is commonly formed at 1 atm by applying sinusoidal pulses at tens of kilohertz and a high voltage of 10–15 kV between two electrodes of varied shapes and sizes spaced about a millimetre apart. The polymer web moves across the plasma area at speeds ranging from tens to hundreds of meters per minute, spending only a few seconds in the discharge. Energy depositions range from 0.1 to a few Joules/cm [6]. Traditional corona treatment of textiles is challenging due to its uniformity and high local energy level caused by lightning discharge [7].

(b) Dielectric Barrier Discharge

In dielectric barrier discharge (DBD) plasma is produced by alternating or pulsing electric fields. DBD reactors typically use planar or cylindrical electrode designs. The electrodes are separated by a dielectric barrier, which can be made of ceramic, quartz, glass, enamel, silicone rubber, Teflon, or plastic. The spacing between the electrodes can vary from 0.1 mm to several centimeters. The dielectric barrier is a key feature of DBDs, preventing the discharge from transitioning to an electric arc and accumulating charge. Their chemical composition and surface characteristics influence charge buildup in dielectric materials. With its flexible electrode shape and dielectric material, this technology is ideal for atmospheric plasma creation and has a wide range of applications [4].

(c) Atmospheric Pressure Glow Discharge

Atmospheric Pressure Glow Discharge (APGD) is a technique that includes applying a frequency of a few Hz to MHz and an alternating current voltage of 1–100 kV between electrodes inserted into a cell containing a gas, commonly used Helium under atmospheric pressure. This approach is distinguished by its relatively lengthy duration, homogeneity, and low-to-moderate areal power densities, preventing surface heating or damage [6]. APGD systems have characteristics similar to low-pressure systems, such as generation method and treatment consistency with plasma, but do not require low pressures to function. The discharge is often generated using helium, which is becoming increasingly scarce and expensive. However, argon and nitrogen can also be utilized in some cases [3]

(d) Atmospheric Pressure Plasma Jet

The most recent surface modification technique, Atmospheric Pressure Plasma Jet (APPJ), is less stiff than a plasma torch but still quite effective at ambient temperatures [6]. These systems process materials at atmospheric pressure, enhancing machine capabilities and lowering costs and loading times [7]. This method uses specific gas mixtures, including helium, to prevent arcing [2]. Advantages of this method over DBD include the ability to manufacture homogenous reactive gasses and apply them to any shaped object. APPJ can only be used on one side of the treated material, facing the plasma [6].

3 Mechanisms of Surface Modification by Plasma Treatment

During plasma treatment on any textile material, the material's surface is attacked by various particles, including electrons, UV photons, ions, and radicals, all with varying energies. Some particles have enough energy to disrupt chemical bonds and cause modification on the fiber surface. Therefore, it is essential to discuss the three mechanisms describing how plasma species can modify any surface during the plasma treatment.

3.1 Surface Etching

Surface etching using plasma involves selectively removing surface material from substrates through physical etching and chemical reactions, forming volatile by-products. Plasma etching commonly uses air, oxygen, nitrogen, and inert gases (argon, helium, etc.) as carrier gases. Surface etching by plasma is also known as dry etching because, unlike other etching procedures, it avoids using aggressive wet chemicals such as acids. During plasma treatment, energetic species collide with the surface, breaking polymer chains and converting them into gaseous by-products, which are evacuated by a vacuum pump. Figure 1 shows a schematic representation of plasma etching [8]. Various factors, including plasma composition, the type of substrate, and operational conditions such as power, gas flow rate, and substrate positioning influence the etching rate. One of the applications of plasma etching is the removal of impurities and contaminants (oils, greases, and oxides) from the substrate surface which is called surface cleaning. Surface cleaning using plasma will improve the adhesion and hydrophilicity of the surface [9].

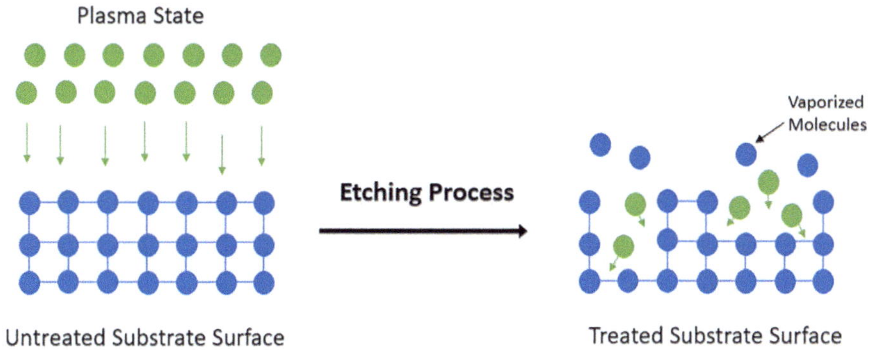

Fig. 1 Plasma surface etching process [8]

3.2 *Surface Activation*

Surface activation by plasma treatment increases surface-free energy by introducing new functional groups onto the treated surface. During plasma formation, gas molecules decompose into active particles with high kinetic energy, enabling them to move freely and bond with the material's surface. This results in functional groups, such as carboxylic, hydroxyl, amine, carbonyl, and nitrile groups, which modify the material's surface characteristics [10].

The introduction of new polar groups enhances fiber surface hydrophilicity and creates specific micro-roughness, improving accessibility for nanoparticles. Increased surface energy accelerates wetting and the finishing process. Surface activation enhances textile fibres' wettability, printability, and adhesion properties [11].

The plasma activation process employs various gases that do not polymerize but dissociate and react with the surface by introducing different functional groups. Argon, Oxygen, Helium, Carbon Dioxide, Ammonia, Nitrogen, and Hydrogen are some gases used for surface activation of textiles. Figure 2 shows a typical surface activation process.

3.3 *Surface Coating*

Surface coating by plasma enables the deposition of an organic monomer directly onto the substrate surface, forming a thin polymer coating through polymerization. The process begins with plasma bombardment to activate the surface, creating reactive bonding sites. Plasma then deposits and grafts molecular fragments while polymerizing them. A key technique is plasma polymerization or vapor deposition, which applies thin, highly cross-linked polymer films to textile surfaces, enhancing their technical properties. Organic monomers such as methane (CH_4), ethane (C_2H_6),

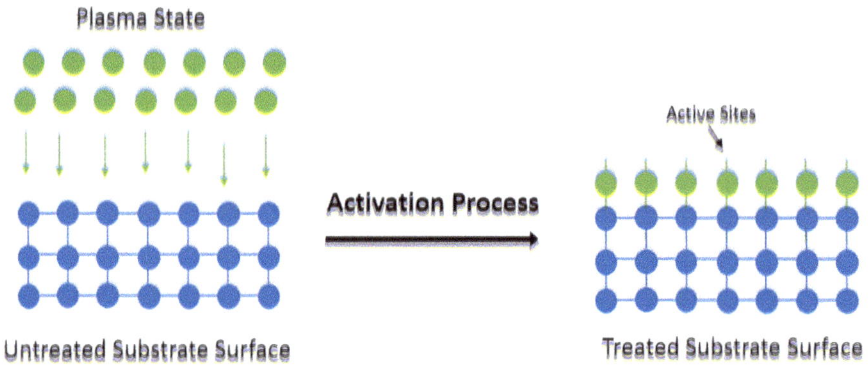

Fig. 2 Plasma surface activation process [8]

tetrafluoroethylene (C_2F_4), or hexafluoropropylene (C_3F_6), etc. in gas form are ionized or dissociated into active species, reacting to form plasma-polymerized films, a process independent of the substrate's nature [12, 13]. Figure 3 schematically presents the plasma surface coating by polymerization.

The deposited thin coatings can exhibit various properties based on the gas and process parameters. These coatings are highly cross-linked and adhere more strongly to the substrate than conventional polymerisation films. As a result, they are widely used in applications such as water repellency, anti-static finishes, antibacterial coatings, gas barrier layers, and UV-protective coatings [14].

Plasma can also activate surfaces and introduce free radicals, which act as initiators for graft-polymerizing vinyl monomers (in liquid or gas form) onto the substrate. This grafting technique tailors surface properties to suit specific applications. A common method involves treating the textile with non-polymerizing gas plasmas (e.g., argon or oxygen) to activate the surface, followed by exposure to a vinyl monomer for grafting. This approach allows for precise surface modification [13].

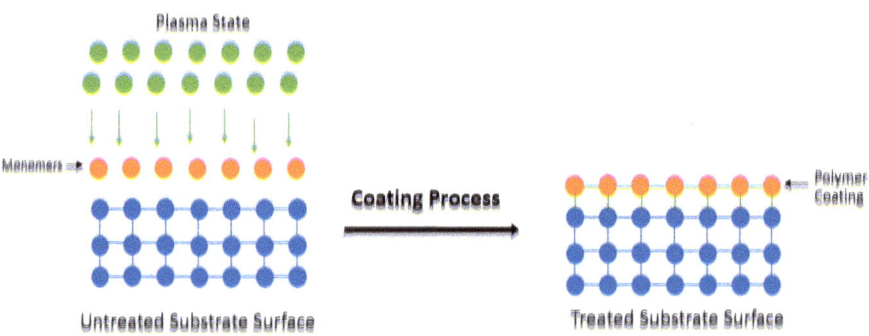

Fig. 3 Plasma surface coating process [8]

4 Techniques of Textile Surface Modification by Plasma Application

4.1 Plasma Induced Graft Polymeristaion on Textile Surface

In the textile, plasma grafting is a very successful technique for altering the surface characteristics of polymers and composites. Properties that involve wettability, adhesion, dyeing efficiency, flame retardancy, anti-microbial resistance, and mechanical strength are all enhanced by this method. Different textile applications, such as sportswear, protective apparel, medical textiles, and high-performance composites, can benefit from plasma grafting. Plasma grafting is a valuable technique for the textile and composite industries since it is environmentally friendly and can enhance the functionality and performance of materials without changing their bulk qualities. Key Steps of the Grafting Technique are:

- Activation: Plasma activates the surface of the polymer or composite, making it highly reactive.
- Grafting: Reactive monomers or polymers are introduced and bonded to the activated surface during the process.
- Post-treatment: The surface properties are enhanced by the newly grafted materials.

Plasma grafting is often used to modify the wettability of polymer and composite materials, making them more hydrophilic or hydrophobic, depending on the functional monomer grafted onto the surface (Water-Repellent Textiles). It can enhance the adhesion of functional coatings, such as flame retardants, UV blockers, or anti-microbial agents, to polymer or composite textiles [15].

4.2 Plasma-Assisted Nanoparticle Application on Textile Surface

Enhancing the functional qualities of textiles is made possible by using plasma technology in the textile industry, particularly for the deposition and modification of nanoparticles on textiles. Without changing the fabric's natural qualities, plasma treatments allow for the exact application of nanoparticles to fabric surfaces, providing various features like anti-microbial, water-repellent, UV-protective, and self-cleaning qualities.

Plasma-assisted nanoparticle synthesis refers to a process where plasma (ionized gas) generates and manipulates nanoparticles. The primary advantage of using plasma is that it can induce reactions at lower temperatures without solvents or harmful chemicals, making it an environmentally friendly method. It is also a dry process, which minimizes water and energy usage, making it more sustainable than traditional methods [16].

The applications of plasma assisted nano-particle treated textiles are listed below.

(a) Medical Textiles: Plasma treatment of textiles with silver or copper nanoparticles can be used to produce anti-microbial fabrics for wound dressings, hospital linens, and surgical garments, helping to prevent infections and promote healing.
(b) Sportswear and Activewear: Silver and zinc oxide nanoparticles can create odor-resistant and antibacterial fabrics, ensuring that sportswear remains fresh even after heavy use.
(c) Outdoor Gear: Fabrics treated with silica or titanium dioxide nanoparticles can become water-resistant, UV-protective, and self-cleaning, making them ideal for outdoor jackets, tents, and other gear exposed to harsh environmental conditions.
(d) Protective Clothing: Plasma treatment of textiles with nanoparticles can also enhance the durability and protective properties of workwear or fire-resistant clothing fabrics.
(e) Fashion Industry: Nanoparticle-treated textiles can offer unique visual effects, such as reflective fabrics or those that change color under UV light, opening up new design possibilities in the fashion industry.

The advantages of plasma-assisted nanoparticle impregnation on textiles are listed below.

(a) Sustainability: Plasma processes are environmentally friendly because they don't require harmful chemicals, solvents, or high temperatures, reducing waste generation and energy consumption. The process is dry and solvent-free, minimizing water and chemical use, which is particularly advantageous for industries aiming to reduce their environmental impact.
(b) Size and Shape Control: Plasma-assisted techniques allow for precise control over nanoparticle size, shape, and distribution. By adjusting factors such as power, pressure, and gas composition, the morphology of the nanoparticles can be carefully controlled during synthesis. This ability to fine-tune the process is crucial for applications that demand nanoparticles with specific characteristics.
(c) High Purity and Uniformity: Plasma processes produce nanoparticles with high purity because they often occur in a vacuum or low-pressure environment, minimizing contamination. The uniform distribution of nanoparticles is essential for ensuring consistent performance across the entire surface of textiles or other materials.
(d) Scalability and Efficiency: Plasma-assisted nanoparticle synthesis can be performed at relatively low temperatures, which allows for rapid processing and large-scale production, making it a scalable option for industrial applications. The process can be easily integrated into existing production lines, particularly in industries like textiles, where nanoparticle deposition is needed on large quantities of fabric.

5 Applications of Plasma in Textile Processing

5.1 Pretreatment

Atmospheric air plasma treatment offers very efficient removal of sizing material from cotton fabric and potentially saves water consumption [9]. Kan et al. [17] compared atmospheric pressure plasma (APP) treatment with conventional wet treatments (desizing, scouring, bleaching) on grey cotton fabric. Using helium and oxygen as gases, APP effectively removed starch-based sizing agents but did not improve fabric whiteness compared to traditional methods. Wang et al. [18] reported that oxygen plasma-treated cotton fabric exhibited notable etching and chemical modification, along with the highest degree of impurity removal, compared to traditional desizing methods such as amylase desizing and water desizing.

Scouring the plasma-treated samples does not remove significantly more sizing material than scoured samples without plasma treatment. For most plasma treatments and durations, the remaining sizing material is more resistant to removal during scouring. Cornelius et al. [19] suggested that the starch not etched by the plasma has undergone cross-linking.

Atmospheric air plasma treatment can also improve the desizing rate of polyvinyl alcohol (PVA) from 100% polyester [20]. Yan et al. [21] reported that the use of caustic soda for desizing PVA sizing material from cotton could be reduced by 60% compared to the traditional desizing process by combining it with atmospheric pressure low-temperature plasma treatment. Bae et al. [22] showed that ion implantation treatments have also proven effective in removing PVA and reducing the desizing time when using hot water. Cornelius et al. [23] reported that the plasma gas mixture significantly affects the removal of PVA from cotton/PET blends, with desizing efficiency increasing in the order $CF_4 > O_2 >$ He.

He/O_2 atmospheric pressure plasma jet pretreatment enhanced the $NaHCO_3$ desizing of cotton fabrics sized with starch phosphate and PVA. The plasma treatment roughened the surface, introduced oxygen-containing polar groups, and increased hydrophilicity, altering surface chemistry and morphology. This significantly improved the percentage desizing ratio (PDR) and reduced desizing time compared to $NaHCO_3$ desizing alone [24].

5.2 Improving Dyeability

Plasma technology can effectively remove natural or synthetic grease and wax from textile fibers while improving dye diffusion, enhancing color intensity, and increasing washing fastness for both natural and synthetic materials. Successful results have been achieved in dyeing polyamide, polyester, and wool fabrics using various dyestuffs, such as acid and disperse dyes. The dyeing performance of plasma-treated fibres is linked to surface chemistry and morphology changes. However, the extent of improved dyeability depends on factors like exposure time, gas mixture composition, and energy applied. Table 1 shows the application of plasma to improve the dyeability of different synthetic dyes.

Table 1 Application of plasma to improve the dyeability of different synthetic dyes

Dye	Fibre	Type of plasma	Carrier gases	References
Reactive dye	Cotton	Corona discharge Plasma	Air	[25, 26]
	Cotton	Glow discharge	Air, Nitrogen	[27]
	Cotton	DBD Plasma	Helium + Oxygen	[28]
	Wool	Glow Discharge	Helium	[29]
	Silk	Low-pressure plasma	Oxygen, Hydrogen, Nitrogen	[30]
	Viscose/cotton	Low-pressure plasma	Air	[31]
	Polyester	Glow discharge	Air, Nitrogen	[27]
	Nylon	Glow discharge	Air, Nitrogen	[27]
	Polyamide	DBD Plasma	Air	[32]
Disperse dye	Cotton	Corona discharge Plasma	Air	[33]
	Polyester	Glow Discharge Plasma	Air	[34]
	Polyester	Atmospheric-pressure plasma	Oxygen	[35]
	Polyester	Corona discharge Plasma	Air	[36]
	Polyester	Low-pressure plasma	Nitrogen, Air	[37, 38]
	Polypropylene	Low-pressure plasma	Oxygen	[39]
Direct dye	Cotton	Corona discharge Plasma	Air	[40]
	Polyamide	DBD plasma	Air	[32, 41, 42]
Basic dye	Polypropylene	DBD plasma	Air, Argon	[43, 44]
	Polypropylene	Low-pressure plasma	Nitrogen	[45]
	Polyamide	Low-pressure plasma	Nitrogen	[45]
	Polyester	Low-pressure plasma	Nitrogen	[45]
Acid Dye	Cotton	DBD plasma	Air + TETA, Helium + Oxygen	[28, 46]
	Wool	Glow discharge	Helium, Air + Helium and Helium + Oxygen	[29, 47]
	Polyester	Low-pressure plasma	Oxygen + Argon	[48]
	Polypropylene	DBD plasma	Argon	[44]
	Polypropylene	Low-pressure plasma	Oxygen	[39]
Vat dye	Polypropylene	DBD plasma	Argon	[49]

Natural dyes are increasingly valued for their non-toxic and eco-friendly properties. However, their application faces challenges due to inherent limitations, such as poor attraction to textile substrates, the need for mordants and fixers, low colour yield, and inadequate color fastness. These issues have hindered their use on an industrial scale. Plasma treatment has emerged as an eco-friendly solution to enhance the dyeing uptake of natural dyes on textiles. Table 2 shows the application of different plasma treatments on the natural dyeing of textiles.

5.3 *Chmeical Finishing*

Plasma surface engineering can modify textile surface properties to meet specific application needs or desired characteristics, such as anti-microbial treatment, self-cleaning, and flame retardancy, without altering the material's bulk properties. This technology offers versatile applications for enhancing textile functionality.

(a) Anti-microbial Finishing

Plasma treatment is very efficient in anti-microbial finishing. During plasma treatment, UV irradiation and free radicals are produced that can react with hydrocarbon bonds, weakening microbial cell walls and disrupting unsaturated bonds of nucleoproteins. Muller et al. [78] found that dielectric barrier discharges significantly reduced microorganisms on cotton fabrics. Shahidi and Ghoranneviss [79] showed that plasma treatment sterilized cotton fabrics inoculated with Staphylococcus aureus. Recently, plasma engineering has been used to bond chitosan derivatives, quaternary ammonium compounds, or silver nanoparticles onto textiles, enhancing anti-microbial properties and reducing risks of odors, infections, and related issues [80, 81] Shahidi et al. [82] highlighted the strong antibacterial activity of silver-sputtered cotton fabrics, unaffected by dyeing processes. Kostic et al. [83] demonstrated that incorporating silver ions into DBD-treated cotton/polyester fabrics created anti-microbial materials effective against a broad spectrum of microbes. Similarly, Sophonvachiraporn et al. [84] achieved comparable results by treating PET fabric with DBD plasma and coating it with chitosan. Senthilkumar and Karthik [85] studied the impact of argon plasma treatment on polyester fabrics, noting its effectiveness against Staphylococcus aureus and Escherichia coli. Malshe et al. [86] used atmospheric pressure glow discharge plasma to graft DADMAC onto nylon/cotton blends, producing durable anti-microbial military fabrics capable of detoxifying harmful microorganisms and preventing odor.

(b) Wrinkle-resistant Finishing

Cotton fibers exhibit poor elasticity and resilience due to the reformation of hydrogen bonds after the external force is applied. Plasma treatment can improve cotton's resilience by creating an etching effect on the fabric surface. This effect allows

Table 2 The application of different plasma treatments on natural dyeing of textiles

Natural dye	Fibre	Type of plasma	Carrier gases	References
Almond shell waste	Wool	Atmospheric pressure plasma	Air + Argon	[50]
Arnebia euchroma	Wool	Low pressure plasma	Oxygen	[51]
Caesalpinia Sappan	Polyester	Low pressure plasma	Oxygen	[52]
Chinese cork tree	Wool	Low pressure plasma	Oxygen, NH_3, CF_4	[53]
Cochineal	Cotton	DBD plasma	Air	[54]
	Wool	Low pressure plasma	Oxygen, NH_3, CF_4	[53, 55]
	Wool	Atmospheric pressure plasma	Air	[56]
	Nylon	Low pressure plasma	Oxygen, NH_3, CF_4	[53]
Cotton pods	Wool	Low pressure plasma	Oxygen	[51, 57]
Curcumin	Polyester	DBD plasma	Air	[58]
	Cotton	Low pressure plasma	Oxygen + Ammonia	[59]
Fruits Of *Sambucus Ebulus L*	Silk	Low pressure plasma	Oxygen	[60]
Grape leaves	Wool	Low pressure plasma	Oxygen	[61]
Green tea	Cotton	Low pressure plasma	Oxygen + Ammonia	[59]
Harmal seed	Wool	Low pressure plasma	Oxygen	[51]
Henna	Polyester	DBD plasma	Air	[62]
	Wool	Low pressure plasma	Oxygen	[63]
Lac	Silk	Low pressure plasma	Oxygen and Argon	[64]
	Polyester	Atmospheric pressure plasma	Air	[65]
Madder	Polyester	DBD plasma	Air	[66]
	Wool	Low pressure plasma	Oxygen, Argon + Cu and Fe Sputtering, NH_3, CF_4	[53, 67–69]
	Wool	Atmospheric pressure plasma	Air	[70]

(continued)

Table 2 (continued)

Natural dye	Fibre	Type of plasma	Carrier gases	References
	Cotton	Low pressure plasma	Oxygen	[71]
Onion skin and pulp	Cotton	Low pressure plasma	Oxygen	[72]
Pomegranate peel	Wool	Low pressure plasma	Oxygen	[73]
Pomegranate rinds	Cotton	Low pressure plasma	Oxygen + Chitosan	[74]
Shrimp shell extract	Wool	Low pressure plasma	Oxygen	[75]
Turmeric	Polyester	Atmospheric pressure plasma	Air	[65]
Weld	Cotton	Low pressure plasma	Oxygen	[71]
	Wool	Low pressure plasma	Argon + Cu and Fe Sputtering	[67, 68]
	Wool	Atmospheric pressure plasma	Air	[76]
Yerba mate	Silk	DBD plasma	Air	[77]
Yarrow	Wool	Low pressure plasma	Oxygen	[63]

finishing agents to penetrate the fibers more effectively, forming additional cross-linkages and increasing the wrinkle recovery angle, enhancing the fabric's overall performance.

Kan et al. [87] demonstrated that atmospheric plasma pretreatment enhances the application of BTCA wrinkle-resistant finishing on cotton fabric without adverse effects. Lam et al. [88] further optimized plasma treatment conditions for improving wrinkle resistance in BTCA-treated cotton fibers and found that treatment speed was the most influential factor, followed by jet-to-substrate distance and oxygen flow rate.

Lao et al. [89] achieved a durable, super hydrophilic finish on cotton fabrics by combining plasma treatment with the attachment of charged silica nanoparticles. The resulting nanoparticle-coated fabrics demonstrated excellent wrinkle-free performance, maintained tensile strength, and exhibited strong abrasion resistance even after 25,000 abrasion cycles. Overall, plasma pretreatment significantly improves the effectiveness of wrinkle-resistant finishes on cotton fabric.

Eyupoglu et al. [90] reported that cellulose acetate-based fabrics underwent a resin-based wrinkle-resistant treatment combined with oxygen plasma pretreatment and showed a significant improvement in the wrinkle-resistance angle for plasma-treated samples. ANOVA analysis confirmed that plasma treatment positively influenced wrinkle resistance. Additionally, plasma treatment increased tensile strength, while the wrinkle-resistant resin reduced it.

(c) Shrink Proofing of Woolen Fabrics

Felting is a significant challenge for wool garments due to the fiber's scale-like, hydrophobic surface. Low-temperature plasma (LTP) treatment has emerged as an environmentally friendly and effective solution to address this issue. Kana et al. [91] demonstrated that LTP treatment imparts shrink-resistant and anti-felting properties to wool fabric by etching the fiber surface, reducing differential friction coefficients, and minimizing shrinkage. Cai et al. [92] further highlighted that combining atmospheric plasma pretreatment with Synthappret BAP finishing significantly enhances shrink resistance and anti-felt performance, surpassing the limitations of the finishing agent alone. Mori et al. [93] reinforced these findings, showing that dielectric barrier discharge (DBD) treatment improves shrink-proofing effects as treatment time increases. Additionally, plasma sputtering treatment not only enhances shrink resistance and anti-felting properties but also transforms the wool surface from hydrophobic to super-hydrophobic [94]. Overall, plasma treatment offers a sustainable and highly effective alternative to traditional methods, improving wool fabric performance while meeting machine washability standards.

Abou-Taleb and El-Sayed [95] investigated AOX-free plasma-mediated coating with keratin on wool to make wool fabrics machine-washable and resistant to felting. Unlike untreated samples, the finished wool samples showed no felt ball formation during washing and demonstrated improved felting resistance, enhanced dyeability, and long-term durability.

(d) Water-resistant Finishing

Plasma treatment can create water-resistant finishes on textiles by depositing hydrophobic coatings or modifying surface properties, enhancing water repellency without compromising fabric breathability. This eco-friendly approach offers a durable and efficient alternative to traditional chemical treatments.

Thin nanostructured coatings were applied to aramid yarns using atmospheric pressure plasma by injecting acetone liquor into argon carrier gas, resulting in a reaction within the plasma state. This process enhanced the yarns' buoyancy and waterproof properties [96]. In a separate study[97], superhydrophobic cotton samples were created using low-pressure plasma-enhanced chemical vapor deposition to deposit hexamethyl disiloxane (HMDSO) coatings on Egyptian cotton. This method retained the fabric's breathability while achieving superhydrophobicity.

(e) Flame-retardant Finishing

Plasma treatment has the potential to enhance the flame retardant finishing. For instance, Raslan et al. [98] demonstrated that DBD atmospheric air plasma treatment, combined with aluminum oxide powder, significantly improved flame retardancy in polyester fabric. Qi et al. [99] also reported that the flame-retardancy of polyester fabrics can be enhanced with flame-retardant and drip-resistant coatings through a two-step process: initial pretreatment with oxygen plasma to generate additional grafting sites, followed by soaking and UV-curing. Nandy et al. [100] showed that using 3-hydroxyphenyl phosphinyl-propanoic acid (3HPP) and DBD plasma, flame-retardant properties were effectively bonded to polyester fabric without compromising strength. So, the plasma treatment enhanced washing fastness by creating a strong chemical bond between the fabric and the FR additive.

Ghoranneviss and Shahidi [101] found that low-pressure nitrogen plasma pretreatment enhanced the flame-retardant properties of cotton fabrics when treated with metallic salts, with no adverse effects from subsequent dyeing. In a follow-up study [102], they further showed that nitrogen plasma treatment followed by nanoclay application increased the flame retardancy of cotton fabric, highlighting the synergistic effects of plasma and chemical treatments.

(f) Self-cleaning Properties

The need to create "self-cleaning" surfaces on textiles is a significant and valuable goal, often referred to as the "lotus effect." This functionality enhances the utility and environmental quality of textiles. Plasma pretreatment can achieve this by significantly increasing water repellency without altering the material's bulk properties.

Qi et al. [103] found that oxygen plasma treatment followed by TiO_2 loading significantly enhanced the self-cleaning properties of PET fibers, improving stain degradation and bacterial activity due to increased active sites and TiO_2 adhesion. Similar results were reported by Shahidi et al. [104], Hashemizad et al. [105] and Baghriche et al. [106]. Mihailovic et al. [106] noted that oxygen and argon plasma-treated PET fabrics with TiO_2 exhibited superior self-cleaning properties, with oxygen plasma proving more effective. Mejia et al. [107, 108] also demonstrated that RF plasma and TiO_2 treatments improved self-cleaning in cotton and nylon fabrics, attributing this to plasma-induced surface roughness enhancing TiO_2 interaction. Akter et al. [109] reported that TiO_2 with plasma treatment efficiently minimizes the time required for self-cleaning defense uniforms (cotton-polyester blended fabric), simplifies maintenance, and ensures environmental sustainability. Saleem et al. [110] showed that the plasma-TiO_2-coated cotton fabric exhibited excellent self-cleaning efficiency, attributed to the strong bonding of nanoparticles on the plasma-functionalized surface.

5.4 *Wastewater Treatment*

One of the biggest concerns of the textile industry is the massive volume of effluents that contain various synthetic dyes, carcinogenic and toxic chemicals. Scientists are exploring conventional treatment methods—chemical, biological, and physical—but these face limitations, such as leaving non-reactive chemical residues, and inorganic solids, and being ineffective against low-concentration, non-biodegradable organic pollutants. To address the limitations of conventional methods, technologists are increasingly focusing on advanced oxidation processes, such as cold plasma technology [111].

Discharge plasma is highly effective for mineralising pollutants and eliminating pathogens in wastewater due to its anti-microbial properties. The process generates oxidizing species like radicals (O^-, OH^-, H^-), molecules (O_3, H_2O_2), UV light, and shockwaves, which degrade pollutants or break them into simpler compounds [112]. Plasma targets and inactivates pathogens by damaging their genetic material, proteins, and cell envelopes. Reactive oxygen and nitrogen species (RONS) are key to this mechanism, penetrating and disrupting viruses' outer protein layer and genetic material, ensuring effective disinfection [113].

The study by Pandiyaraj et al. [114] demonstrated that combining non-thermal atmospheric pressure plasma treatment with copper-doped cerium oxide nanoparticles synergistically degrades reactive black-5 dye (RB-5) in textile effluents. This method generated reactive species ($OH\cdot$, H_2O_2, O) that achieved 77.4% degradation efficiency in 25 min, decolorizing the solution. The treated effluent was non-toxic to E. coli and S. aureus, making it safe for agricultural use. This approach is highly effective for treating heavily polluted industrial textile wastewater.

Kusumandari et al. [115] used plasma corona discharge to treat batik wastewater containing synthetic dyes like naphthol and procion, achieving 55% degradation efficiency after 40 min at 11.5 kV. Reactive species (O_3, OH^+, N_2O^+, NO) generated by the plasma oxidized the dyes, reducing OH group intensity (confirmed by FTIR) due to oxidative damage. The treatment also improved wastewater properties such as pH, conductivity, Total Dissolved Solids (TDS), and dissolved oxygen, demonstrating its effectiveness for industrial dye degradation.

6 Examples of Industrial Scale Applications

A real world application of industrial plasma for textile treatment has emerged both continuous and semicontineous roll to roll system as a transformative technology in the textile industry, offering innovative solutions for enhancing the surface functional properties. It involves the use of ionized gases to modify textile surfaces at the nanoscale, enabling improved dyeability, adhesion, hydrophilicity and functionalization without altering the bulk properties of the material. Both low-pressure plasma and atmospheric pressure plasma treatments are commonly employed industrially. LPP provides precise control over surface modification through uniform discharges in vacuum environments, while APP enables roll-to-roll processing for large-scale applications. For instance, Grabher Group GmbH's Plasma-Nanocoater-Bag (Austria) employs a low-pressure plasma reactor with a volume of 12 cubic meters for the permanent modification of textile surfaces (Fig. 4).

This system features a roll-to-roll setup with a 17-m long plasma passage, capable of treating fabric up to 1.63 m in width. The reactor uses a variety of gases, including ammonia, acetylene, and ethylene, to create plasma environments that

Fig. 4 Semi-contineous roll to roll Nano-Plasma-coater-Bag of Grabher Group GmbH, Austria [116]

can deposit nanoporous coatings. These coatings are crucial for enabling substrate-independent dyeing, where the dyeing process becomes less dependent on the chemical nature of the textile fibers. The process also involves the deposition of functional groups onto the textile surfaces, which significantly enhances surface adhesion, hydrophilicity or hydrophobicity etc. Beyond dyeing, commercialy plasma technology offers broader applications in textile surface engineering. It can be used to deposit ultra-thin nanocoatings for water repellency, enhance adhesion for composite materials, and create biocompatible surfaces for medical textiles. For example, Plasma-Enhanced Chemical Vapor Deposition (PECVD) enables the formation of conformal barrier coatings that protect textiles from environmental contaminants while maintaining their flexibility and breathability. Companies like Plasmalex specialize in both LPP and APP technologies, providing turnkey solutions tailored to industrial needs. Their systems incorporate advanced features such as multi-zone treatment and precise control over up to 89 plasma parameters, ensuring consistent quality across diverse substrates. Their plasma polymerization technology is used to deposit polymer films (nano plasma coatings) onto textile substrates under both roll to roll process for atmospheric pressure and vacuume systems, creating tailored functional surfaces (Fig. 5). This treatment can improve properties like adhesion, hydrophobicity, biocompatibility, and chemical resistance without altering the textile's bulk properties.

On the other hand Sefar AG (Switzerland) utilizes one of the world's largest atmospheric pressure plasma systems for their high-performance filtration solutions

Fig. 5 Plasmalex roll-to-roll vacuum plasma technology [117]

[118]. This system, equipped with a high voltage power supply, is integrated into their inline textile processing, capable of treating materials up to 4.0 m wide at speeds up to 25 m per minute.

These industrial applications underscore the transformative potential of plasma technology in textile processing, offering solutions that are not only technically superior but also environmentally sustainable and economically viable paving the way for future innovations in plasma based textile technology.

7 Conclusion

Plasma technology represents a significant advancement in sustainable textile manufacturing, offering a viable alternative to traditional, environmentally damaging processes. The ability to precisely control and modify textile surfaces at the molecular level allows for the creation of high-performance fabrics with enhanced properties, such as improved dyeability, water resistance, and antimicrobial capabilities. Plasma treatment's dry, energy-efficient nature reduces water usage, minimises chemical waste, and lowers carbon emissions, addressing key environmental concerns associated with conventional textile processing. The versatility of plasma technology, demonstrated through various applications like pretreatment, dyeing, and finishing, positions it as the future green technology in the textile industry.

Despite the demonstrated benefits, the widespread adoption of plasma technology faces challenges, including cost-effectiveness, process optimization, and the scalability of equipment. Overcoming these obstacles requires ongoing research and development focused on optimising plasma parameters, exploring hybrid processing techniques, and creating new functional coatings. As industrial applications of plasma technology continue to expand, supported by advancements in both low-pressure and atmospheric pressure systems, the textile industry can move towards more sustainable practices. By addressing the existing challenges and further exploring the potential of plasma treatment, the industry can unlock new possibilities for creating innovative, high-quality textiles while minimising its environmental footprint.

References

1. A. Haji, H. Barani, Introduction of plasma technology, in *Advances in Plasma Treatment of Textile Surfaces* (2024), pp. 1–12. https://doi.org/10.1016/B978-0-443-19079-7.00002-6
2. R.A. Jelil, A review of low-temperature plasma treatment of textile materials. J Mater Sci **50**. https://doi.org/10.1007/s10853-015-9152-4
3. M.R.S. McCoustra, R.R. Mather, Plasma modification of textiles: understanding the mechanisms involved. Text. Prog. **50**(4), 185–229 (2018). https://doi.org/10.1080/00405167.2019.1637115
4. K.H. Kale, A.N. Desai, Atmospheric pressure plasma treatment of textiles using non-polymerising gases (2011)
5. M. Domonkos, P. Ticha, Low-temperature atmospheric pressure plasma treatment in the polymer and textile industry. IEEE Trans. Plasma Sci. **51**(7), 1671–1681 (2023). https://doi.org/10.1109/TPS.2023.3235266
6. A brief review on plasma treatment of textile materials. www.austinpublishinggroup.com
7. A. Hassabo, E. El-Sayed, Recent advances in the application of plasma in textile finishing (a review). J. Text. Color. Polym. Sci. (2021). https://doi.org/10.21608/jtcps.2021.67798.1050
8. H. Ssekasamba et al., Review of plasma surface engineering technology toward sustainable textile materials. Mater. Circ. Econ. **6**(1) (2024). https://doi.org/10.1007/s42824-024-00114-z
9. N.V. Bhat, An Netravali, av Gore, mp Sathianarayanan, ga Arolkar, and rr Deshmukh, "Surface modification of cotton fabrics using plasma technology." Text. Res. J. **81**(10), 1014–1026 (2011). https://doi.org/10.1177/0040517510397574
10. A. Vesel, R. Zaplotnik, G. Primc, M. Mozetič, Evolution of the surface wettability of PET polymer upon treatment with an atmospheric-pressure plasma jet. Polymers (Basel) **12**(1) (2020). https://doi.org/10.3390/polym12010087
11. R.M. Thurston, J.D. Clay, M.D. Schulte, Effect of atmospheric plasma treatment on polymer surface energy and adhesion. J. Plast. Film Sheeting **23**(1), 63–78 (2007). https://doi.org/10.1177/8756087907078698
12. S. Gaur, G. Vergason, Plasma polymerization: theory and practice, in *43rd Annual Technical Conference Proceedings* (Denver, 2000). https://vergason.com/wp-content/uploads/2018/11/Plasma-Polymerization-Theory-and-Practice.pdf
13. G. Nageswaran, L. Jothi, S. Jagannathan, Plasma assisted polymer modifications, in *Non-Thermal Plasma Technology for Polymeric Materials: Applications in Composites, Nanostructured Materials, and Biomedical Fields* (Elsevier, 2018), pp. 95–127. https://doi.org/10.1016/B978-0-12-813152-7.00004-4
14. D. Hegemann, Plasma polymerization and its applications in textiles, Indian J. Fibre Text. Res. **31**, 99–115 (2006). https://nopr.niscpr.res.in/bitstream/123456789/24497/1/IJFTR%2031%281%29%2099-115.pdf
15. D. Grujić, A. Savić, L. Topalić-Trivunović, J. Milanović, A. Haji, Plasma surface modification of textiles for application of natural dyes, in *Advances in Plasma Treatment of Textile Surfaces* (2024), pp. 323–365. https://doi.org/10.1016/B978-0-443-19079-7.00006-3
16. U.S. Meda, V.G. Soundarya, H. Madhu, N. Bhat, Nano-engineered textiles: development and applications. Mater. Sci. Eng. B **296**, 116636 (2023). https://doi.org/10.1016/J.MSEB.2023.116636
17. C.W. Kan, C.F. Lam, C.K. Chan, S.P. Ng, Using atmospheric pressure plasma treatment for treating grey cotton fabric. Carbohydr. Polym. **102**(1), 167–173 (2014). https://doi.org/10.1016/j.carbpol.2013.11.015
18. X. Wang et al., The application of atmospheric plasma for cotton fabric desizing. Fibers Polym. **20**(11), 2334–2341 (2019). https://doi.org/10.1007/s12221-019-9330-0
19. C. Cornelius, M. McCord, M. Bourham, P. Hauser, Desizing of starch sized cotton fabrics with atmospheric pressure plasma. Cellulose **24**(12), 5685–5695 (2017). https://doi.org/10.1007/s10570-017-1509-1

20. L.M. Li, Y.Q. Li, J.Q. Liu, Study of PVA on polyester desizing with atmospheric pressure dielectric barrier discharge plasma, in *Advanced Materials Research* (2013), pp. 11–15. https://doi.org/10.4028/www.scientific.net/AMR.681.11
21. J. Yan et al., Application of atmospheric pressure low temperature plasma in cotton fabric desizing. J. Phys. Conf. Ser. (2021). https://doi.org/10.1088/1742-6596/1790/1/012060. IOP Publishing Ltd
22. P.H. Bae et al., Size removal on polyester fabrics by plasma source ion implantation device. Chemosphere **63**(6), 1041–1047 (2006). https://doi.org/10.1016/j.chemosphere.2005.09.055
23. C. Cornelius, M. McCord, M. Bourham, P. Hauser, Desizing of PVA sized pet/cotton fabrics with atmospheric pressure plasma. Cellulose **25**(1), 869–881 (2018). https://doi.org/10.1007/s10570-017-1586-1
24. X. Li, Y. Qiu, The effect of plasma pre-treatment on NaHCO 3 desizing of blended sizes on cotton fabrics. Appl. Surf. Sci. **258**(11), 4939–4944 (2012). https://doi.org/10.1016/j.apsusc.2012.01.124
25. A. Patiño, C. Canal, C. Rodríguez, G. Caballero, A. Navarro, J.M. Canal, Surface and bulk cotton fibre modifications: plasma and cationization. Influence on dyeing with reactive dye. Cellulose **18**(4), 1073–1083 (2011). https://doi.org/10.1007/s10570-011-9554-7
26. M.E. Yazdanshenas, Dyeability improvement of cationized corona discharge treated cotton fabric. Asian J. Chem. **20**(5), 3543–3548 (2008). https://www.researchgate.net/publication/270284957
27. T.S.E.F. Karunarathne et al., Development of universal fabric dyeing and adhesion through RF glow discharge plasma treatment. Vacuum **204**, 111394 (2022). https://doi.org/10.1016/j.vacuum.2022.111394
28. J. Correia et al., Surface functionalization of greige cotton knitted fabric through plasma and cationization for dyeing with reactive and acid dyes. Cellulose **28**(15), 9971–9990 (2021). https://doi.org/10.1007/s10570-021-04143-8
29. M. Naebe et al., Effects of plasma treatment of wool on the uptake of sulfonated dyes with different hydrophobic properties. Text. Res. J. **80**(4), 312–324 (2010). https://doi.org/10.1177/0040517509338308
30. Y. Iriyama, T. Mochizuki, M. Watanabe, M. Utada, Plasma treatment of silk fabrics for better dyeability. J. Photopolym. Sci. Technol. **15**(2), 299–306 (2002). https://doi.org/10.2494/photopolymer.15.299
31. S. Md, M. Kabir, S.R. Das, Improvement of the levelness of viscose-cotton blended fabric dyed with reactive dye using low-temperature air plasma. Cellulose **31**(12), 7749–7764 (2024). https://doi.org/10.1007/s10570-024-06048-8
32. A. Pedro, F. Ribeiro, N. Carneiro, Polyamide 6.6 modified by dbd plasma treatment for anionic dyeing processes, in *Textile Dyeing* (InTech, 2011). https://doi.org/10.5772/21755
33. M. Gorenšek, M. Gorjanc, V. Bukošek, J. Kovač, P. Jovančić, D. Mihailović, Functionalization of PET fabrics by corona and nano silver. Text. Res. J. **80**(3), 253–262 (2010). https://doi.org/10.1177/0040517509105275
34. K. Gotoh, A. Yasukawa, Atmospheric pressure plasma modification of polyester fabric for improvement of textile-specific properties. Text. Res. J. **81**(4), 368–378 (2011). https://doi.org/10.1177/0040517510387207
35. M.M. Kamel, M.M. El Zawahry, H. Helmy, M.A. Eid, Improvements in the dyeability of polyester fabrics by atmospheric pressure oxygen plasma treatment. J. Text. Inst. **102**(3), 220–231 (2011). https://doi.org/10.1080/00405001003672366
36. W. Xu, X. Liu, Surface modification of polyester fabric by corona discharge irradiation. Eur. Polym. J. **39**(1), 199–202 (2003). https://doi.org/10.1016/S0014-3057(02)00169-6
37. Kh. El-Nagar, M.A. Saudy, A.I. Eatah, M.M. Masoud, DC pseudo plasma discharge treatment of polyester textile surface for disperse dyeing. J. Text. Inst. **97**(2), 111–117 (2006). https://doi.org/10.1533/joti.2005.0169
38. M. Lehocký, A. Mráček, Improvement of dye adsorption on synthetic polyester fibers by low temperature plasma pre-treatment. Czech J. Phys. **56**(S2), B1277–B1282 (2006). https://doi.org/10.1007/s10582-006-0362-5

39. A. Haji, R.S. Rahbar, A. Ebrahimzade, Improving the dyeability of polypropylene fibers using nanoclay and plasma treatment. Fibers and Polymers **22**(5), 1304–1313 (2021). https://doi.org/10.1007/s12221-021-0101-3
40. N. Carneiro et al., Dyeability of corona-treated fabrics. Color. Technol. **117**(5), 298–302 (2001). https://doi.org/10.1111/j.1478-4408.2001.tb00079.x
41. F.R. Oliveira, A. Zille, A.P. Souto, Dyeing mechanism and optimization of polyamide 6,6 functionalized with double barrier discharge (DBD) plasma in air. Appl. Surf. Sci. **293**, 177–186 (2014). https://doi.org/10.1016/j.apsusc.2013.12.126
42. A.P. Souto, F.R. Oliveira, M. Fernandes, N. Carneiro, Poliamid Boyama İşleminde DBD Plazma Modifikasyonunun Etkisi, in *Tekstil ve Mühendis* (2012), pp. 20–25. https://doi.org/10.7216/130075992012198505
43. N. Yaman, E. Özdoğan, N. Seventekin, Effect of surrounded air atmospheric plasma treatment on polypropylene dyeability using cationic dyestuffs. Fibers Polym. **14**(9), 1472–1477 (2013). https://doi.org/10.1007/s12221-013-1472-x
44. N. Yaman, E. Özdoğan, N. Seventekin, H. Ayhan, Plasma treatment of polypropylene fabric for improved dyeability with soluble textile dyestuff. Appl. Surf. Sci. **255**(15), 6764–6770 (2009). https://doi.org/10.1016/j.apsusc.2008.10.121
45. F. Ferrero, C. Tonin, R. Peila, F.R. Pollone, Improving the dyeability of synthetic fabrics with basic dyes using *in situ* plasma polymerisation of acrylic acid. Color. Technol. **120**(1), 30–34 (2004). https://doi.org/10.1111/j.1478-4408.2004.tb00203.x
46. H.A. Karahan, E. Özdoğan, A. Demir, H. Ayhan, N. Seventekin, Effects of atmospheric plasma treatment on the dyeability of cotton fabrics by acid dyes. Color. Technol. **124**(2), 106–110 (2008). https://doi.org/10.1111/j.1478-4408.2008.00129.x
47. Z. Cai, Y. Qiu, Dyeing properties of wool fabrics treated with atmospheric pressure plasmas. J. Appl. Polym. Sci. **109**(2), 1257–1261 (2008). https://doi.org/10.1002/app.25865
48. M.M. Hossain, A.S. Herrmann, D. Hegemann, Plasma deposition of amine-embedded nanoporous ultrathin films on polyester enables substrate independent surface dyeing. Plasma Process. Polym. **4**(S1), S1068–S1074 (2007). https://doi.org/10.1002/ppap.200732407
49. N. Yaman, E. Özdoğan, N. Seventekin, Atmospheric plasma treatment of polypropylene fabric for improved dyeability with insoluble textile dyestuff. Fibers Polym. **12**(1), 35–41 (2011). https://doi.org/10.1007/s12221-011-0035-2
50. Ö. Erdem İşmal, E. Özdoğan, L. Yıldırım, An alternative natural dye, almond shell waste: effects of plasma and mordants on dyeing properties. Colo. Technol. **129**(6), 431–437 (2013). https://doi.org/10.1111/cote.12047
51. A. Haji, Application of D-optimal design in the analysis and modelling of dyeing of plasma-treated wool with three natural dyes. Color. Technol. **136**(2), 137–146 (2020). https://doi.org/10.1111/cote.12445
52. Y. Park, K. Koo, S. Kim, J. Choe, Improving the colorfastness of poly(ethylene terephthalate) fabrics with the natural dye of *Caesalpinia sappan* L. Wood extract and the effect of chitosan and low-temperature plasma. J. Appl. Polym. Sci. **109**(1), 160–166 (2008). https://doi.org/10.1002/app.27899
53. T. Wakida, S. Cho, S. Choi, S. Tokino, M. Lee, Effect of low temperature plasma treatment on color of wool and nylon 6 fabrics dyed with natural dyes. Text. Res. J. **68**(11), 848–853 (1998). https://doi.org/10.1177/004051759806801110
54. H.M. Ahmed, K.A. Ahmed, H.M. Mashaly, A.A. El-Halwagy, Treatment of cotton fabric with dielectric barrier discharge (DBD) plasma and printing with cochineal natural dye. Indian J. Sci. Technol. **10**(10), 1–10 (2017). https://doi.org/10.17485/ijst/2017/v10i10/100973
55. T. Sajed, A. Haji, M.K. Mehrizi, M. Nasiri Boroumand, Modification of wool protein fiber with plasma and dendrimer: effects on dyeing with cochineal. Int J Biol Macromol **107**, 642–653 (2018). https://doi.org/10.1016/j.ijbiomac.2017.09.038
56. A. Quispe-Quispe, F. Lozano, V. Machaca-Machaca, J. Quispe-Marcatoma, Textile characteristics, medullation, and colorimetry of wool fiber dyed with dactylopius coccus using atmospheric pressure plasma jet (APPJ). Appl. Sci. **15**(1), 421 (2025). https://doi.org/10.3390/app15010421

57. A. Haji, M.K. Mehrizi, J. Sharifzadeh, Dyeing of wool with aqueous extract of cotton pods improved by plasma treatment and chitosan: Optimization using response surface methodology. Fibers Polym. **17**(9), 1480–1488 (2016). https://doi.org/10.1007/s12221-016-6457-0
58. A. Kerkeni, N. Behary, A. Perwuelz, D. Gupta, Dyeing of woven polyester fabric with curcumin: effect of dye concentrations and surface pre-activation using air atmospheric plasma and ultraviolet excimer treatment. Color. Technol. **128**(3), 223–229 (2012). https://doi.org/10.1111/j.1478-4408.2012.00367.x
59. M. Gorjanc, M. Mozetič, A. Vesel, R. Zaplotnik, Natural dyeing and UV protection of plasma treated cotton. Eur. Phys. J. D **72**(3), 41 (2018). https://doi.org/10.1140/epjd/e2017-80680-9
60. H. Dayioglu, D. Kut, N. Merdan, S. Canbolat, The effect of dyeing properties of fixing agent and plasma treatment on silk fabric dyed with natural dye extract obtained from Sambucus Ebulus L. plant. Proc. Soc. Behav. Sci. **195**, 1609–1617 (2015). https://doi.org/10.1016/j.sbspro.2015.06.201
61. A. Haji, F.K. Bizhaem, Optimization of oxygen plasma treatment to improve the dyeing of wool with grape leaves Sayyed Sadroddin Qavamnia (2016). https://www.researchgate.net/publication/307631895
62. S. Shahidi, M. Ghoranneviss, J. Wiener, Improving synthetic and natural dyeability of polyester fabrics by dielectric barrier discharge. J. Plast. Film Sheeting **31**(3), 286–308 (2015). https://doi.org/10.1177/8756087914567013
63. A. Haji, Natural dyeing of wool with henna and yarrow enhanced by plasma treatment and optimized with response surface methodology. J. Text. Inst. **111**(4), 467–475 (2020). https://doi.org/10.1080/00405000.2019.1642710
64. K. Boonla, S. Saikrasun, Influence of silk surface modification via plasma treatments on adsorption kinetics of lac dyeing on silk. Text. Res. J. **83**(3), 288–297 (2013). https://doi.org/10.1177/0040517512458344
65. P.S. Vankar, A. Gangwar, Natural dyeing mediated by atmospheric air pressure plasma treatment of polyester. Pigm. Resin Technol. **53**(5), 569–575 (2024). https://doi.org/10.1108/PRT-07-2022-0093
66. H. Dave et al., Use of dielectric barrier discharge in air for surface modification of polyester substrate to confer durable wettability and enhance dye uptake with natural dye eco-alizarin. Compos. Interfaces **19**(3–4), 219–229 (2012). https://doi.org/10.1080/15685543.2012.702594
67. Z. Motaghi, S. Shahidi, J. Wiener, Application of low temperature plasma on dye ability of wool with madder (2009). www.SID.ir
68. Z. Motaghi, S. Shahidi, The effect of plasma sputtering on dye ability of the polyester/wool blends fabrics. J Text Sci Eng **02**(03) (2012). https://doi.org/10.4172/2165-8064.1000112
69. H. Barani, H. Maleki, Plasma and ultrasonic process in dyeing of wool fibers with madder in presence of lecithin. J. Dispers. Sci. Technol. **32**(8), 1191–1199 (2011). https://doi.org/10.1080/01932691.2010.505525
70. C. Eyupoglu, S. Eyupoglu, N. Merdan, Z. Omerogullari Basyigit, Natural dyeing of air plasma-treated wool fabric with <scp> *Rubia tinctorum* </scp> L. and prediction of dyeing properties using an artificial neural network. Color. Technol. **140**(1), 91–102 (2024). https://doi.org/10.1111/cote.12700
71. A. Haji, A. Haji, Dyeing of cotton fabric with natural dyes improved by Mordants and Plasma treatment (2019). https://www.researchgate.net/publication/333669251
72. C. Chen, W.-Y. Chang (2007) Antimicrobial activity of cotton fabric pretreated by microwave plasma and dyed with onion skin and onion pulp extractions (2007)
73. J. Peran, S. Ercegović Ražić, A. Sutlović, T. Ivanković, M.I. Glogar, Oxygen plasma pretreatment improves dyeing and antimicrobial properties of wool fabric dyed with natural extract from pomegranate peel. Color. Technol. **136**(2), 177–187 (2020). https://doi.org/10.1111/cote.12464
74. A. Haji, Improved natural dyeing of cotton by plasma treatment and chitosan coating. Optimization by response surface methodology (2017)

75. M. Molakarimi, M. Khajeh Mehrizi, A. Haji, Effect of plasma treatment and grafting of β-cyclodextrin on color properties of wool fabric dyed with Shrimp shell extract. J. Text. Inst. **107**(10), 1314–1321 (2016). https://doi.org/10.1080/00405000.2015.1102459
76. A. Talebian, S. Habibi, P. Neshat, Green dyeing of weld on corona discharge treated wool fabric. J. Text. Inst. **112**(1), 144–151 (2021). https://doi.org/10.1080/00405000.2020.1744236
77. N.T. Vila, A. Ferreira, M.G. da Silva, M. Fernandes, S. Fiori, Surface modification of silk by (DBD) dielectric barrier discharge treatment for dyeing with natural dye yerba mate (Ilex paraguariensis). Proc. Eng. **200**, 170–177 (2017). https://doi.org/10.1016/j.proeng.2017.07.025
78. S. Müller, R.-J. Zahn, T. Koburger, K.-D. Weltmann, Smell reduction and disinfection of textile materials by dielectric barrier discharges. Nat Sci (Irvine) **02**(09), 1044–1048 (2010). https://doi.org/10.4236/ns.2010.29128
79. S. Shahidi, M. Ghoranneviss, Sterilization of cotton fabrics using plasma treatment. Plasma Sci. Technol. **15**(10), 1031–1033 (2013). https://doi.org/10.1088/1009-0630/15/10/13
80. M.M. Said et al., Multifunctional hydroxyapatite/silver nanoparticles/cotton gauze for antimicrobial and biomedical applications. Nanomaterials **11**(2), 429 (2021). https://doi.org/10.3390/nano11020429
81. J. Mosebolatan Jabar, Antimicrobial functional textiles, in *Textiles for Functional Applications,* (IntechOpen, 2021). https://doi.org/10.5772/intechopen.97806
82. S. Shahidi, S. Rezaee, E. Hezavehi, Effect of dyeing on antibacterial efficiency of silver coated cotton fabrics. Eur. Phys. J. Appl. Phys. **66**(1), 10801 (2014). https://doi.org/10.1051/epjap/2014130566
83. M. Kostic, N. Radic, B. Obradovic, S. Dimitrijevic, M. Kuraica, P. Skundric, Antimicrobial textile prepared by silver deposition on dielectric barrier discharge treated cotton/polyester fabric. Chem. Ind. Chem. Eng. Q. **14**(4), 219–221 (2008). https://doi.org/10.2298/CICEQ0804219K
84. P. Sophonvachiraporn, R. Rujiravanit, T. Sreethawong, S. Tokura, S. Chavadej, Surface characterization and antimicrobial activity of chitosan-deposited DBD plasma-modified woven PET surface. Plasma Chem. Plasma Process. **31**(1), 233–249 (2011). https://doi.org/10.1007/s11090-010-9276-x
85. P. Senthilkumar, T. Karthik, Effect of argon plasma treatment variables on wettability and antibacterial properties of polyester fabrics. J. Inst. Eng. (India) Ser. E **97**(1), 19–29 (2016). https://doi.org/10.1007/s40034-015-0074-0
86. P. Malshe, M. Mazloumpour, A. El-Shafei, P. Hauser, Functional military textile: plasma-induced graft polymerization of DADMAC for antimicrobial treatment on nylon-cotton blend fabric. Plasma Chem. Plasma Process. **32**(4), 833–843 (2012). https://doi.org/10.1007/s11090-012-9380-1
87. C.W. Kan, Y.L. Lam, C.W.M. Yuen, K.S. Chen, Using plasma treatment for enhancing conventional flame-retardant finishing of cotton fabric, in *Asia-Pacific International Symposium on the Basics and Applications of Plasma Technology [APSPT]* (2012)
88. Y.L. Lam, C.W. Kan, C.W.M. Yuen, Effect of plasma pretreatment on the wrinkle-resistance properties of cotton fibers treated with a 1,2,3,4-butanetetracarboxylic acid–sodium hypophosulfite system with titanium dioxide as a cocatalyst. J. Appl. Polym. Sci. **120**(3), 1403–1410 (2011). https://doi.org/10.1002/app.32960
89. L. Lao, L. Fu, G. Qi, E.P. Giannelis, J. Fan, Superhydrophilic wrinkle-free cotton fabrics via plasma and nanofluid treatment. ACS Appl. Mater. Interfaces **9**(43), 38109–38116 (2017). https://doi.org/10.1021/acsami.7b09545
90. S. Eyupoglu et al., The effect of oxygen plasma treatment on wrinkle resistance of cellulose acetate based fabric. J Natl Fibers **19**(17) (2022). https://doi.org/10.1080/15440478.2022.2131686
91. C.W. Kana, K. Chan, M. Yuen, Low-temperature plasma treatment of wool fabric for its industrial use (2004)
92. Z. Cai, Y. Qiu, Effect on the anti-felt properties of atmospheric pressure plasma treated wool. J. Appl. Polym. Sci. **107**(2), 1142–1146 (2008). https://doi.org/10.1002/app.26647

93. M. Mori, V. von Arnim, A. Dinkelmann, M. Matsudaira, T. Wakida, Modification of wool fibers by atmospheric pressure plasma treatment. J. Text. Inst. **102**(6), 534–539 (2011). https://doi.org/10.1080/00405000.2010.498173
94. S.M. Borghei, S. Shahidi, M. Ghoranneviss, Z. Abdolahi, Investigations into the anti-felting properties of sputtered wool using plasma treatment. Plasma Sci. Technol **15**(1), 37–42 (2013). https://doi.org/10.1088/1009-0630/15/1/07
95. M. Abou-Taleb, H. El-Sayed, Durable machine-washable wool via AOX-free plasma-mediated coating with keratin. J. Natl. Fibers **21**(1) (2024). https://doi.org/10.1080/15440478.2024.2408626
96. F. Sohbatzadeh, E. Shakerinasab, S. Mirzanejhad, Surface modification of aramid yarn by atmospheric pressure plasma: Reinforcement and floating properties. Polym. Test. **117**, 107836 (2023). https://doi.org/10.1016/j.polymertesting.2022.107836
97. M. Prado et al., Hexamethyldisiloxane coating by plasma to create a superhydrophobic surface for fabric masks. J. Market. Res. **17**, 913–924 (2022). https://doi.org/10.1016/j.jmrt.2022.01.003
98. W.M. Raslan, U.S. Rashed, H. El-Sayad, A.A. El-Halwagy, Ultraviolet protection, flame retardancy and antibacterial properties of treated polyester fabric using plasma-nano technology. Mater. Sci. Appl. **02**(10), 1432–1442 (2011). https://doi.org/10.4236/msa.2011.210194
99. L. Qi et al., Durable flame retardant and dip-resistant coating of polyester fabrics by plasma surface treatment and UV-curing. Prog. Org. Coat. **172**, 107066 (2022). https://doi.org/10.1016/j.porgcoat.2022.107066
100. S. Nandy, T.V. Sreekumar, S. Palaskar, N. Mehra, Plasma-assisted incorporation of flame-retardant chemicals for improved flame retardancy of polyester fabrics. Polym. Eng. Sci. **64**(11), 5411–5425 (2024). https://doi.org/10.1002/pen.26924
101. M. Ghoranneviss, S. Shahidi, Flame retardant properties of plasma pretreated/metallic salt loaded cotton fabric before and after direct dyeing. J. Fusion Energy **33**(2), 119–124 (2014). https://doi.org/10.1007/s10894-013-9642-9
102. S. Shahidi, M. Ghoranneviss, Effect of plasma pretreatment followed by Nanoclay loading on flame retardant properties of cotton fabric. J. Fusion Energy **33**(1), 88–95 (2014). https://doi.org/10.1007/s10894-013-9645-6
103. K. Qi, J.H. Xin, W.A. Daoud, C.L. Mak, Functionalizing polyester fiber with a self-cleaning property using anatase TiO_2 and low-temperature plasma treatment. Int. J. Appl. Ceram. Technol. **4**(6), 554–563 (2007). https://doi.org/10.1111/j.1744-7402.2007.02168.x
104. S. Shahidi, M. Ahmadi, A. Rashidi, M. Ghoranneviss, Effect of plasma treatment on self-cleaning of textile fabric using titanium dioxide. Micro Nano Lett. **10**(8), 408–413 (2015). https://doi.org/10.1049/mnl.2015.0053
105. S. Hashemizad, A. Haji, S. Saeed Mireshghi, C. Author, Environmentally friendly plasma pretreatment for preparation of self-cleaning polyester fabric with enhanced deposition of TiO2 nanoparticlesnet Environmentally friendly plasma pretreatment for preparation of self-cleaning polyester fabric with enhanced deposition of TiO2 nanoparticles. Artic. J. Biodivers. Environ. Sci. **5**(3), 220–226 (2014). http://www.innspub.net
106. D. Mihailović et al., Improved properties of oxygen and argon RF plasma-activated polyester fabrics loaded with TiO_2 nanoparticles. ACS Appl. Mater. Interfaces **2**(6), 1700–1706 (2010). https://doi.org/10.1021/am100209n
107. M.I. Mejía et al., Innovative UVC Light (185 nm) and radio-frequency-plasma pretreatment of nylon surfaces at atmospheric pressure and their implications in photocatalytic processes. ACS Appl. Mater. Interfaces **1**(10), 2190–2198 (2009). https://doi.org/10.1021/am900348u
108. M.I. Mejía et al., Self-cleaning modified TiO2–cotton pretreated by UVC-light (185nm) and RF-plasma in vacuum and also under atmospheric pressure. Appl. Catal. B **91**(1–2), 481–488 (2009). https://doi.org/10.1016/j.apcatb.2009.06.017
109. S. Akter et al., Application of low-temperature air plasma for the enhancement of defense fabric's self-cleaning property. ACS Omega **9**(39), 41053–41066 (2024). https://doi.org/10.1021/acsomega.4c06893

110. M. Saleem, M.Y. Naz, S. Shukrullah, S. Ali, S.T.A. Hamdani, Ultrasonic biosynthesis of TiO2 nanoparticles for improved self-cleaning and wettability coating of DBD plasma pre-treated cotton fabric. Appl. Phys. A **127**(8), 608 (2021). https://doi.org/10.1007/s00339-021-04767-4
111. P. Gururani et al., Cold plasma technology: advanced and sustainable approach for wastewater treatment. Environ. Sci. Pollut. Res. **28**(46), 65062–65082 (2021). https://doi.org/10.1007/s11356-021-16741-x
112. W.S. Abdul-Majeed, E. Karunakaran, C.A. Biggs, W.B. Zimmerman, Development of wastewater treatment system based on cascade dielectric barrier discharge plasma atomizers. J. Environ. Sci. Health Part A **50**(12), 1249–1258 (2015). https://doi.org/10.1080/10934529.2015.1055150
113. A. Filipić, I. Gutierrez-Aguirre, G. Primc, M. Mozetič, D. Dobnik, Cold plasma, a new hope in the field of virus inactivation. Trends Biotechnol. **38**(11), 1278–1291 (2020). https://doi.org/10.1016/j.tibtech.2020.04.003
114. K. Navaneetha Pandiyaraj, D. Vasu, M.C. Ramkumar, R.R. Deshmukh, R. Ghobeira, Improved degradation of textile effluents via the synergetic effects of Cu-CeO2 catalysis and non-thermal atmospheric pressure plasma treatment. Sep Purif Technol. **258**, 118037 (2021). https://doi.org/10.1016/j.seppur.2020.118037
115. K. Kusumandari, T.E. Saraswati, S.W. Wulandari, A. Qusnudin, Atmospheric air plasma corona discharge for dye degradation of Indonesian batik wastewater. Indian J. Phys. **96**(4), 1001–1008 (2022). https://doi.org/10.1007/s12648-021-02082-5
116. https://www.grabher-group.company/textile-finishing
117. https://www.plasmalex.com/roll-to-roll/
118. H. Mohammad, H. Dirk, Substrate independent dyeing of synthetic textiles treated with low-pressure plasmas, in *Textile Dyeing* (InTech, 2011), pp 173–194. https://doi.org/10.5772/21797. (14 Dec 2011)

Chapter 9
Sustainable Practices and Circularity in Fashion and Textiles Manufacturing

Mazed Islam, Md Shamsuzzaman, Raju Ahmed, Md. Abdullah Al. Mamun, and Abu Sadat Muhammad Sayem

Abstract The recent emergence of sustainable practices and circularity with numerous innovations is recognized as a new horizon of the fashion industry, which can help transform the fashion business to achieve sustainability goals. This chapter presents emerging sustainable practices in manufacturing operations of fashion and textiles (e.g., yarn, knitting, weaving, apparel, printing, washing, packaging), associated benefits, key challenges, and stakeholder roles in fostering circular economy broadly. Drawing upon the manufacturing operational stages, this study highlights the nature and transformative impact of emerging sustainable practices (e.g., circular design principles, bio-based materials, close loop system, sustainable dyeing and finishing, digital printing zero-waste pattern, ethical sourcing) and benefits for diverse stakeholders. It provides insights into the practical and theoretical applications of sustainable practices contributing to the broader understanding of circularity, and opportunities for wider stakeholders to enable a more sustainable fashion and textiles business. Readers can get an overview of the contemporary innovations in fashion circularity.

Keywords Fashion and textiles · Fashion manufacturing · Sustainable practices · Circularity

M. Islam (✉)
Department of Fashion and Textiles, University of Southampton, Southampton, UK
e-mail: m.m.islam@soton.ac.uk

M. Shamsuzzaman
Department of Textile Engineering, World University of Bangladesh, Dhaka, Bangladesh

R. Ahmed · Md. A. Al. Mamun
Department of Materials, The University of Manchester, Manchester, UK

A. S. M. Sayem
Fashion Institute, Manchester Metropolitan University, Manchester, UK

A. S. Muhammad Sayem and P. Goswami (eds.), *SDG 12 and Global Fashion Textiles Production*, SDGs and Textiles, https://doi.org/10.1007/978-981-96-9051-0_9

1 Introduction

The global fashion industry, valued in the trillions of US dollars, has profound impacts on society, economy, and environment [1]. With rapid expansion, concerns over its environmental footprint and labour practices have intensified. Sustainable manufacturing in textiles has emerged as a key response, aiming to reduce environmental harm, ensuring ethical production. This shift is crucial as the industry faces increasing pressure over its resource consumption and waste generation [2]. Sustainability in fashion is gaining momentum due to the urgent need for environmental responsibility and resource efficiency. Innovations in sustainable materials are transforming textile production, providing eco-friendly alternatives, and promoting circularity and ethical manufacturing. The fashion sector is a major contributor to global environmental degradation, consuming large amounts of water, energy, and raw materials, while also generating pollution from microplastics, toxic dyes, and waste [3]. The traditional "take, make, dispose" model worsens these problems, with a significant portion of textile waste ending up in landfills or being incinerated [4].

In response to these challenges, sustainable practices and circularity have become centre to fashion's future. Circularity focuses on extending the life cycle of products and materials through practices like recycling and upcycling, which help reduce waste and increase resource efficiency. Unlike the linear model, circularity aims to keep materials in use as long as possible, promoting regeneration and reducing reliance on new resources. Sustainable practices such as using renewable materials, implementing energy-efficient production, and adopting zero-waste design principles not only lessen environmental impact but also promote fair labor conditions [5]. The growing importance of sustainability and circularity is further driven by increasing regulatory pressures and consumer demand for ethical products. Governments are introducing stricter environmental regulations, while consumers are becoming more conscious of social and environmental consequences of their purchases. Fashion brands, are recognizing the long-term advantages of sustainable practices too, such as cost savings, enhanced brand reputation, and access to new markets. Transitioning to a circular model is not just necessary for mitigating environmental and social impacts, but also an essential strategy for ensuring the fashion industry's sustainable future [6].

This book chapter aims to provide an overview of emerging sustainable practices in the manufacturing of fashion and textiles to guide managers in making strategic decision to adopt the best practices.

2 Research Methodology

The research is based on a comprehensive literature review to delve into the state of the art of academic studies, integrated by analyzing relevant studies to allow mapping of sustainable practices in fashion and textiles linked with manufacturing operations and areas (e.g., fiber, yarn, knitting, weaving, apparel, printing, washing, packaging). To conduct the literature review, we used the following keywords "sustainable practices," "fashion and textiles," "circularity," "fashion manufacturing," on the most relevant academic databases and scientific journals including Science Direct, Springer, Wiley Online Library, Proquest, and Emerald. The search was limited to scientific literature published between 2019 to 2024. Papers and articles written only in English Language were included in the review. A total of 33 pertinent papers were included in this study, and sustainable practices were mapped against various manufacturing operations of fashion and textiles along with key features and environmental benefits as presented in resutls and discussion section.

3 Results and Discussion

3.1 Sustainable Practices in Fashion and Textiles Manufacturing

3.1.1 Materials Innovations

Recent innovations in textile materials are driving sustainable practices with a strong focus on bio-based, biodegradable, and recycled materials. Bio-fabricated textiles, such as mycelium leather and algae-based fabrics, provide sustainable alternatives to conventional textiles and synthetic leathers by naturally decomposing, minimizing long-term waste [7]. Plant-based fibers like banana fiber, Piñatex (pineapple leaves), and apple leather also offer eco-friendly alternatives to cotton, requiring fewer resources and being biodegradable or recyclable [8].

The use of recycled polyester derived from plastic bottles or post-consumer textile waste reduces reliance on virgin polyester. Besides, upcycling and closed-loop systems transform discarded textiles into new fabrics, extend material life cycles, reduce waste, and promote a circular economy [9]. Smart textiles, incorporating wearable electronics and temperature-responsive materials, enhance garment functionality and extend garment life [10]

3.1.2 Textile Fiber Production

Textile fiber production largely depends on water usage, energy consumption, and chemical waste. Organic fibers, such as organic cotton, are cultivated without synthetic pesticides or fertilizers, improving soil health, and enhancing biodiversity. Regenerative agriculture, focusing on soil restoration, helps sequester carbon, making fiber crops more sustainable [11]. Recycled and upcycled fibers reduce reliance on virgin materials and help divert waste from landfills. Closed-loop systems, such as lyocell (Tencel), use non-toxic solvents to recycle fibers and water, minimizing chemical pollution and water consumption [12]. Alternative fibers such as hemp, bamboo, and nettle require fewer resources than conventional cotton and offer a sustainable alternative to synthetic fibers like polyester. Biodegradable fibers like wool, organic cotton, and silk are also crucial, as they naturally decompose, unlike synthetic fibers, which can persist for hundreds of years.

3.1.3 Textile Yarn Manufacturing

Sustainable yarn manufacturing uses eco-friendly and biodegradable raw materials, such as organic cotton, recycled fibers, and biopolymers. Organic cotton yarn avoids synthetic pesticides, preserving soil health and water quality, while recycled yarns reduce reliance on virgin materials [9]. Innovations in spinning processes, like compact and vortex spinning, enhance energy efficiency by reducing consumption and emissions. Closed-loop systems recycle by-products, further reducing waste [13].

3.1.4 Textile Knitting

Sustainable practices in knit fabric manufacturing are essential for reducing environmental impact and promoting resource efficiency. The use of eco-friendly raw materials, such as organic cotton, bamboo, recycled polyester, and biodegradable fibers, is crucial in minimizing resource depletion and pollution [14]. Energy-efficient knitting machines, including modern circular and flat machines, reduce electricity consumption by operating at high speeds with minimal energy use. Incorporating renewable energy sources like solar or wind power in factories further lowers the carbon footprint [15]. Waste minimization strategies, including precision knitting and design optimization, reduce fabric scraps and enhance material efficiency. Recycling post-industrial knit fabric waste into new products supports circular economy principles [16].

3.1.5 Textile Weaving

For sustainable practices in weaving, this sector is implementing new technologies with the combination of man, machines, and materials to support transition to circularity such as waste reduction and energy savings [17]. The use of recyclable or biodegradable warp and weft materials supports circular economy principles [18]. Water and chemical management are crucial, with innovations like waterless sizing and the use of bio-based sizing agents reducing both water and chemical usage while preserving yarn integrity [19]. Minimizing the energy consumption is another concern and for this, machine manufacturers are introducing new energy efficient technologies such as versatile and interactive movable creel, rapid pattern changes, minimal setup times by utilising the automatic machines including drawing-in, denting, knotting and quick style change systems etc. Moreover, energy-efficient looms like air-jet and rapier looms, reduce power consumption and operate with precision, minimizing yarn waster [18]. Waste reduction strategies, such as smart weaving techniques that optimize fabric design and reduce selvage waste, also contribute to sustainability. Social sustainability practices, including fair wages, worker safety, and community engagement, align environmental efforts with ethical standards, fostering a more sustainable and responsible textile industry [19].

3.1.6 Textile Dyeing/Coloration

The sustainable practices in the textile dyeing industry are replacing traditional methods that are water and chemical-intensive, creating wastewater and sludge [19]. Emerging innovations like waterless supercritical CO_2 dyeing are replacing water with carbon dioxide in its supercritical state, while foam dyeing reduces water usage by up to 90%, helping to prevent wastewater contamination and lower chemical consumption [21].

Natural dyes, derived from plants, insects, and minerals, are low-impact dyes, safer, biodegradable, and require less energy than synthetic dyes [20]. Closed-loop systems in dyeing operations recycle water and chemicals, decreasing both freshwater consumption and toxic wastewater discharge. Besides, certifications like OEKO-TEX® and Global Recycled Standard (GRS), are guiding the textile dyeing industry toward a more sustainable future by reducing water, energy, and chemical use.

3.1.7 Textile Printing/Embroidery

Digital printing technology offers eco-friendly alternative to traditional screen printing, reducing the use of water, energy, and chemicals. Digital printing applies ink directly to fabric using inkjet technology, allowing for high customization with minimal material waste [21]. Water-based inks and solvent-free dyes are gaining popularity over traditional plastisol inks and solvent-based dyes. These alternatives contain fewer volatile organic compounds (VOCs), improving air quality and worker safety. Additionally, natural and plant-based dyes are being used more widely, offering biodegradable, non-toxic alternatives to synthetic dyes [22]. Sustainable embroidery practices also minimize electricity consumption, use of recycled threads, and bio-based fabrics that reduce waste and support circularity in production. Closed-loop systems are increasingly adopted to recycle materials such as fabric scraps, inks, and threads, minimizing landfill waste.

3.1.8 Fashion Sewing Operations and Apparel Making

Key sustainable practices in garment industry are focusing on minimizing waste, reducing energy consumption, and fostering circularity while ensuring ethical labor conditions. Zero-waste pattern making, for example, optimizes fabric usage to minimize offcuts, significantly reducing textile waste. Additionally, modular design and design for disassembly are explored to facilitate garment recycling and upcycling, supporting a circular economy [23]. Energy-efficient machinery and automation in the sewing process reduce energy consumption, while the use of renewable energy sources like solar and wind power further decreases carbon emissions [24]. Digital garment prototyping and 3D knitting technologies also minimize material waste and overproduction by reducing the need for traditional sample-making [25]. Circular fashion such as eco-friendly product is promoted through designs aimed at longevity, repair, and recycling, with take-back programs encouraging garment reuse among end consumers [26]. Ethical labour practices and supply chain transparency, ensured by certifications like Fair Trade and blockchain technology, further drive sustainability [26].

3.1.9 Garment Washing and Finishing

Garment washing and finishing processes are associated with significant environmental impacts, including excessive water use, chemical waste, and high energy consumption. However, emerging sustainable practices including waterless or low-water technologies, such as laser fading, ozone washing, significantly reduce water consumption, eliminate harmful chemicals, and safer alternatives. Ozone gas efficiently cleans and bleaches garments, while laser technology replaces water-intensive stonewashing, particularly in denim production [27]. Enzymatic washing, which uses enzymes to create fading effects, requires less water and fewer chemicals, providing a gentler and more sustainable alternative for both denim and non-denim fabrics [28].

3.1.10 Fashion Trimmings and Accessories

A significant development is the use of biodegradable materials in fashion accessories. Items traditionally made from plastic or metal, such as buttons and buckles, are now making from renewable and biodegradable materials like coconut shells, wood, and recycled plastic. Zippers are also being produced from recycled or biodegradable components, offering sustainable alternatives. Recycled and upcycled materials are increasingly used in manufacturing accessories like belts and bags, with reclaimed metals for buckles and clasps, promoting the circular economy [29]. Digital printing has replaced traditional labelling techniques, reducing ink, chemicals, and water usage.

3.1.11 E-textiles

Sustainable practices in e-textiles is increasingly an area of research and innovation, taking into consideration the environmental and social impacts of integrating textiles with electronics. Material innovations within this sector are a cornerstone, where bio-based and recycled alternatives have gained a lot of popularity. Textiles bio fabricated from algae or cellulose, for example, produce biodegradable and renewable options, thus reducing the demand for synthetic materials [10, 29]. Recycled polymers and conductive threads produced from post-consumer waste further enhance the circularity of e-textiles, minimizing waste and conserving resources. Graphene-based or carbon nanotube conductive ink has advanced alternatives, where copper and silver are two major heavy metals, hence even reducing further reliance on non-renewable resources [9]. Energy-efficient production methods, such as supercritical CO_2 dyeing, additive manufacturing, and 3D printing of conductive patterns, have greatly reduced the use of water, energy, and chemicals, setting a benchmark for sustainable manufacturing practices [25]. Similarly, integrating electronic components by laser-assisted methods requires less energy compared to traditional soldering, which helps to conserve energy and reduce emissions [19]. The principles of circular design also benefit the domain of e-textiles, for example, modularity practice for easy disassembly and recycling of components for extended life cycles and more efficient resource use [23]. Closed-loop systems address the metals and fiber recovery of the end-of-life management of e-textiles, therefore reducing their contribution to waste and pollution. Biodegradable electronics and organic polymer-based circuits are being developed to further reduce the ecological impact of electronic components [29]. Ethical practices involving transparent supply chains, fair labor standards, and certifications like Fair Trade will be in higher demand with the inclusion of not just environmental but also social aspects of sustainability [26]. Blockchain technology is increasingly being applied for enhanced traceability to check on sustainable practices that will build trust among consumers. These developments mark a comprehensive beginning of designing environmentally responsible, socially equitable, economically viable e-textile systems in harmony with global sustainable imperatives.

3.2 *Mapping Key Sustainable Practices in Fashion Manufacturing*

Manufacturers of fashion and textiles are emphasizing sustainable innovation and practices in various manufacturing stages which provides environmental and economic benefits. In industrial production, uses of raw materials are largely dependent on bio-based fabric, organic fibers, and recyclable materials highlighting sustainable material innovation. These are promoting biodegradability, decreasing the uses of resources, and increasing circular economy practices, Closed loop systems, and smart/wearables integration are optimizing the utilization of raw materials that extend the usefulness of a garment. Organic or regenerative agriculture or farming that are influencing the cultivation of hemp and bamboo as an alternative to cotton fibers, reduces the amount of water, and chemicals used in the fiber production process. Yarn and fabric manufacturers are also utilizing energy-efficient technologies, renewable energy sources, and waste-reduction techniques. In the dyeing and printing processes, researchers are attempting to use waterless technologies such as closed-loop systems, nature-based dyes, and supercritical CO_2 approaches. Further, advancements in automated patternmaking, the introduction of modular systems for designs, and digital prototyping are cutting waste and emphasizing circular economy practices in the garment industry. To save energy, chemicals, and water in garment washing and finishing, numerous technologies like enzymatic treatment, laser, and ozone finishing are employed. Lastly, incorporating 3R's (reuse, recycle, and repurpose) promotes circular design principles, and uses of biodegradable materials are rising in garment making and production of trims and accessories. Together addressing these practices in multiple departments, fashion and textile industries can incorporate innovative practices, exercise closed-loop systems, and transition from linear to circular practices. These are reducing waste, preserving resources, cutting carbon emissions, and promoting sustainable practices. The Table 1 provides the mapping summary of sustainable practices identified.

Table 1 Sustainable innovation and practices in fashion and textile manufacturing

Areas	Sustainable innovation	Specific practices	Environmental benefit
Materials innovation	Bio-fabricated textiles	Bio-based materials such as mushroom leather and algae-based fabrics	Biodegradable, reduces long-term waste, sustainable alternative to synthetic textiles
	Plant-based fibers	Plant based banana fiber, Piñatex, and apple leather as alternatives to cotton	Lower resource consumption, biodegradable, and recyclable
	Recycled and upcycled materials	Recycled polyester from plastic bottles, post-consumer waste, and upcycling of textiles into new fabrics	Reduces demand for virgin polyester, reduces textile waste
	Closed-loop systems	Materials designed for continuous recycling like circulose (fiber from discarded textiles)	Promotes circularity, reduces resource waste
	Smart textiles	Wearable technologies electronics, temperature-responsive materials	Enhances garment functionality, extends garment life
Fiber production	Organic fiber production	Cultivation of fibers like organic cotton without synthetic pesticides or fertilizers	Reduces chemical runoff, protects biodiversity, improves soil health, and enhances water quality
	Regenerative agriculture	Agricultural practices focusing on soil health restoration, such as crop rotation and cropping	Sequesters carbon in soil improves soil fertility and reduces the need for synthetic fertilizers
	Recycled fibers	Recycled polyester fiber from PET bottles and firber from textile waste	Reduces reliance on virgin materials, diverts waste from landfills
	Closed-loop fiber systems	Systems like lyocell (Tencel) use non-toxic solvents to recycle water and fibers, minimizing waste and pollution	Reduces water use, and chemical discharge, and creates a circular production system
	Alternative fibers (Hemp, Bamboo, Nettle)	Plant-based fibers like hemp, bamboo, and nettle, requiring fewer resources (water, pesticides) than cotton	Less water and pesticide use, sustainable alternatives to cotton and polyester

(continued)

Table 1 (continued)

Areas	Sustainable innovation	Specific practices	Environmental benefit
	Biodegradable fibers	Natural fibers such as wool, organic cotton, and silk decompose naturally	Reduces landfill waste, promotes eco-friendlier disposal, and avoids the environmental burden
	Water-efficient practices	Adoption of methods to reduce water consumption, (e.g., drip irrigation, rainwater harvesting in cultivation	Reduces water usage, especially for cotton
Yarn manufacturing unit	Organic cotton yarn	Organic cotton avoids synthetic pesticides and fertilizers, protecting soil health and water quality	Preserves biodiversity, soil health, and reduces water pollution
	Recycled yarn	Yarn produced from post-consumer or industrial waste or recycled fibers	Reduce demand for virgin fibers, and minimizes environmental impact
	Energy-efficient spinning processes	Compact, vortex, and rotor spinning techniques that reduce energy consumption in manufacturing	Enhances energy efficiency processes
	Closed-loop systems	Systems that recycle by-products, including water and chemicals in the production process	Reuse waste, conserves resource, lowers chemical discharge into the environment
	Renewable energy sources	Solar and wind power to decrease dependence on fossil fuels in yarn production facilities	Reduces carbon footprint, promotes sustainable energy use, and decreases fossil fuel dependence
	Automation and digital monitoring	Integration of digital tools to monitor precision in manufacturing, reducing waste and energy usage	Increases efficiency, reduces waste, and lowers energy consumption
Knit fabric manufacturing	Eco-friendly raw materials	Organic cotton, bamboo, recycled polyester, and biodegradable fibers in knitting production	Reduces resource depletion and promotes sustainable fabric production
	Energy-efficient knitting	Adoption of energy-efficient circular and flat knitting machines	Reduces energy consumption, operational costs, greenhouse gas emissions

(continued)

Table 1 (continued)

Areas	Sustainable innovation	Specific practices	Environmental benefit
	Renewable energy integration	Integration of solar or wind energy into textile factories for knitting operations	Reduces carbon footprint by utilizing renewable energy sources, decreasing dependence on fossil fuels
	Waste reduces via design optimization	Precision knitting and optimization of designs increase material efficiency	Enhances fabric utilization and improves resource efficiency in manufacturing
	Recycling post-industrial knit waste	Recycling post-industrial knit fabric waste into new products to support a circular economy	Reduces textile waste, promotes a circular economy
Woven fabric manufacturing	Eco-friendly raw materials	Adoption of organic cotton, hemp to reduce the need for water, pesticides, and virgin resources	Reduced environmental impact, less reliance on virgin resources
	Recycled yarns	Yarns produced from textile waste or post-consumer products	Reduces waste, conserves resources, and lowers energy consumption
	Energy-efficient looms	Development of looms such as air-jet, rapier looms that consume less power, improve precision, minimize waste	Reduced energy consumption, less material waste
	Renewable energy in weaving	Incorporating renewable energy sources solar, wind power in weaving	Reduced carbon footprint, greater energy sustainability
	Waterless sizing and bio-based agents	Waterless sizing processes and the use of bio-based sizing agents to reduce water and chemical consumption while maintaining yarn integrity	Lower water usage, reduced chemical impact, improved sustainability
	Smart weaving techniques	Optimizing fabric design, minimizing selvage waste, and enabling the reuse of production scraps	Reduced waste generation, and increased material efficiency

(continued)

Table 1 (continued)

Areas	Sustainable innovation	Specific practices	Environmental benefit
Dyeing and coloration	Waterless dyeing technologies	Technologies (supercritical CO2) dyeing replace water with carbon dioxide for dyeing fabrics	Eliminates water usage, reduces pollution, and minimizes wastewater
	Foam dyeing	A water-reducing process where dye is applied in foam form, cutting water usage by up to 90%	Significant reduction in water usage and chemical waste
	Low-impact dyes	Biodegradable dyes, require less energy to fix and are safer compared to conventional synthetic dyes	Reduced environmental and health impacts, less energy consumption
	Natural dyes	Dyes are sourced from plants, insects, and minerals, offering eco-friendly alternatives to synthetic dyes	Reduced environmental harm, biodegradable, and renewable
	Closed-loop systems	Recycling of water and chemicals within dyeing process to minimize freshwater use, toxic wastewater	Decreases freshwater consumption, reduces toxic discharges
Printing and embroidery	Digital printing	Digital printing uses inkjet technology to apply color directly to fabric, eliminating the need for large quantities of water, energy, and chemicals	Significant reduction in water, waste, and chemical runoff
	Water-based inks and solvent-free dyes	Water-based inks and solvent-free dyes are eco-friendly alternatives to plastisol inks and solvent-based dyes, with fewer volatile organic compounds	Reduced VOCs, safer for the environment and workers, improved air quality
	Natural and plant-based dyes	Use of natural and plant-based dyes as biodegradable, non-toxic alternatives to synthetic dyes in textile printing	Biodegradable, non-toxic, and environmentally safer

(continued)

Table 1 (continued)

Areas	Sustainable innovation	Specific practices	Environmental benefit
	Energy-efficient embroidery machinery	Use of energy-efficient machinery in embroidery processes to reduce electricity consumption in stitching	Reduced energy consumption, increased operational efficiency
	Recycled threads and bio-based fabrics	Increasing use of recycled threads and bio-based fabrics in embroidery to reduce waste and support circularity	Waste reduction, supports circular economy, and promotes sustainability
	Closed-loop production systems	Implementation of closed-loop systems that recycle waste materials, fabric scraps, inks, threads in textile printing and embroidery processes	Reduced landfill waste, improved resource efficiency, and waste reduction
Garment sewing and finishing	Zero-waste pattern making	Optimization of fabric usage in pattern making to minimize textile offcuts and scraps, reducing overall waste	Reduces fabric waste, minimizes environmental impact
	Modular design and design for disassembly	Designing garments that can be easily disassembled for recycling or upcycling	Promotes circularity, enhances recycling, and reduces waste.
	Energy-efficient machinery and automation	Use of advanced, energy-efficient machinery and automation to improve precision and reduce energy consumption in garment production	Reduced energy consumption, improved operational efficiency
	Digital garment prototyping and 3D knitting	Use of digital prototyping and 3D knitting technology to reduce the need for traditional sample making, cutting material waste and overproduction	Reduced material waste, improved efficiency, and faster prototyping
	Sustainable materials (recycled, organic, plant-based)	Use of recycled polyester, organic cotton, and plant-based fibers like hemp and Tencel for garment production	Lower environmental impact, reduced reliance on virgin materials

(continued)

Table 1 (continued)

Areas	Sustainable innovation	Specific practices	Environmental benefit
	Bio-fabrication techniques	Creation of leather alternatives using materials such as fungi or algae, reducing reliance on animal products and synthetic materials	Reduced environmental impact, innovative solutions to material scarcity
	Circular fashion and take-back programs	Design garments for longevity, repair, and recycling; take-back programs allow consumers to return old garments for reuse or remanufacturing	Supports circular economy, reduces demand for new materials
	Ethical labor practices and certifications	Ensuring fair wages and environmentally responsible production through certifications like Fair Trade and GOTS	Improved worker welfare, responsible and sustainable production
	Transparency/ blockchain in supply chains	Using blockchain technology to enhance transparency in supply chains and ensure ethical sourcing of materials and labor	Greater transparency, responsible sourcing, and ethical production
Garment washing/finishing unit	Waterless or low-water technologies (ozone washing)	Ozone washing uses ozone gas to clean and bleach garments, reducing water consumption and eliminating harmful chemicals	Reduces water usage, eliminates harmful chemicals, and improves efficiency
	Laser technology for denim distressing	Laser technology replaces traditional stonewashing in denim finishing, using laser beams to create faded and distressed effects without water or chemicals	Reduces water and chemical usage and minimizes fabric damage
	Enzymatic washing	Enzymatic washing uses enzymes to break down fibers and create fading effects, requiring less water, chemicals	Gentle on fabrics, reduces water and chemical use, and is effective for both denim and non-denim garments

(continued)

Table 1 (continued)

Areas	Sustainable innovation	Specific practices	Environmental benefit
	Energy-efficient machinery	Energy-efficient machinery, such as tumble dryers and washing machines, in garment washing processes	Reduces energy consumption and lowers carbon emissions
	Closed-loop water systems	Closed-loop systems recycle and treat water, reducing freshwater consumption to enable reuse of treated water in subsequent washing cycles	Reduces water waste, minimizes freshwater consumption, and improves efficiency
	Filtration and reverse osmosis for wastewater	Wastewater treatment processes, filtration and reverse osmosis treat, recycle wastewater for reuse in washing, reduced consumption	Minimizes water waste, enables reuse of treated water, and reduces discharge of pollutants
	Low-impact, biodegradable chemicals (ZDHC)	The use of low-impact, biodegradable chemicals in garment washing, supported by initiatives like the Zero Discharge of Hazardous Chemicals	Reduces environmental and health risks associated with hazardous chemicals
	Repair and upcycling programs	Garment repair and upcycling programs encourage extending garment lixsfe and reducing waste	Reduces waste, promotes circularity, and extends garment life
Trimmings and accessories unit	Biodegradable and eco-friendly materials	Fashion trimmings (buttons, buckles, and zippers), made from biodegradable, renewable materials (coconut shells, wood, coral, and recycled plastics)	Reduces environmental impact, biodegrades naturally, and is safer for the environment
	Sustainable labels and tags	Use of paper-based, plant-based materials like hemp and organic cotton for labels, and digital printing replacing traditional label production	Reduces plastic usage, lowers ink, chemical, and water consumption

(continued)

Table 1 (continued)

Areas	Sustainable innovation	Specific practices	Environmental benefit
	Recycled and upcycled materials	Use of recycled polyester, nylon, and reclaimed metals for fashion accessories belts, bags, and clasps	Supports circular economy, reduces demand for new raw materials, and reduces waste
	Circular design principles	Encouraging the creation of accessories designed for disassembly, repair, and recycling to promote longevity and reduce waste	Supports circular economy, reduces landfill waste, and promotes sustainability
E-textiles	Bio-based conductive materials	Use of bio-based materials like algae or cellulose for conductive threads	Biodegradable, renewable, reduces reliance on synthetic and petroleum-based materials
	Recycled conductive threads	Conductive threads made from recycled PET or post-consumer textile waste	Reduces textile waste, minimizes raw material extraction, supports circular economy
	Graphene and carbon nanotube alternatives	Use of graphene-based conductive inks instead of heavy metals like silver or copper	Reduces reliance on non-renewable and non-recyclable metals
	Additive manufacturing	Printing conductive pathways directly onto textiles using 3D printing or screen-printing techniques	Minimizes waste, reduces water and chemical usage
	Laser-assisted electronics integration	Embedding electronics using laser processes that require less energy than traditional soldering	Lowers energy consumption, minimizes waste
	Modular design	Designing e-textiles for easy disassembly of electronic components for reuse or recycling	Extends product life cycle, facilitates recycling
	Closed-loop systems	Recovery of metals and fibers from discarded e-textiles	Reduces landfill waste, promotes resource efficiency
	Biodegradable electronics	Use of components that naturally degrade after disposal, such as organic polymer circuits	Reduces environmental impact of e-waste

(continued)

Table 1 (continued)

Areas	Sustainable innovation	Specific practices	Environmental benefit
	Take-back programs	Programs for consumers to return used e-textiles for recycling or repurposing	Promotes material recovery, reduces landfill dependency
	Transparent supply chains	Blockchain technology to ensure ethical sourcing of raw materials and traceability of production	Ensures ethical production, builds consumer trust

4 Challenges, Stakeholder Roles and Future Directions to Adopt Sustainable Practices in Fashion Manufacturing

The adoption of sustainable practices in fashion and textile manufacturing includes multiple challenges, stemming from both technological, managerial and systemic factors. Firstly, economic barriers to sustainable materials innovation and optimization. For example, organic cotton, eco-friendly dyes, biodegradable textiles are typically more expensive due to the higher costs of production, certification, and lower economies of scale. Mass market brands and small entrepreneurs struggle in absorbing additional expenses without increasing product prices [20]. There exist, technological limitations in key practice areas like material recycling and waste management. Current textile recycling methods often cannot fully recover the quality of original fibers, leading to inefficiencies and complexities in processing mixed material garments. Moreover, the lack of standardization in materials and processes complicates efforts to develop closed-loop systems where products can be continuously recycled without loss of quality [31]. The fragmented and globalized supply chain of the fashion industry also hinders sustainable practices. Fashion manufacturing involves multiple tiers of production across various countries, making it difficult to ensure transparency, traceability, and uniform compliance with sustainability standards. This lack of coordination often results in "greenwashing," where companies make unsubstantiated claims about sustainable practices, further eroding consumer trust [32].

However, sustainable practices in fashion and textile manufacturing are strongly influenced by policy, industry standards, and multi-stakeholder collaboration. Policymakers play a key role in establishing regulatory frameworks, enforcing environmental laws, and ensuring transparency. Initiatives like the European Union's Green Deal and Extended Producer Responsibility (EPR) laws hold brands accountable for their product lifecycles, from production to disposal, driving sustainable practices. Industry standards and certifications, such as the Global Organic Textile Standard (GOTS), OEKO-TEX, and the Global Recycled Standard (GRS), offer manufacturers clear guidelines on materials, labor practices, and environmental impacts [33]. These certifications ensure the credibility of sustainability claims and help consumers make informed choices, thereby boosting demand for sustainable products [30]. Collaboration among stakeholders, including manufacturers, brands, NGOs, and consumers, is vital for advancing sustainability in the fashion industry. Initiatives like the 'Ellen MacArthur Foundation's Circular Fashion' and the 'Fashion for Good' platform facilitate the sharing of knowledge, resources, and innovations, accelerating circularity. Multi-stakeholder partnerships also enable the development of infrastructure, such as textile recycling facilities, which individual companies may struggle to establish alone. These collaborative efforts enhance the implementation of sustainable practices and create a unified voice for systemic change, promoting long-term environmental and social sustainability.

Looking forward, future directions for sustainable fashion involve significant investments in technology innovation, particularly in the development of scalable textile recycling methods, biodegradable materials, and digital manufacturing techniques such as 3D printing, and 3D wholegarment knitting. The implementation of circular economy principles will require improvements in waste management systems and new business models, such as leasing, take-back schemes, and product-as-a-service models. Additionally, stronger policy frameworks and collaborative industry efforts are essential to harmonize sustainability standards, drive systemic change, and incentivize investments in green technologies and products [32]. These efforts must also be paired with consumer education to foster demand for more sustainable products [33]. Figure 1 serves as a roadmap, starting with identifying current challenges, proposing strategic solutions, and setting long-term goals for a more sustainable fashion industry.

5 Conclusion

Sustainable practices in fashion and textiles manufacturing are crucial in reducing the industry's significant environmental and social impacts. Innovations in materials, like bio-based textiles, recycled fibers, and upcycled fabrics, are leading the way toward more sustainable production processes. Closed-loop systems and circular economy principles, which focus on recycling and reusing materials, are essential for reducing waste and resource depletion. Sustainable dyeing methods, waterless technologies, and the use of natural dyes are minimizing water and chemical use, further contributing to eco-friendly practices. Ethical sourcing, supported by certifications like GOTS and Fair Trade, ensures that production aligns with social and environmental responsibility. Collaboration among policymakers, industry stakeholders, and consumers accelerates the adoption of these practices, creating a more sustainable and circular fashion industry. Ultimately, integrating these sustainable practices will help drive the fashion manufacturing to be environmental and socially sustainable, shaping a responsible future for both the industry and the planet. However, implementing sustainable practices in fashion manufacturing includes multiple challenges, stemming from both technological, managerial and systemic factors, for

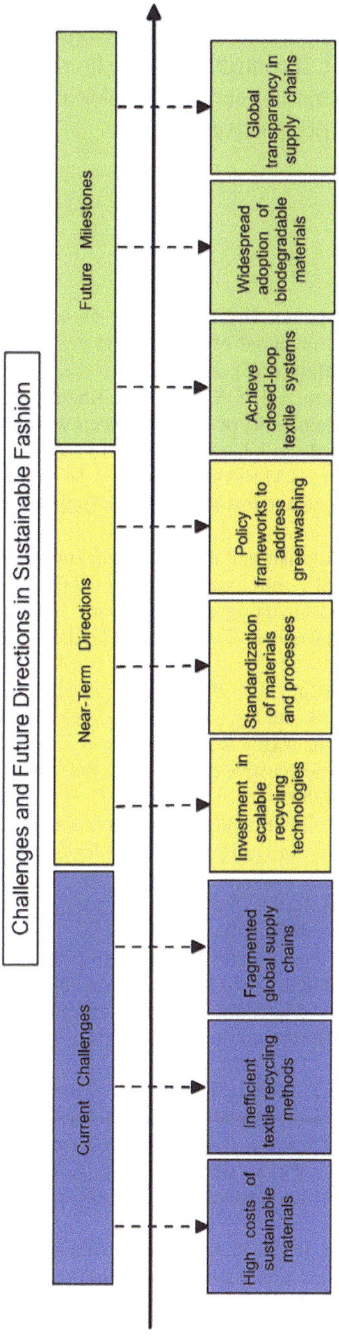

Fig. 1 Challenges and future directions to adopt sustainable practices in fashion manufacturing

which collaborative efforts among key stakeholders are critical to be more sustainable. This study provides a summary of mapping and categorisation of a wide range of sustainable practices in fashion manufacturing with environmental benefits and highlighting areas for improvement. It contributes to the literature on sustainable fashion and textiles manufacturing operations and has broader implications for stakeholders from theoretical and practical perspectives.

References

1. M. Shamsuzzaman, M.A. Al. Mamun, H.R.U. Hasan, R. Hassan, A. Zulkernine, M.A.R. Atik, M. Islam, Fashion circularity: potential of reusing and recycling remnant fabric to create sustainable products. Sustainability **17**(5), 2010 (2025)
2. M. Shamsuzzaman, M.A. Kashem, A.S.M. Sayem, A.M. Khan, S.M. Shamsuddin, M.M. Islam, Quantifying environmental sustainability of denim garments washing factories through effluent analysis: a case study in Bangladesh. J. Clean. Prod. 1–13 (2021)
3. M. Shamsuzzaman, M.M. Islam, H.M Rakib Ul Hasan, A.M. Khan, A.S.M. Sayem, Mapping environmental sustainability of knitted textile production facilities. J. Clean. Prod. **405**, 136900 (2023)
4. M.M.R. Khan, M.M. Islam, Materials and manufacturing environmental sustainability evaluation of apparel product: knitted T-shirt case study. Tex. Cloth. Sus. **1**, 1–12 (2015)
5. M. Shamsuzzaman, et al., Waste management in textile industry ed. by M.M. Rahman, M. Mashud, M.M. Rahman, *Advanced Technology in Textiles. Textile Science and Clothing Technology* (Springer, 2023)
6. M.C.S.D. Abreu, F.N.H. Ferreira, J.F. Proenca, D. Ceglia, Collaboration in achieving sustainable solutions in the textile industry. J. Bus. Ind. Mark. **36**(9), 1614–1626 (2021)
7. G. Marzocchi, Decarbonizing the textile & fashion industry: material-driven guidelines in an ecological dimension-oriented perspective exploring the field of biomaterials (online) (2021), https://hdl.handle.net/10589/212599
8. S.S. Muthu, M. Ramchandani, Environmental footprint, toxicity, and life cycle assessment within the vegan leather industry. In *Vegan Alternatives for Leather* (Cham: Springer Nature, 2024), pp. 37–61
9. L.T. Korley, T.H. Epps III., B.A. Helms, A.J. Ryan, Toward polymer upcycling—adding value and tackling circularity. Science **373**(6550), 66–69 (2021)
10. J. Tabor, K. Chatterjee, T.K. Ghosh, Smart textile-based personal thermal comfort systems: current status and potential solutions. Adv. Mater. Technol. **5**(5), 1901155 (2020)
11. P.M. Stathatou, L. Corbin, J.C. Meredith, A. Garmulewicz, Biomaterials and regenerative agriculture: a methodological framework to enable circular transitions. Sustainability **15**(19) (2023)
12. C. Arana, I.B. Franco, A. Joshi, J. Sedhai, SDG 15 life on land: a review of sustainable fashion design processes: upcycling waste organic yarns, in *Actioning the Global Goals for Local Impact: Towards Sustainability Science, Policy, Education and Practice*, pp. 247–264 (2020)
13. G. Dong, D. Zhang, J. Yue, L. Wu, L. Han, Z. Zhang, X. Qian, Progress of air flow field and fiber motion analysis in textile manufacturing process: a review. Text. Res. J. **94**(13–14), 1581–1612 (2024)
14. M.M.K. Akter, U.N. Haq, M.M. Islam, M.A. Uddin, Textile-apparel manufacturing and material waste management in the circular economy: a conceptual model to achieve sustainable development goal (SDG) 12 for Bangladesh. Clean. Environ. Syst. **4**, 100070 (2022)
15. A.K. Aliyana, G. Stylios, A review on the progress in Core-Spun Yarns (CSYs) based textile TENGs for real-time energy generation, capture and sensing. Adv. Sci. **10**(29) (2023)

16. J. Scott, R. Kaiser, D. Ozkan, A. Hoenerloh, A. Agraviador, E. Elsacker, B. Bridgens, Knitted cultivation: textiling a multi-kingdom bio architecture, in *Structures and Architecture A Viable Urban Perspective?*. CRC Press, pp. 3–10 (2022)
17. M. Shamsuzzaman, M. Islam, M.A.A. Mamun, R. Rayyaan, K. Sowrov, S. Islam, A.S.M. Sayem, Fashion and textile waste management in the circular economy: a systematic review. Clean. Waste Syst. 100268 (2025)
18. M.S. Hoque, L.M. Degenstein, P.I. Dolez, Innovations in woven textiles, in *Sustainable Innovations in the Textile Industry* (Woodhead Publishing, 2024), pp. 125–163
19. C. Var, S. Palamutcu, Sustainable approaches in textile-sizing process, in *Sustainable Manufacturing Practices in the Textiles and Fashion Sector* (Cham: Springer, Nature, 2024), pp. 55–74
20. E. Khalil, J. Sarkar, M.M. Rahman, M. Shamsuzzaman, D. Das, Advanced technology in textile dyeing, in *Advanced Technology in Textiles: Fibre to Apparel* (Springer Nature, 2023), pp. 97–138
21. T. Saha, Z. Uddin, M.N. Islam, M. Shamsuzzaman, A. Al Tahsin, M.D. Islam, Assessing the effectiveness and environmental sustainability of reactive dyes for their structural diversity. Text. Leather. Rev. **5**(322), 103–119 (2022)
22. E. Khalil, J. Sarkar, M.M. Rahman, M. Shamsuzzaman, D. Das, Innovative textile printing technology, in *Advanced Technology in Textiles: Fibre to Apparel* (Springer Nature, 2023), pp. 139–159
23. E. Gayathiri, P. Prakash, K. Selvam, M.K. Awasthi, R. Gobinath, R.R. Karri, M.G. Ragunathan, J. Jayanthi, V. Mani, M.A. Poudineh, S.W. Chang, Plant microbe based remediation approaches in dye removal: a review. Bioengineered **13**(3), 7798–7828 (2022)
24. E. D'Itria, F. Vacca, Shaping sustainable solutions in fashion through design-led strategies, approaches, and practices. Discov. Sustain. **5**(1), 409 (2024)
25. M. Shamsuzzaman, M. Mashud, M.M. Rahman, M.M. Rahman, E. Hoq, D. Das, Management and maintenance of textile machinery, in *Advanced Technology in Textiles: Fibre to Apparel* (Springer Nature, 2023), pp. 31–63
26. M. Sobuj, A.M. Khan, M.A. Habib, M.M. Islam, Factors influencing eco-friendly apparel purchase behavior of Bangladeshi young consumers: case study. Res. J. Text. Appar. **25**(2), 139–157 (2021)
27. N. Friedman, J. Ormiston, Blockchain as a sustainability-oriented innovation?: Opportunities for and resistance to blockchain technology as a driver of sustainability in global food supply chains. Technol. Forecast. Soc. Chang. **175** (2022)
28. Z.A. Awal, M. Shamsuzzaman, D. Das, Impact of laser intensities at various DPI and pixel time on the properties of denim garments. J. Adv. Res. Mater. Sci. **77**(1), 1–13 (2021)
29. M.G. Uddin, M.M. Islam, M.R. Islam, Effects of reductive stripping of reactive dyes on the quality of cotton fabric. Fash. Text. **2**, 1–12 (2015)
30. A. Patti, D. Acierno, Towards the sustainability of the plastic industry through biopolymers: properties and potential applications to the textiles world. Polymers **14**(4), 692 (2022)
31. K. Baker, L.S. McNeill, A five-stage framework for sustainability certification for fashion brands: can a standardised approach work in the fashion industry? Int. J. Fash. Des. Technol. Educ. 1–15 (2024)
32. M.M. Islam, P. Perry, S. Gill, Mapping environmentally sustainable practices in textiles, apparel and fashion industries: a systematic literature review. J. Fash. Mark. Manag.: Int. J. **25**(2), 331–353 (2021)
33. M. Islam, M. Shamsuzzaman, H.R.U. Hasan, M.A.R. Atik, Environmental sustainability of fashion product made from post-consumer waste: impact across the life cycle. Sustainability **17**(5), 1917 (2025)

Chapter 10
How Greenwashing Negates Sustainable Production and Consumption of Fashion Textiles

Md Nazmul Haque, Chunmin Lang, and Chuanlan Liu

Abstract The official designation of SDG 12 is "Ensure sustainable consumption and production patterns." This chapter analyzes the setting and circumstances of the fashion business via a literature study to attain the objectives of SDG 12. After an in-depth analysis of greenwashing in the fashion sector and its emergence within the fashion industry, evidence and reports of greenwashing, its adverse impacts on fashion consumers, existing regulations, and various mitigation strategies, it is concluded that greenwashing in the fashion industry poses significant threats to achieve SDG 12. Considerable evidence of greenwashing legislation from developed countries compared with developing countries has been observed, and developing countries must keep pace with the tempo of reforming regulations in the fashion sector. Developed countries' primary focus is on consumption while developing nations are renowned for their role as manufacturing centers. Both parties must implement legislation to prevent greenwashing in the fashion industry, as fashion brands rely on complex supply chains. Furthermore, developed countries must establish comprehensive regulations and can observe each other's regulations to build more effective regulations. Additionally, informing consumers about greenwashing and their readiness to advocate for and comprehend sustainability was imperative, as it is crucial for sustainable production and consumption. For the customer, sustainability may represent a preference rather than an obligatory need. Conversely, when fashion businesses fail to comply with rules, they face fines. Educating consumers is essential to prevent greenwashing in the fashion sector when regulations are organized in this way. A few preventive measures have been provided to curb greenwashing phenomena.

Keywords Greenwashing · SDG · Eco-lexicon · Legislation · Fashion consumer · Fashion industry

---

M. N. Haque (✉) · C. Lang · C. Liu
Department of Textiles, Apparel Design, and Merchandising, College of Agriculture, Louisiana State University, Baton Rouge, USA
e-mail: mhaqu18@lsu.edu

A. S. Muhammad Sayem and P. Goswami (eds.), *SDG 12 and Global Fashion Textiles Production*, SDGs and Textiles, https://doi.org/10.1007/978-981-96-9051-0_10

1 Introduction

The United Nations established 17 Sustainable Development Goals in 2015. One of these goals, "responsible consumption and production," is Sustainable Development Goal 12 (SDG 12 or Global Goal 12). Ensuring sustainable consumption and production patterns is the official form of SDG 12 [1]. The purpose of SDG 12 is to guarantee the efficient utilization of resources, enhance energy efficiency and sustainable infrastructure, provide access to essential services, generate green and respectable employment opportunities, and ensure a superior quality of life for all. Eleven targets will be accomplished by 2030, and 13 indicators will be employed to assess progress toward the targets in SDG 12 [2]. The three implementation targets are [3] to assist developing nations in enhancing their scientific and technological capabilities, to create and apply instruments for monitoring sustainable development outcomes, and to eliminate market distortions, such as fossil fuel subsidies, that promote excessive consumption [1].

The fashion industry, characterized by its highly globalized supply chain, is the third largest polluter, behind the food and construction sectors. It constituted 10% of global greenhouse gas emissions, releasing 1.2 billion tons of carbon dioxide each year, above the total emissions of the shipping and aviation industries combined. If the present trend continues, the industry's greenhouse gas emissions are projected to increase by about 50% by 2030 [4]. In 2000, less than 50 billion garments were manufactured globally. In 2019, the figure exceeded 130.6 billion [5]. Less than 1% of materials used in clothing manufacture are repurposed into new garments, resulting in a waste of $100 billion in material value [6]. The United States and China account for the highest textile production and consumption levels globally. The textile sector has a mere 0.3% circularity [7].

The fashion industry poses serious threats to achieving SDG 12. Aside from statistical data, research also showed that the top ten fast fashion companies, including Inditex, Fast Retailing, H&M, Next, C&A, Lojas Renner, Bestseller, Gap, Mango, and ASOS, have participated in anti-consumerist washing, a kind of greenwashing, by not complying with the 2030 agenda of Sustainable Development Goal 12 goals [8]. The fashion industry is a prominent perpetrator of greenwashing, mainly attributable to the significant environmental damage it inflicts [5]. Greenwashing is a broad term used to describe actions that deceive consumers about an organization's environmental performance. It refers to a scenario when there is a discrepancy between what an organization communicates and what it performs in terms of fulfilling its environmental responsibilities [9].

This chapter offers a detailed perspective on greenwashing within the fashion industry. The first section of this chapter outlines the origins of greenwashing and its entrance into the fashion industry. The next subsequent section consecutively examines the mechanism of greenwashing, its detrimental effects on fashion consumers, current legislation, and several prevention measures. This will assist policymakers, researchers, and fashion sector consumers in comprehending the adverse impacts of greenwashing on attaining Sustainable Development Goal 12.

2 Greenwashing in Fashion Textiles

2.1 *Greenwashing Historical Background*

The issue of greenwashing can be dated back to the 1960s. The nuclear power sector of the American electrical conglomerate Westinghouse was a forerunner in greenwashing. Due to the anti-nuclear movement of the 1960s, Westinghouse's advertisements obscured the persistent ecological hazard presented by radioactive waste. In the mid-1980s, Chevron, an oil firm, invested significantly in television and print commercials to advocate its environmental stewardship. During the campaign, Chevron contravened the Clean Air and Water Acts and released oil into wildlife refuges [10]. In 1986, Jay Westerveld coined the phrase "greenwashing," followed by a sequence of events shown in Fig. 1. Ultimately, in 1999, the official meaning of "greenwashing" was included in the Oxford English Dictionary. The term "greenwashing" was used in the 1980s to characterize exaggerated corporate environmental claims. Three decades later, the process has become far more complex [10].

The phenomenon of greenwashing has been extensively examined in various academic disciplines and has been the subject of deliberation among numerous governmental and non-governmental organizations. Multiple interrelated definitions have emerged, all sharing a conceptual foundation, highlighting various fundamental aspects of the notion or its manifestations [11]. Lyon and Montgomery [9] argue that the definition of greenwashing is challenging due to its complex and multifaceted characteristics [12]. The most common negative remarks associated with greenwashing definitions are misleading consumers [9, 13–17]. Other negative comments related to the greenwashing definitions including environmental whitewash and brainwash of consumer [18], withhold of negative information [19], poor environmental but positive communication [14], deflect attention and unfriendly activities [20], symbolic information without substantive action [21], disseminating disinformation [22, 23], falsely promoting and deceptive information [24], decoupling

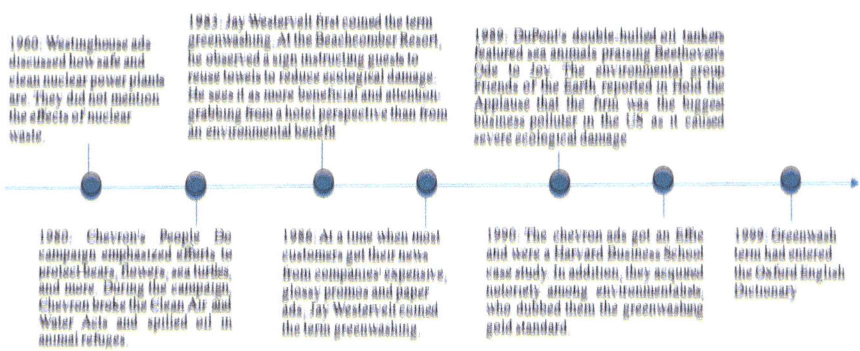

Fig. 1 Greenwashing definition evolution. *Source* Prepared by authors of this paper

behavior [25], merely symbolic [26], misleading communication [9], false perception [13], claims that are not true [27], statement of sustainability but not in action [28], disclose negative but disseminating positive [16], deceitful advertising [17], hiding less than stellar environmental records [29], misleading environmental claim [12], pretense of environmental beneficial [30], misleading perception and deceiving of public [31], misleading information and unsubstantiated claim to deceive [32], unjustified appropriation of environmental virtue [33], and false or deceptive picture [34].

2.2 Greenwashing Historical Path of Introduction to the Textile Industry

The fashion industry's initial introduction of greenwashing differs slightly from the definition's origin. The fast fashion concept originated in the 1970s when merchants started outsourcing manufacturing to countries, especially in Asia, where labor costs were much cheaper than in Western nations [35]. In 1983, Benetton initiated a campaign of advertisements explicitly celebrating diversity and inclusiveness. Although Benetton garnered public attention via advertising, the advertisements seemed to prioritize shock value above promoting apparel [36]. Levi's and Nike were subjected to scrutiny and public outcry from 1980 to 1990 due to sweatshops and discriminatory labor rights in their supply chains [37]. When Zara arrived in New York at the beginning of the 1990s, the phrase "fast fashion" was first used. Fast fashion is a business model that offers inexpensive, stylish clothing via a swift supply chain, characterized by the capacity to produce contemporary items bi-weekly, adherence to transient trends, rapid inventory turnover, mass production, and short lead times [38].

Fast fashion, misleading product environmental benefits, and marketing initiatives were the dominant factors in greenwashing in the fashion industry from 1990 to the present. For instance, Patagonia was an early proponent of organic cotton use in 1996 [39] and has produced fleece from recycled bottles since the 1990s [40]. This is a notable transition from the industry's traditional cotton practices, which are condemned for their environmental repercussions stemming from pesticide use and excessive water consumption [39]. 13% of Patagonia's polyester is derived from virgin petroleum sources [39]. During laundry, an individual fleece jacket may release up to 250,000 synthetic fibers [41]. Zara, a Spanish apparel brand, was among the first to introduce new merchandise to its stores on a bi-weekly basis in the early 2000s [5].

Marks & Spencer (M&S) initiated Plan A in 2007, including 180 goals to fulfill by 2015, by collaborating with BP (British Petroleum), has been accountable for many environmental catastrophes, including the release of over one million liters of oil in Alaska in 2006 [42]. Researcher Bertola, et al. [43] stated that abstaining from corporate social responsibility (CSR) practices is seen as greenwashing. The

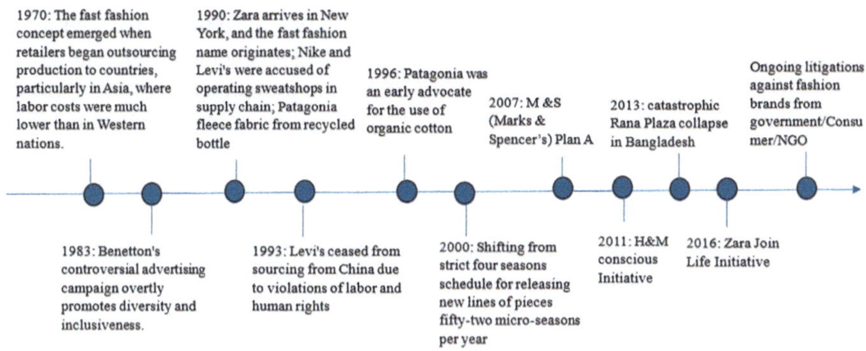

Fig. 2 Greenwashing introduction path in the fashion industry with notable events. *Source* Prepared by authors of this paper

catastrophic Rana Plaza collapse in Bangladesh on April 24, 2013, which claimed the lives of 1,134 garment workers, persistent questions and concerns over corporate social responsibility (CSR) policies in the textile and apparel sector remain. Furthermore, after the collapse of the Rana Plaza building, U.S. corporations neglected to attend planned meetings on compensation for the families of the Bangladeshi women who perished. Walmart and GAP have been associated with a misleading history of contracts, inaccurately asserting their involvement in CSR measures after the Rana Plaza disaster [44].

Several marketing initiatives that benchmark for greenwashing in the fashion industry have been noticed. For instance, H&M Conscious (2011) and Zara Join Life (2016) used public relations strategies to win over environmentally aware consumers. Deposit the unwanted clothing in the designated containers inside the shop. This initiative was criticized for being a pull strategy to lure consumers into the shops with the promise of discount coupons. In that initiative, fast fashion brands generated more than mere collections [45]. Consumer groups, environmental protection organizations, and the government have accused various fashion brands of greenwashing. The majority of these allegations are still pending [46] (Fig. 2).

3 Mechanisms of Greenwashing in Fashion

3.1 Misleading Labels and Certifications

A recent report by the Changing Markets Foundation identifies ten principal certifications, labels, and voluntary industry initiatives within the fashion sector: Higg Index and SAC, Bluesign, OEKO-TEX, Cradle to Cradle, Ellen MacArthur Foundation, Textile Exchange, The Microfiber Consortium, ZDHC, WRAP, and the EU Eco-label. Regrettably, many of these initiatives function as sustainability facades for businesses, facilitating global greenwashing, characterized by insufficient transparency, limited accountability, and compromised independence inherent in the programs [47]. Up to 96% of their clients submit fraudulent claims [48]. Furthermore, they use ambiguous terminology, jargon (e.g., eco-, organic, chemical-free, and sustainable), and misleading marketing strategies in their routine practices. It leads to the misinterpretation of environmentally relevant terminology and creates a disconnect between consumer expectations and corporate communications [49]. The use of ambiguous wording further exacerbates confusion. The absence of standardization may advantage the manufacturer by allowing a particular interpretation of issues, sometimes referred to as "greenwash." As a relative beginner in sustainability, the fashion industry seems to have appropriated marketing terminology from the environmental movement [50].

Thomas [50] assembled a compendium of prevalent eco-centric terminology and an eco-lexicon for scholars, branding professionals, editors, authors, and the general populace to elucidate the meanings of ideas and their accurate application (Fig. 3). From this, Thomas [50] created a vocabulary, twelve entries of which Peirson-Smith and Evans [51] used in their study on the customer perception gap about fashion businesses that advocate sustainable practices within the Asian setting. Study findings revealed the green language lexicon as too complex, ill-defined, and overwhelming for fashion users. Furthermore, an analysis of advertising materials in Hong Kong shows that clients of environmentally conscious fashion firms express dissatisfaction with vague, ambiguous, or inconsistent messaging. Moreover, compelling research indicates that this approach is unlikely to enhance awareness, engagement, and action or influence brand loyalty because eco-lexicon is misleading [52].

Furthermore, Ranasinghe and Jayasooriya [53] conducted a comprehensive study to examine the attributes of textile eco-labels currently utilized worldwide, explicitly focusing on eco-labeling or certification. According to the "Eco-label Index," which is a comprehensive database of eco-labels, it is challenging to track the entire life cycle of a product from a single eco-label because many countries export the yarn and raw materials required for the manufacturing process [53].

3.2 *False Advertising and Marketing Initiative*

The demand for ecologically favorable products among consumers has led to businesses' frequent use of greenwashing. TerraChoice Environmental Marketing reported that in 2009, 98% of the items making green claims, irrespective of the sector, were engaging in greenwashing. From 2006 to 2009, the increase in green advertising in prominent publications was from 3.5% to 10% [54, 55]. Sailer, et al. [56] documented instances of greenwashing on Instagram during the black Friday period. To identify postings that were explicitly associated with Black Friday 2020, Sailer, et al. [56] analyzed the Instagram accounts of the 76 highest-ranked firms. The investigation identified 115 instances of greenwashing in all relevant Black Friday advertisements from 39 sustainable fashion companies. Furthermore, Castillo-Abdul,

Most Common Lexicon used by fashion brand to appear sustainable. Most cases, meaning appear to be exaggerated from the literal meaning

Eco is short of ecology. The narrative of the interaction between humans and their environment. Because of its non-judgmental and possibly non-factual approximation, the term is frequently used despite its deliberate flexibility.

Common application of ecological terminology is the use of **hyphens**. The hyphen is frequently employed to prefix the terms "eco" and "green," as seen by phrases such as "light-green environmentalist," "eco-power pinup," "eco-narcissism," "eco-razz," and "eco-ethical"

Ethical has a quasi-religious connotation, with "ethical" referring to actions and behaviors governed by philosophical principles determined by their impact on others. By implication, the term ethical is problematic because it may be perceived as judgmental, in part due to its perceived overlap with the word morals.

Fair trade based on a philosophy and method of trading with developing countries enables the producers and manufacturers of products to receive an equitable wage or a living wage. Once restricted to food and handicrafts, fair trade now includes cotton and other fibers

Green, not independently verified, the term alludes to a product or service positively associated with the environment. Green is one of the oldest terms in English; it alludes to a 1960s political movement but has fallen out of use. In general writing about environmental issues, the term may appear hyperbolic or be employed simply

Natural as terms are interchangeable with organic and are most applied to fibers. Since the term refers to the origin of the thread's staples or fiber, the term 'natural' does not apply. In addition, it does not imply that no chemicals (herbicides, pesticides, or fungicides) were used in the propagation and growth of these plants

Organic pertains precisely to fibers and fabrics, referring explicitly to the organic standard established in the United Kingdom in 1973 by Soil Association Certification Limited (OSA). Organic is a relatively novel concept in the study of fashion and textiles, having originated in agriculture and the food industry. Due to its interchangeability with natural, it conveys the same meaning

Recycle, when a garment is no longer useful to its original owner, it is recycled and may have a second existence as secondhand/vintage apparel [or as new] [or as new]. "

Sustainability is not synonymous with environmentalism due to its broader scope and inclusive nature.

Upcycling occurs when the value of abandoned clothing is increased through alteration or customization. Similarly, it may be redeployed by being altered, deconstructed, dyed, or customized and resold into the apparel system

Fig. 3 Most common and exaggerated lexicon used by the fashion brand

et al. [57] looked into the corporate social responsibility communication for male luxury fashion brands on Instagram, Facebook, and TikTok revealed that the utilization of pressing and persistent problems such as the Ukraine war and the endangered Asian tiger species was employed to captivate consumer attention, alongside intermittent sustainable messages.

The 2007 introduction of Diesel's "Global Warming Ready" (GWR) campaign marked a significant departure from traditional brand messaging. The GWR advertisement, which depicted appealing models in apocalyptic scenarios (such as an inundated Rio de Janeiro or a Paris overtaken by tropical vegetation), was swiftly criticized for its ability to attract attention to the dire consequences of climate change [58]. The findings of the study conducted by Glozer and Morsing [58] indicated that Diesel, an immoral and hypocritical corporation, aims to profit from the world's suffering. This collection of works underscores the perception of hypocrisy as incongruity, which has led to evaluations of greenwashing. This perspective portrays [58]

Fast fashion firms implement numerous initiatives to demonstrate their commitment to sustainability. Nevertheless, the initiatives were motivated by financial gain and consumption in the name of sustainability. For example, Zara's "Join Life" marketing tagline encourages consumers to "bring the clothes you no longer want and deposit them in the containers located in the stores." Initially, the company attracted consumers to its stores, and subsequently, it emphasized the disposability of its apparel [59]. The H&M Conscious initiative has been perceived as a public relations strategy to attract environmentally conscious consumers. This initiative has been criticized as a ploy to entice consumers to increase their spending through discount vouchers, even though H&M has promoted the disposal of second-hand apparel in its stores as an environmentally responsible business practice [60]. According to the Guardian, H&M's claim of accumulating 1,000 tons of used apparel is solely a form of greenwashing, as they produce the same volume in just two days [61].

3.3 Overstatements and Deceptions of Eco-friendly Innovations

Jones [61] asserts that Zara's collaboration with "Lenzing," innovators in Tencel and Modal production, to produce and use sustainable materials was seen as a manifestation of greenwashing. LanzaTech's technology can capture CO_2 from industrial, agricultural, or residential waste operations. A fermentation process transforms it into ethanol, an essential element in the manufacture of polymers like PET, which is used in the creation of polyester thread. Lanzanol is later transformed into low-carbon monoethylene glycol (MEG) by India Glycols Limited, which is then processed into low-carbon polyester yarn used by Zara. The resulting fabrics, however, are not wholly comprised of trapped carbon. The final polyester consists of 20% MEG provided by Lanzatech, with the remaining 80% originating from Purified Terephthalic Acid (PTA) [62].

The fashion revolution underscored that H&M's assurance of accumulating 1,000 tons of used clothing through its take-back initiative could not be fully recycled due to the variations in textile combinations among each garment. Therefore, H&M should have been transparent in their marketing communications and not mislead consumers, as such a technology does not exist in the market [45, 60].

In 2022, the FTC demonstrated its commitment to addressing misleading environmental claims by filing lawsuits against Kohl's and Walmart for "misrepresenting numerous rayon textile products as bamboo." Both firms faced allegations of making misleading environmental assertions, namely that ecologically sound methods manufactured the "bamboo" fabrics. The clothes comprised rayon derived from bamboo via a process that requires harmful chemicals and produces dangerous pollutants. According to the FTC, rayon production utilizes ecologically toxic substances like sodium hydroxide, carbon disulfide, carbonyl sulfide, ethylene oxide, methanol, methyl chloride, propylene oxide, and toluene [5].

Boohoo's "sustainable collection" incorporates recycled materials such as polyester to become a more sustainable brand. It was discovered that certain items were constructed from acrylic, a form of plastic. Although the brand suggests that this material may be recycled, inadequate evidence supports this claim [63, 64]. Furthermore, chemicals such as DDT and PCBs may adhere to plastic materials exposed to water during disposal and laundry, rendering wastewater treatment facilities incapable of eliminating all synthetic fibers [41]. The development of cancer is known to be at risk due to oxidative damage, DNA damage, and alterations in gene activity that microplastics can cause, as indicated by research. An infant's initial excretion, meconium, and human breast milk have both been discovered to contain microplastics [65].

4 Negative Effects of Greenwashing on Fashion Consumers

Fashion consumers are adversely affected by greenwashing when they make environmentally eco-friendly purchases. Several recent studies have examined the impact of greenwashing, and Table 1 illustrates the absence of academic research in the field from consumers in various countries.

5 Regulatory and Industry Responses to Combat Greenwashing

5.1 Brief Summary of Nation-Wise Existing Legislation

Developed countries have implemented comprehensive legislation and initiatives to combat greenwashing in the fashion industry. Nevertheless, there was not much evidence from developing nations. For example, China has no specific laws addressing greenwashing in the fashion industry. They have additional laws that are classified to prevent greenwashing in the fashion industry. Developing nations are renowned for their function as manufacturing centers, while developed nations primarily concentrate on consumption [76]. To prevent greenwashing in the fashion industry, both parties must implement legislation. Furthermore, the extensive and decentralized nature of the fast fashion supply chain network makes it challenging for fashion brands to oversee overall compliance effectively [77]. Table 2 discusses some legislation from developed and developing countries to curb greenwashing in the fashion industry.

Table 1 Research focuses on the greenwashing effect on fashion consumers

Research sample (fashion consumers)	Findings	*Source*
Chinese consumers	• The results indicate that customers' views on greenwashing adversely influence their propensity to buy eco-friendly items. Moreover, perceived betrayal partly buffers this link, whereas environmental responsibility exacerbates the negative impact of greenwashing views on green purchase intentions	[66]
	• Consumers' perception of greenwashing within the fast fashion industry adversely impacts their intention to acquire eco-friendly products; (2) This perception indirectly diminishes purchase intention by heightening consumers' risk perceptions, encompassing both financial and environmental risks; and (3) Impulsive buying behavior in the fast fashion sector amplifies the positive correlation between greenwashing perceptions and financial risk perceptions, functioning as a moderating variable	[67]
	• The study's findings suggest a moderating effect of green clothing type (green home-in wear vs green home-out wear). Unlike green home-out clothing, the relationship between expected moral benefit and perceived greenwashing for green home-in wear had a reduced negative effect on purchase intentions	[68]
Australian consumers	• The results confirm the significant positive influence of environmental concern and confidence in the sustainable clothing brand; nevertheless, perceived greenwashing hinders the acceptance of sustainable fashion	[69]
	• Cynical persons reject sustainability assertions, judging them as ambiguous or unfounded, thereby constituting greenwashing. Nonetheless, conspicuous consumption diminishes this skepticism	[70]

(continued)

Table 1 (continued)

Research sample (fashion consumers)	Findings	*Source*
German consumers	• The study examined consumer vulnerability to greenwashing influenced by product color and pricing indicators. Due to deceptive marketing, consumers often misidentify things as eco-friendly when the signals correspond with their preconceived notions. Precise categorization improves when indicators align with the actual product condition. Even seasoned eco-conscious consumers may be misled by superficial indicators	[71]
Slovakian consumers	• The influence of greenwashing on brand impression was evaluated among both educated and ignorant consumers. Ignorant consumers saw a fast fashion company as more valuable after witnessing false or unfounded advertisements, underscoring the efficacy of greenwashing on the ignorant. Informed consumers with environmental information opposed such techniques, demonstrating education's significance in combating greenwashing and fostering openness	[72]
Korean consumers	• The linkage of "eco-friendly fabric" with terms indicative of "eco-friendly" is the paramount component in establishing a sustainable fashion brand image. Furthermore, terms like "marketing" and "campaign" were positioned in the top 10 for H&M and Zara, suggesting a potential opportunistic use of greenwashing	[73]
UAE consumers	• Consumers believe that fast fashion companies in the UAE do not provide sufficient proof to support their altruistic assertions, leading to increased caution and skepticism when buying eco-friendly items	[45]
Indonesian consumers	• Greenwashing has a substantial positive impact on negative green word of mouth and green skepticism. Meanwhile, negative green word of mouth is significantly positively impacted by green skepticism	[74]
Portuguese consumers	• Greenwashing techniques negatively impact customers' brand perceptions and purchasing intentions to a greater extent in the luxury industry compared to the fast fashion sector	[75]

Table 2 Legislation of some countries to curb greenwashing in the fashion sector

Country	Legislation for greenwashing in the fashion sector
USA	• In the United States, enforcement actions against "greenwashing" are primarily carried out by the Federal Trade Commission (FTC) and the Securities and Exchange Commission (SEC). The FTC regulates environmental claims following the Federal Trade Commission Act and, depending on the industry, under specific sector regulations, such as the Textile Fiber Products Identification Act for the textile sector. The SEC classifies these issues as violations of the Securities Act and the Exchange Act [78] • To aid marketers in avoiding deceptive environmental statements to consumers, the FTC released the first edition of "Guides for the Use of Environmental Marketing Claims" (Green Guides) [5, 78–81] in 1992. The Green Guides were amended in 1996, 1998, and 2012 to provide guidance on (1) overarching concepts applicable to all environmental marketing claims, (2) consumer interpretation of specific claims, (3) strategies for marketers to engage consumers [5] • The latest version from 2012, in addition to regular modifications, including additional sections on carbon offsets, "green" certifications and seals, as well as assertions about renewable energy and materials that validate these claims. Furthermore, it examines how marketers might substantiate their assertions to prevent misleading representations [5] • The Green Guides specify that a claim such as "Eco-friendly: made with recycled materials" is not deceptive if (1) the phrase "made with recycled materials" is clear and prominent, and (2) the marketer can verify that the entire product or package, excluding minor incidental components, consists of recycled material. In conjunction with requirements (1) and (2), the claim is not deceptive if (3) "manufacturing the product with recycled materials amplifies its overall environmental advantages" and (4) the context of the advertising does not imply any other misleading statements [5] • If the FTC determines that an advertisement has contravened the Green Guidance, it will generally issue a cease-and-desist order to the offender. If the practice of greenwashing persists, the FTC may impose a fine or a jail sentence of up to one year on the offender [81]

(continued)

Table 2 (continued)

Country	Legislation for greenwashing in the fashion sector
United Kingdom	• The U.K. Competition & Markets Authority (CMA) is the primary competition regulator and has diligently provided extensive suggestions for companies to avoid greenwashing [5, 78, 80]. In September 2021, the CMA published its Green Claims Code (GCC), providing guidance to firms on complying with existing consumer protection laws when making environmental claims for their products or services [5]. • In September 2024, the CMA published a practical Guidance entitled "Complying with Consumer Law When Making Environmental Claims in the Fashion Retail Sector" (the "Guidance"), relevant to all businesses that assert environmental claims concerning apparel, footwear, fashion accessories, or related services such as packaging, delivery, and returns. The CMA asserts that the Guidance is relevant for manufacturers and suppliers (including third-party branded suppliers), wholesalers, and distributors, and may also be used by other consumer-facing businesses. The Guidance includes a company's sustainability claims on its products, services, processes, branding, or overall operations and specifies 11 requirements for enterprises, including the need that their environmental statements [78]: (1) are explicit and precise, (2) do not exclude critical information, (3) avoid ambiguous terminology, (4) do not feature deceptive imagery and icons, (5) include comparisons that are transparent, equitable, and equivalent, and (6) clearly delineate any actions a consumer must undertake for the claim to be realized.
French	• The primary elements of France's policy to combat greenwashing consist of the Climate and Resilience Law and the Anti-Waste and Circular Economy Law [5]. • The 2021 Resilience Law mandates that ads be visible or accessible to consumers at the point of purchase, ensuring they are trustworthy and readily comprehensible to safeguard French consumers and address this developing attitude. Furthermore, assertions on the environmental impact of a product must be accurate throughout the whole life cycle of the garment and consider the environmental consequences of the used manufacturing technique [5] • The Climate and Resilience Law, enacted on August 22, 2021, governs assertions about carbon neutrality. Advertisers must disclose on their website a summary report detailing the carbon footprint of the promoted product or service, the planned reduction trajectory, and the compensation methods [78] • The 2021 Resilience Law further expanded penalties for deceptive consumer actions under the FCC. The FCC formerly stipulated that participation in deceptive business activities might result in a maximum of two imprisonment and a fine of €300,000. It stipulated that the penalties may be augmented to 10% of the turnover or 50% of the expenses spent on advertising or the conduct constituting the infraction. The 2021 Resilience Law revised the penalties under the FCC to permit fines to escalate to eighty percent of the expenses expended for the objectionable marketing technique if it pertains to environmental concerns [5]

(continued)

Table 2 (continued)

Country	Legislation for greenwashing in the fashion sector
Italy	• Italian law regulates green claims under the Consumer Code ("Legislative Decree n. 206 of 2005") since it seeks to sway consumers' purchase decisions towards more sustainable alternatives [78] • The Italian Competition Authority (ICA) can investigate and issue penalties for unfair commercial activities with green claims. The Advertising Self-Discipline Institute (IAP) forbids both direct and indirect environmental assertions that lack true, pertinent, and scientifically verified evidence [78] • Upon violations of the Consumer Code, the ICA may mandate the cessation of the unfair practice and impose an administrative fine (between 5,000 Euros and 10,000,000 Euros), considering the severity and duration of the violation as well as the economic and asset circumstances of the professional [78]
Germany	• Greenwashing and its legal ramifications are not now explicitly governed by German law. The matter pertains to the "Gesetz gegen den unlauteren Wettbewerb" ("UWG"), which regulates German competition law [78] • The UWG asserts that "any individual engaging in a deceptive commercial practice likely to induce a consumer or other market participant to make a commercial decision they would not have otherwise made is acting unfairly" [78] • The principal criteria is whether advertising assertions are "deceptive" to customers or other market participants in their purchase decisions. A violation of the UWG may result in orders for claim retraction and cease-and-desist demands [78] • The German Federal Court of Justice (FCJ) has determined that the phrase "climate neutral" may deceive consumers if it lacks further information elucidating its foundation [78]
Australia	• In December 2023, the Australian Competition and Consumer Commission (ACCC) released Guidance about the applicability of the Australian Consumer Law ("ACL") to environmental and sustainability assertions. "Making Environmental Claims: A Guide for Business" outlines the ACCC's views on optimal methods for companies to ensure their environmental and sustainability claims are not misleading [78] • The guideline comprises eight principles, accompanied by examples, to assist companies in understanding what the ACCC considers acceptable or unacceptable. The examples suggest that the ACCC requires more detail in environmental claims and will not accept vague assertions or representations that are only partly accurate [78]
China	• China has not enacted a distinct legislation to regulate greenwashing marketing methods; nevertheless, it has tackled such acts via its advertising law, consumer protection law, trademark law, and anti-unfair competition law [82] • Advertising is deemed 'false' if it meets any of the following conditions: (1) The products or services are non-existent; (2) The information regarding the products or services is inconsistent with the actual circumstances concerning performance, function, origin, purpose, quality, specifications, ingredients, price, producer, validity period, sales, awards, or any guarantees associated with the products or services, and such discrepancies significantly impact purchasing decisions; (3) Utilizing scientific advancements, statistics, research findings, abstracts, quotations, and other unverifiable or fabricated information as evidentiary material; (4) Asserting false effects of product or service utilization; or (5) Other situations wherein consumers are deceived or misled by false or misleading content [82]

(continued)

Table 2 (continued)

Country	Legislation for greenwashing in the fashion sector
Canada	• The federal Competition Act, Textile Labelling Act, and Consumer Packaging and Labelling Act in Canada ban false or misleading assertions [82] • The Competition Act forbids claims on a product's performance, effectiveness, or longevity without sufficient and appropriate testing. Likewise, Canada's Trademarks Act prohibits the dissemination of substantially false and misleading assertions about the character, quality, amount, composition, geographical origin, or method of manufacturing, production, or performance of products or services [82]
Singapore	• The Consumer Protection (Fair Trading) Act (CPFTA) protects customers against unfair business practices, such as fraud, misleading representations, and the exploitation of consumers' incapacity to understand transaction consequences. Discontented customers may lodge a complaint with the Competition and Consumer Commission of Singapore [82] • The Misrepresentation Act allows a customer to seek damages from the merchant for a misrepresentation that affected a business transaction, in accordance with contractual responsibilities [82] • The Singapore Code of Advertising Practice mandates that all ads must be lawful, respectable, honest, and truthful [82]

5.2 *Brief Summary of Some Industry Initiatives*

Some initiatives are discussed in Table 3 to curb greenwashing in the fashion industry. The fashion transparency index is an industry-based initiative, GRI is renowned for standardized sustainability reporting, make the label count to authenticate sustainability claim, and the GOTS certificate is renowned for comprehensive sustainable apparel production.

6 Preventive Measures

6.1 *Educating Consumers and Better Access to Learn About Greenwashing*

Even environmentally conscious consumers acknowledge acquiring inexpensive fast-fashion items to satisfy their fashion preferences [89]. For example, despite their environmental concerns regarding the activities of fast fashion brands [74, 90, 91], generation Z exhibits a significant addiction to fast fashion [92]. In addition, consumers struggle to distinguish between legitimate and fraudulent environmental claims [89]. It is challenging to identify greenwashing, even among professional consumers who are well-informed about the industry and greenwashing. The accusing process is even more difficult for average consumers who are oblivious to or have limited knowledge about the phenomenon [12].

Table 3 Some initiatives to curb greenwashing

Initiatives	Description
Fashion transparency index	• The Fashion Transparency Index assesses firms' public disclosure of human rights and environmental concerns across five key domains, using 258 metrics. Key topics encompass: (1) Policies and Commitments, (2) Governance, (3) Supply Chain Traceability, (4) Know, Show, and Fix, (5) Spotlight Issues, which include: (a) Decent Work, addressing forced labor, living wages, purchasing practices, unionization, and collective bargaining, (b) Gender and Racial Equality. (c) Sustainable sourcing and materials, (d) Overconsumption, business models, waste, and circularity, (e) Water and chemicals, (f) Climate change, fossil fuels, and biodiversity [83] • All previous editions of the Fashion Transparency Index are available for review and comparison, year-over-year [83]
Global reporting initiative (GRI)	• The GRI Standards encapsulate global best practices for public reporting on diverse economic, environmental, and social consequences. An organization's beneficial or detrimental impacts on sustainable development are shown via sustainability reporting that complies with the Standards [84, 85]. • The Standards are a modular framework consisting of three interconnected series: Universal Standards, Sector Standards, and Topic Standards. Organizations may use the GRI Standards to produce a sustainability report that complies with the Standards or to provide information tailored for certain users or purposes by leveraging particular Standards (or segments of its content) [85] • For instance, Nike (USA), Inditex (Spain), and Ralph Lauren (USA) are among the prominent fashion companies that incorporate GRI indicators into their sustainability reports [84]
Global organic textile standard (GOTS)	• GOTS is a standard governing the processing of organic fibers across the whole supply chain, from harvesting raw materials to selling the completed product to consumers. Numerous manufacturers globally adhere to it, resulting in the widespread acceptance of GOTS-certified goods worldwide [86]. GOTS certification is founded on many distinctive and unique essential aspects that underscore its significance [87] through-(a) Compliance support, (b) Organic fibers, (c) Due diligence, (d) Third-party certification, (e) Human rights and social criteria, (f) Environmental criteria, (g) All processing stages, (h) Traceability • Scope Certificates provide details on the extent of a company's GOTS certification. This encompasses the certified items, the manufacturing locations of these products, and the value chain partners engaged in the production process. These certifications guarantee that the whole value chain for a GOTS-certified product is transparent and that all pertinent stakeholders are dedicated to GOTS' sustainability and ethical manufacturing criteria [87] • A Transaction Certificate encompasses all information from the preceding Certificate and the new transaction facts and is issued subsequent to the transfer of material ownership. The Certification Body may, therefore, reconcile volumes and verify the mass balances reported by the Certified Entities. The Transaction Certificate will be granted only after verifying that the Entity has sufficient input material to provide the output specified in the Certificate. This establishes a clear record of the product's progression through the value chain, facilitating the verification of its organic certification and ensuring compliance with GOTS' stringent standards for sustainability and ethical manufacturing [87]
Make the label count	• Make the Label Count is an international coalition of organizations representing diverse environmental groups and natural fiber producers. They share a commonality: It strives to guarantee that the sustainability claims made by textile manufacturers in the European Union are credible and equitable [88] • The Commission is investigating the potential of the Product Environmental Footprint (PEF) methodology to verify green claims and inform ecodesign requirements. The Commission recognizes that the PEF fails to take into account critical impacts, including the discharge of microplastics [88]

Environmental impact could be enhanced by providing consumers with comprehensive information on end-of-life recycling, care, and repair through RFID technology. It is also possible to scan it with any RFID-enabled smartphone. However, RFID identifiers themselves are not particularly sustainable. Most of RFID's carbon footprint is attributed to the materials, including a silicon chip, a copper or aluminum antenna, and a PET-based (Polyethylene Terephthalate) laminate layer [93]. For specific information, the Quick Response (QR) code [94] and more details through blockchain Digital Product Passport (DPP) [95] are available for consumers to access information about fashion products to curb greenwashing. However, QR codes possess limited information, and DPP has the potential for data breaches and privacy concerns. Furthermore, not many fashion brands use it for every product; consumer willingness to use it during shopping is also essential.

To the consumer, sustainability may be a choice, not a mandatory obligation. Conversely, when fashion brands fail to adhere to the regulations, they will incur penalties. It is imperative to educate consumers to prevent greenwashing in the fashion industry when laws and regulations is structured in this manner.

Research is essential to disseminating information about the manifestation of greenwashing and educating consumers regarding sustainability. An alternative approach may be necessary to inform consumers and comprehend the significance of sustainability, necessitating additional research in this context. For instance, consumers engage in retail activities to enhance their shopping experiences. Without disrupting the purchasing experience, it is imperative to investigate how consumers can more readily access information regarding the sustainability of clothing products.

6.2 Collaboration in Improving Legislation

The guidelines established by the United States, the United Kingdom, and France assist companies in determining whether or not they have violated consumer protection laws [5]. Nevertheless, the laws of each country contained some exhaustive, creative, and robust regulations that other countries could emulate.

The governments of the United States, the United Kingdom, and France have underscored the need of validating environmental assertions with evidence. The degree to which this requires third-party assistance remains unclear. Businesses must justify their statements with rigorous, trustworthy, relevant, and current evidence, as articulated by the U.K. GCC. The GCC does not require firms to substantiate their claims with third-party research. The GCC provides many instances of proof, including published research or studies that a corporation has commissioned or undertaken. Nonetheless, this fails to meet the explicit criterion for third-party verification. The United States Green Guide stipulates that claims must be substantiated by evidence derived from tests, analyses, research, or studies conducted and assessed objectively by qualified individuals, which are widely recognized in the profession as producing accurate and reliable results. This does not explicitly require third-party examination; rather, the stipulation that competent personnel validate statements provides a more stringent threshold for corporations to verify their assertions [5].

The Green Guides in the United States were first produced in 1992 to provide a nationwide standard for environmental claims in marketing. Subsequent revisions occurred in 1996, 1998, and 2012. Nonetheless, the Green Guides may be amended to provide clear definitions and verification criteria for often used terms in fashion marketing, such as "sustainable," [79, 80] "carbon," and "recyclable." The FTC might accomplish these goals by amending the Green Guides to align with the principles of the United Kingdom, the European Union, and California legislation [80]. In 2021, the Competition and Markets Authority ("CMA") in the United Kingdom published a detailed guidance for firms on the responsible formulation of green claims. This guide defines "sustainability claims" as assertions indicating that a product is manufactured, a service is provided, or a firm operates in accordance with the concepts of sustainability, sustainable consumption, or sustainable development. This may include assertions pertaining to environmental issues, climate change, biodiversity, animal welfare, labor welfare, or corporate social responsibility [79]. The Council of the European Union ("CEU") offers the FTC an enhanced framework for delineating "sustainability." The CEU Directive, a provisional agreement, was promulgated in 2023 by the CEU, which characterizes "sustainability label" as "any voluntary trust mark, quality mark, or equivalent, whether public or private, intended to distinguish and promote a product, process, or business regarding its environmental or social attributes or both" [80]. A sophisticated regulatory regime for the use of the recycling symbol and the advertisement of the recyclability of products is enforced by California's SB 343 for recycling claims. SB 343 mandates that individuals utilizing the terms "recyclable" or the "chasing arrows symbol" must furnish proof that the product or packaging in question is made from materials recognized by recycling facilities serving a minimum of sixty percent of California's population, and that these materials can be effectively sorted into specified streams for recycling operations by large-scale transfer or processing facilities that cater to at least sixty percent of California's recycling initiatives. The Voluntary Carbon Market Disclosures Act (AB 1305) of California ("VCMDA") establishes a regulatory framework for carbon claims. Any qualified entity that claims its product is "carbon neutral" or makes a similar assertion regarding its non-contribution to greenhouse gas emissions must, according to the VCMDA, disclose all information pertaining to the verification of the claim's accuracy or actual achievement, along with the methodology employed to measure interim progress toward that goal [80].

Moreover, the FTC need to replicate the measures taken by other countries and undertake more comprehensive investigations to combat greenwashing in the fashion sector. The CMA executed its first worldwide and industry-wide internet "sweep" from November 9 to November 27, 2020, aimed at detecting "misleading environmental claims." The CMA investigated over five hundred "randomly selected websites" for deceptive environmental claims. In March 2021, the Netherlands Authority for Consumers and Markets ("ACM"), in conjunction with the CMA, undertook an assessment of 170 enterprises in the Netherlands to detect any deceptive sustainability claims [79]. In October and November 2022, the ACCC performed an online investigation in Australia to detect deceptive environmental and sustainability marketing assertions in an assessment of 247 enterprises [78].

7 Conclusion

The comprehensive research of this chapter unequivocally demonstrates that greenwashing in the fashion sector is a substantial obstacle to attaining Sustainable Development Goal 12. The trend of greenwashing, alongside legislative actions from various nations, NGOs, and consumer and environmental protection organizations, provides substantial evidence of greenwashing within the fashion industry. Moreover, current law has significant limits in addressing greenwashing within the fashion business. The USA FTC must amend the green recommendations, and there is a substantial need for stringent rules from emerging nations.

Recommendations for enhancing current regulations and preventative actions have been presented. Without stringent rules and consumer awareness fostered by sustainability education, mitigating greenwashing in the fashion sector is challenging, which poses substantial challenges to achieving sustainability. This chapter presented concise data and information about greenwashing in the fashion industry, enabling policymakers, academics, and consumers to comprehend how greenwashing poses a danger to sustainable production and consumption within this sector.

References

1. General Assembly, Resolution adopted by the General Assembly on 6 July 2017 (2017). https://ggim.un.org/documents/a_res_71_313.pdf
2. S. Tracker, Measuring progress towards the sustainable development goals (2018), https://sdg-tracker.org/inequalitytargets
3. J. Bartram, C. Brocklehurst, D. Bradley, M. Muller, B. Evans, Policy review of the means of implementation targets and indicators for the sustainable development goal for water and sanitation. NPJ Clean Water **1**(1), 1–5 (2018)
4. N. Shukla, Fast fashion pollution and climate change (2022), https://earth.org/fast-fashion-pollution-and-climate-change/
5. E. Sammons, "Green" is the new black: enforcing consumer protection laws against greenwashing in the fashion industry. Emory Int'l. L. Rev. **38**, 191 (2024)
6. Geneva Environmental Network, Environmental Sustainability in the Fashion Industry, (2024) https://www.genevaenvironmentnetwork.org/resources/updates/sustainable-fashion/
7. CGR, A circular textile industry is within reach (2024), https://www.circularity-gap.world/textiles
8. B. Garcia-Ortega, J. Galan-Cubillo, F.J. Llorens-Montes, B. de-Miguel-Molina, Sufficient consumption as a missing link toward sustainability: the case of fast fashion. J. Clean. Prod. **399**, 136678 (2023)
9. T.P. Lyon, A.W. Montgomery, The means and end of greenwash. Organ. Environ. **28**(2), 223–249 (2015)

10. B. Watson, The troubling evolution of corporate greenwashing (2016), https://www.theguardian.com/sustainable-business/2016/aug/20/greenwashing-environmentalism-lies-companies
11. N. Nemes, S.J. Scanlan, P. Smith, T. Smith, M. Aronczyk, S. Hill, D. Stabinsky, An integrated framework to assess greenwashing. Sustainability **14**(8), 4431 (2022)
12. S.V. De Freitas Netto, M.F.F. Sobral, A.R.B. Ribeiro, G.R.D.L. Soares, Concepts and forms of greenwashing: a systematic review. Environ. Sci. Eur. **32**, 1–12 (2020)
13. B. Parguel, F. Benoît-Moreau, F. Larceneux, How sustainability ratings might deter 'greenwashing': a closer look at ethical corporate communication. J. Bus. Ethics **102**, 15–28 (2011)
14. M.A. Delmas, V.C. Burbano, The drivers of greenwashing. Calif. Manage. Rev. **54**(1), 64–87 (2011)
15. TerraChoice, The sins of greenwashing: home and family edition (2010), https://www.twosides.info/wp-content/uploads/2018/05/Terrachoice_The_Sins_of_Greenwashing_-_Home_and_Family_Edition_2010.pdf
16. E. Tateishi, Craving gains and claiming "green" by cutting greens? An exploratory analysis of Greenfield housing developments in Iskandar Malaysia. J. Urban Aff. **40**(3), 370–393 (2018)
17. A. Corcione, What is greenwashing (2020), https://www.businessnewsdaily.com/10946-greenwashing.html
18. CorpWatch. "Greenwash Fact Sheet." (2023), https://www.corpwatch.org/article/greenwash-fact-sheet
19. T.P. Lyon, J.W. Maxwell, Greenwash: corporate environmental disclosure under threat of audit. J. Econ. Manag. Strat. **20**(1), 3–41 (2011)
20. C. Marquis, M.W. Toffel, *The globalization of corporate environmental disclosure: accountability or greenwashing?* (Harvard Business School, Boston, 2011)
21. K. Walker, F. Wan, The harm of symbolic actions and green-washing: corporate actions and communications on environmental performance and their financial implications. J. Bus. Ethics **109**, 227–242 (2012)
22. L.M. Baum, It's not easy being green… or is it? A content analysis of environmental claims in magazine advertisements from the United States and United Kingdom. Environ. Commun.: J. Nat. Cult. **6**(4), 423–440 (2012)
23. J. Pearsall, Concise Oxford English Dictionary, 10th ed. Oxford, UK (2003)
24. S.O. Idowu, N. Capaldi, L. Zu, A.D. Gupta, *Encyclopedia of corporate social responsibility* (Springer, Berlin, 2013)
25. R. Guo, L. Tao, L. Yan, P. Gao, The effect path of greenwashing brand trust in Chinese microbiological industry from decoupling view. Indian J. **10**(7), 1827–1831 (2014)
26. F. Bowen, J.A. Aragon-Correa, Greenwashing in corporate environmentalism research and practice: the importance of what we say and do. Organ. Environ. **27**(2), 107–112 (2014)
27. L.G. Vaqué, Directive 2005/29/EC on unfair commercial practices and its application to food-related consumer protection. European Food and Feed Law Review 210-221 (2015)
28. A. Siano, A. Vollero, F. Conte, S. Amabile, "More than words": expanding the taxonomy of greenwashing after the Volkswagen scandal. J. Bus. Res. **71**, 27–37 (2017)
29. M. De Ferrer, What is greenwashing and why is it a problem? (2020), https://www.euronews.com/green/2020/09/09/what-is-greenwashing-and-why-is-it-a-problem
30. Ethical Consumer, What is greenwashing? (2023), https://www.ethicalconsumer.org/transport-travel/what-greenwashing
31. Ecolife, What is Greenwashing? (2022), https://ecolife.com/dictionary/greenwashing/
32. A. Hayes, What is Greenwashing? How it works, examples, and statistics (2022), https://www.investopedia.com/terms/g/greenwashing.asp
33. Sourcewatch Greenwashing. The Center for Media and Democracy, https://www.sourcewatch.org/index.php/Greenwashing
34. EPA, Top green home terms (2010), https://sor.epa.gov/sor_internet/registry/termreg/searchandretrieve/glossariesandkeywordlists/search.do?details=&glossaryName=Top%20Green%20Home%20Terms

35. D. Kelleher, Fast fashion (2025), https://www.britannica.com/art/fast-fashion
36. K. Collins, Benetton & fashioning controversy (2019), https://blogs.library.duke.edu/rubenstein/2019/04/08/benetton-fashioning-controversy/
37. D.J. Doorey, The transparent supply chain: from resistance to implementation at Nike and Levi-Strauss. J. Bus. Ethics **103**, 587–603 (2011)
38. G.P. Cachon, R. Swinney, The value of fast fashion: quick response, enhanced design, and strategic consumer behavior. Manag. Sci. **57**(4), 778–795 (2011)
39. B. Oliwia, Navigating branding frontiers: de-influencing, accessibility, and anti-greenwashing at Patagonia (2024), https://www.diva-portal.org/smash/record.jsf?pid=diva2%3A1880723&dswid=-2321
40. E. Woodhouse, Patagonia's decision to give away its brand proves greenwashing won't save us (2022), https://www.newstatesman.com/thestaggers/2022/09/patagonia-decision-greenwashing-climate-change#:~:text=20%20September%202022-,Patagonia's%20decision%20to%20give%20away%20its%20brand%20proves%20greenwashing%20won,real%20climate%20action%20is%20expensive
41. M.C. O'Connor, Patagonia's new study finds fleece jackets are a serious pollutant (2016), https://www.outsideonline.com/outdoor-gear/gear-news/patagonias-new-study-finds-fleece-jackets-are-serious-pollutant/
42. B. Miller, Co-operative, M&S and Waitrose accused of greenwash over close links to 'unethical' companies (2012), https://theecologist.org/2012/feb/07/co-operative-ms-and-waitrose-accused-greenwash-over-close-links-unethical-companies
43. P. Bertola, C. Colombi, V.M. Iannilli, F. Vacca, From cultural branding to cultural empowerment through social innovation: I was a sari—a design-driven Indian case study. Fash. Pract. **12**(2), 245–263 (2020)
44. V. Crinis, Corporate social responsibility, human rights and clothing workers in Bangladesh and Malaysia. Asian Stud. Rev. **43**(2), 295–312 (2019)
45. S. Munir, V. Mohan, Consumer perceptions of greenwashing: lessons learned from the fashion sector in the UAE. Asian J. Bus. Ethics **11**(1), 1–44 (2022)
46. Truthinadvertising.org, Companies accused of greenwashing (2024), https://truthinadvertising.org/articles/companies-accused-greenwashing/
47. Changing Markets Foundation, Changing markets foundation, to greenwash license: how certification schemes and voluntary initiatives are fueling fossil fashion (2022), https://changingmarkets.org/portfolio/fossil-fashion/
48. ECO-AGE, Eco-Age SRL: fashion industry addicted to greenwashing and fossil-fuel based synthetics: new report from changing markets (2021), https://eco-age.com/resources/synthetics-anonymous-changing-markets-report/
49. G. Kaner, Greenwashing: how difficult it is to be transparent to the consumer—H&M case study, in *Green Marketing in Emerging Markets: Strategic and Operational perspectives*, pp. 203–226 (2021)
50. S. Thomas, From "green blur" to ecofashion: fashioning an eco-lexicon. Fash. Theory **12**(4), 525–539 (2008)
51. A. Peirson-Smith, S. Evans, Fashioning green words and eco language: an examination of the user perception gap for fashion brands promoting sustainable practices. Fash. Pract. **9**(3), 373–397 (2017)
52. S. Evans, A. Peirson-Smith, The sustainability word challenge: Exploring consumer interpretations of frequently used words to promote sustainable fashion brand behaviors and imagery. J. Fash. Mark. Manag.: Int. J. **22**(2), 252–269 (2018)
53. L. Ranasinghe, V.M. Jayasooriya, Ecolabelling in textile industry: a review. Resour. Environ. Sustain. **6**, 100037 (2021)
54. R. Dahl, Green washing: do you know what you're buying? (2010), https://ehp.niehs.nih.gov/doi/full/https://doi.org/10.1289/ehp.118-a246
55. L. Alizadeh, M.C. Liscio, P. Sospiro, The phenomenon of greenwashing in the fashion industry: a conceptual framework. Sustain. Chem. Pharm. **37**, 101416 (2024)

56. A. Sailer, H. Wilfing, E. Straus, Greenwashing and bluewashing in black Friday-related sustainable fashion marketing on Instagram. Sustainability **14**(3), 1494 (2022)
57. B. Castillo-Abdul, E. Ortega Fernandez, L.M. Romero-Rodriguez, Corporate social responsibility communication of male luxury fashion brands: analysis on Instagram, Facebook and TikTok. Manag. Decis. **62**(2), 471–491 (2024)
58. S. Glozer, M. Morsing, Helpful hypocrisy? Investigating 'double-talk' and irony in CSR marketing communications. J. Bus. Res. **114**, 363–375 (2020)
59. N. Macchiato, The ultimate guide to greenwashing (2019), https://www.natimacchiato.com/the-ultimate-guide-to-greenwashing
60. R. Dalton, The cooler sustainability gets, the more fast-fashion will greenwash you (2018), https://wellmadeclothes.com/articles/AsSustainabilityGetsCoolerFastFashionWillGreenwash/
61. L. Jones, Best Of 2018: are you being greenwashed? (2018), .https://wellmadeclothes.com/
62. F. Leach, Euroviews. Clothes made from 'carbon emissions': why Zara's new line is just more greenwashing (2022), https://www.euronews.com/green/2022/06/26/clothes-made-from-carbon-emissions-why-zaras-new-line-is-just-more-greenwashing
63. A. Badhwar, S. Islam, C.S.L. Tan, T. Panwar, S. Wigley, R. Nayak, Unraveling green marketing and greenwashing: a systematic review in the context of the fashion and textiles industry. Sustainability **16**(7), 2738 (2024)
64. S. Rauturier, Greenwashing examples: 8 notorious fast fashion claims and campaigns (2022), https://goodonyou.eco/greenwashing-examples/
65. S. Dutchen, Microplastics everywhere. The tiny particles are even in our bodies. What might this mean for our health? (2023), https://magazine.hms.harvard.edu/articles/microplastics-everywhere#:~:text=Studies%20in%20cell%20cultures%2C%20marine,meconium%2C%20an%20infant's%20first%20stool.
66. Y. Sun, B. Shi, Impact of greenwashing perception on consumers' green purchasing intentions: a moderated mediation model. Sustainability **14**(19), 12119 (2022)
67. X. Lu, T. Sheng, X. Zhou, C. Shen, B. Fang, How does young consumers' greenwashing perception impact their green purchase intention in the fast fashion industry? An analysis from the perspective of perceived risk theory. Sustainability **14**(20), 13473 (2022)
68. G. Zhang, L. Wang, H. Meng, Environmental knowledge level and consumer behavior regarding green fashion: a moderated mediation model. Asia Pac. J. Mark. Logist. **36**(4), 977–992 (2024)
69. V. Apaolaza, M.C. Policarpo, P. Hartmann, M.R. Paredes, C. D'Souza, Sustainable clothing: why conspicuous consumption and greenwashing matter. Bus. Strateg. Environ. **32**(6), 3766–3782 (2023)
70. M.C. Policarpo, V. Apaolaza, P. Hartmann, M.R. Paredes, C. D'Souza, Social cynicism, greenwashing, and trust in green clothing brands. Int. J. Consum. Stud. **47**(5), 1950–1961 (2023)
71. L. Ende, M.A. Reinhard, L. Göritz, Detecting greenwashing! The influence of product colour and product price on consumers' detection accuracy of faked bio-fashion. J. Consum. Policy **46**(2), 155–189 (2023)
72. M. Marko, A. Kusá, Greenwashing and the nature of education in relation to consumer trust in fast fashion marketing communication. Commun. Today **14**(1), 86 (2023)
73. Y. Kim, K.W. Oh, Which consumer associations can build a sustainable fashion brand image? Evidence from fast fashion brands. Sustainability **12**(5), 1703 (2020)
74. R. Promalessy, T. Handriana, How does greenwashing affect green word of mouth through green skepticism? Empirical research for fast fashion business. Cogent Bus. Manag. **11**(1), 2389467 (2024)
75. V.C. Von Unruh, The effects of greenwashing on consumer behavior in the fashion industry: a study comparing the luxury fashion and fast fashion industries-literature review part 1 (Master's thesis, Universidade NOVA de Lisboa (Portugal)) (2023)
76. A. Desore, S.A. Narula, An overview on corporate response towards sustainability issues in textile industry. Environ. Dev. Sustain. **20**, 1439–1459 (2018)

77. B. Wren, Sustainable supply chain management in the fast fashion Industry: a comparative study of current efforts and best practices to address the climate crisis. Clean. Logist. Supply Chain. **4**, 100032 (2022)
78. S. Hoffmann, et al., Navigating the evolving era of greenwashing regulations in the fashion industry, https://www.whitecase.com/insight-alert/navigating-evolving-era-greenwashing-regulations-fashion-industry
79. K.A. West, Goodbye to greenwashing in the fashion industry: greater enforcement and guidelines. NCL Rev. **101**, 841 (2022)
80. E. Bank, Fashion's greenwashing problem and how to better protect consumers. Rutgers Univ. L. Rev. Comment. **76**, 125 (2023)
81. B. Ballan, J.J. Czarnezki, Disclosure, greenwashing, and the future of ESG litigation. Wash. Lee Law Rev. **81**(2), 545 (2024)
82. C. Bey, I. Liang, S. Uhera, D. Smith, Greenwashing: exploring the risks of misleading environmental marketing in China, Canada, France, Singapore and the UK (2023), https://gowlingwlg.com/en/insights-resources/articles/2023/the-regulation-of-greenwashing
83. Fashion Revolution, Fashion transparency index 2023 (2023), https://www.fashionrevolution.org/fashion-transparency-index/
84. etOso, Sustainability in fashion companies: what and how to report (2024), https://etoso.io/resources/articles/sustainability-in-fashion-companies-what-and-how-to-report
85. GRI, How to use the GRI standards (2024), https://www.globalreporting.org/how-to-use-the-gri-standards/gri-standards-english-language/
86. S.S. Muthu, *Sustainability in the Textile Industry*. Springer (2017)
87. GOTS, Key features. Global organic textile standard (GOTS) (2024), https://global-standard.org/the-standard/gots-key-features
88. #MakeTheLabelCount, Delivering environmental policy through fair comparison of natural and synthetic in PEF (2022) https://www.makethelabelcount.org/globalassets/make-the-label-count/documents/gd4505-mtlc-pef-executive-summary-final.pdf
89. J. Fernandes, S. Segev, J.K. Leopold, When consumers learn to spot deception in advertising: testing a literacy intervention to combat greenwashing. Int. J. Advert. **39**(7), 1115–1149 (2020)
90. B. Um, C. Santos, S. Abdirahman, A, Power, Z. Uddin, EcoFashion scanner: bridging the Gen Z and Millennial'Green Gap'by facilitating sustainable fashion consumption behaviours, in *Extended Abstracts of the CHI Conference on Human Factors in Computing Systems*, pp. 1–6 (2024)
91. D. Zimand-Sheiner, S. Lissitsa, Generation Z-factors predicting decline in purchase intentions after receiving negative environmental information: fast fashion brand SHEIN as a case study. J. Retail. Consum. Serv. **81**, 103999 (2024)
92. W. Wojdyla, T. Chi, Decoding the fashion quotient: an empirical study of key factors influencing US generation Z's purchase intention toward fast fashion. Sustainability **16**(12), 5116 (2024)
93. Checkpoint RFID and retail sustainability: how RFID labels can support green initiatives, https://checkpointsystems.com/apparel-labeling-solutions/rfid-and-retail-sustainability/
94. Checkpoint Demystifying QR Codes and how they can help you with customer engagement, https://checkpointsystems.com/blog/demystifying-qr-codes/
95. Checkpoint Digital product passport guide: 6 things every fashion retailer needs to know, https://checkpointsystems.com/blog/digital-product-passport-what-you-need-to-know/

Chapter 11
Climate Change, Decent Work and Sustainable Production: An Insight from Sri Lankan Apparel Industry

Devathanthrige Janaka Chamara Harshana Senadeera, Abu Sadat Muhammad Sayem, Walter Leal, Grace Farhat, and Haruna Musa Moda

Abstract In the face of escalating climate risks, this chapter rigorously examines the integration of climate resilience into workplace safety within Sri Lanka's apparel industry, an essential sector of the national economy. With Sri Lanka ranked among the most vulnerable countries to climate change, the apparel industry faces significant threats from increased temperatures, extreme weather events, and associated economic disruptions. These challenges necessitate the implementation of robust climate resilience strategies to safeguard both the workforce and production processes. The chapter discusses the current state of health and safety within the industry, highlighting historical inadequacies and the pressing need for improvements in regulatory frameworks and awareness initiatives. It delves into strategic adaptations that include enhancing workplace health and safety, leveraging technological innovations, and fostering multi-actor partnerships to strengthen sector-wide

---

D. J. C. H. Senadeera (✉) · A. S. M. Sayem
Faculty of Arts and Humanities, Manchester Metropolitan University, Manchester, United Kingdom
e-mail: DEVATHANTHRIGE.J.SENADEERA@stu.mmu.ac.uk

A. S. M. Sayem
e-mail: ASM.Sayem@mmu.ac.uk

W. Leal
FTZ-Hamburg University of Applied Sciences, Hamburg, Germany
e-mail: walter.leal2@haw-hamburg.de

G. Farhat
Faculty of Health and Education, Manchester Metropolitan University, Manchester, United Kingdom
e-mail: G.Farhat@mmu.ac.uk

H. M. Moda
Department of Environmental Health and Safety, University of Doha for Science and Technology, Doha, Qatar
e-mail: haruna.moda@udst.edu.qa

A. S. Muhammad Sayem and P. Goswami (eds.), *SDG 12 and Global Fashion Textiles Production*, SDGs and Textiles, https://doi.org/10.1007/978-981-96-9051-0_11

resilience and sustainability. By aligning these efforts with the Sustainable Development Goals, particularly SDG 8 (Decent Work and Economic Growth), SDG 12 (Sustianble Consumption and Porduction) and SDG 13 (Climate Action), the chapter underscores the global relevance of Sri Lanka's initiatives, proposing a model for integrating environmental considerations with economic and social development goals. This analysis not only illuminates the path towards a resilient apparel industry in Sri Lanka but also offers insights into the broader implications for global supply chains and international trade in the context of escalating climate risks.

1 Introduction

Climate change, characterised by the steady increase in global temperatures due to the accumulation of greenhouse gases such as carbon dioxide and methane, presents profound challenges across the globe. Data from the National Oceanic and Atmospheric Administration [1] highlights a marked increase in the frequency, intensity, and duration of various natural hazards, a consequence of a 1 ºC temperature rise over the past 47 years [1, 2]. This trend is visibly affecting natural systems and human activities, evidenced by increased occurrences of natural disasters like wildfires, extreme heat, and significant flooding events. The year 2023 stands out as the warmest year on record, underscoring the accelerated pace of climate change, which has seen a tripling in the rate of temperature increase per decade since 1982 [1–3].

2 Overview of Global Climate Change Impacts

The significance of climate change is recognised globally, as evidenced by recent reports from the Intergovernmental Panel on Climate Change [4], the World Meteorological Organization [5], and other entities. The IPCC [7] highlights that human activities have caused global warming over the past 200 years, resulting in a 1.1 °C temperature rise above pre-industrial levels, leading to more frequent and hazardous weather events. Meanwhile, the WMO [5] reported that 2023 experienced record highs in global temperatures, sea levels, and extreme weather events, underscoring the urgent need for action. Additionally, Canada's historic 2023 wildfire season, which burned approximately 7.8 million hectares and released nearly four times more carbon dioxide than the entire global aviation sector in 2022, exemplifies the direct impacts of climate change [8]. These events reflect the growing concern and commitment from leaders, organisations, and citizens to tackling the climate crisis.

Furthermore, the broader definitions and causes of climate change, incorporating both natural and human-induced factors, are comprehensively described by the Intergovernmental Panel on Climate Change [4]. The IPCC defines climate change as any significant, long-term alteration in climate patterns, and emphasises that while natural processes do contribute to climate variability, human activities, especially since the mid-20th century, have become the primary drivers of recent global warming. These human activities include the burning of fossil fuels, deforestation, and various industrial processes that have led to increased greenhouse gas emissions [4, 7].

The health and safety implications of climate change are both profound and interrelated. The World Health Organization [9] has highlighted that climate change significantly intensifies health risks, resulting in a rise in heat-related illnesses, vector-borne diseases, and water-borne diseases, which underscores the urgent need for coordinated global action to address these health challenges. Simultaneously, evolving climate conditions necessitate adaptations in workplace safety. The International Labour Organization [10] stresses that the changing climate not only impacts public health but also introduces new challenges for occupational safety. This makes it essential to revise and strengthen safety protocols across various sectors to safeguard workers from the emerging risks associated with climate change [10].

3 Significance of the Apparel Industry

The global apparel industry plays a vital and interconnected role in the world economy, serving as a significant driver of employment, trade, and innovation. According to the World Trade Organization [11], the industry's trade reached $492 billion in 2022, underscoring its importance in fostering international trade and economic integration. Beyond mere figures, the sector provides livelihoods for millions, particularly in developing nations such as Bangladesh, Vietnam India and Sri Lanka, where it has catalysed industrial growth, created countless job opportunities, and contributed substantially to GDP growth and foreign exchange earnings [12, 16].

Moreover, the apparel industry has witnessed significant transformations in the last few decades, with clothing production doubling between 2004 and 2019, primarily fueled by increased global consumption and emerging middle classes in developed economies [13]. This growth has significantly impacted the environment and society, intensifying issues like carbon emissions, air pollution, and water contamination [12, 13]

However, alongside its positive contributions, the apparel industry faces significant challenges and negative impacts. One of the most pressing concerns is its environmental footprint, with the United Nations [12, 42] reporting that the industry is responsible for approximately 8–10% of global carbon emissions, making it one of the most polluting industries worldwide. Moreover, it generates nearly 20% of global wastewater, largely from water-intensive processes like dyeing and finishing textiles, contributing to pollution and impacting water resources. Moreover, McKinsey [14] analysis also shows that clothing brands are making public sustainability pledges, aiming to reduce their Scope 1, 2, and 3 emissions by 2030. Despite these efforts, only about 37% of brands are on track to meet their decarbonisation goals. Regulatory pressures, such as the European Union's new sustainability directives, are further pushing brands to accelerate their sustainability initiatives, encouraging the adoption of greener materials and more energy-efficient production methods. Social issues are also prevalent, as industry is often associated with poor working conditions, low wages, and exploitation, especially in developing countries (15). The collapse of the Rana Plaza factory in Bangladesh in 2013 serves as a tragic example of the unsafe working conditions that many garment workers face. Initiatives like the Sustainable Apparel Coalition's Higg Index are working towards encouraging sustainability and ethical practices in the industry, but there remains a considerable need to balance economic growth with environmental sustainability and social responsibility [16].

4 Sri Lanka's Vulnerability to Climate Change

Sri Lanka, a tropical island strategically located in the Indian Ocean, faces heightened risks from climate-related hazards due to its geographical and climatic conditions. The susceptibility of the island to increasing frequencies, intensities, and durations of natural hazards such as storms, floods, and heat waves is well-documented by numerous international assessments [16, 17]. The Global Climate Risk Index, which analyses historical data on weather-related losses, has consistently ranked Sri Lanka among the nations most affected by these events. Specifically, Sri Lanka was ranked 4th in 2016, 2nd in 2017, and 6th in 2018, which underscores its ongoing and severe vulnerability to climate-related disasters [18]. Despite a decrease in the death toll in 2020, the economic repercussions have been significant, with the highest losses recorded in 2018 amounting to approximately USD 3.63 billion, which is 1.24% of the national GDP [18].

Further analysis from the ND-GAIN Index sheds light on the nation's vulnerability and readiness to adapt to climate change. This index assesses exposure to climatic impacts, the sensitivity of sectors to these hazards, and the availability of resources for effective adaptation. In 2022, Sri Lanka scored a vulnerability score of 0.466, placing it 123rd in terms of vulnerability and 107th in readiness with a score of 0.379 [19]. These rankings reflect the critical need for enhanced adaptation strategies and resilience-building measures to cope with imminent climate challenges.

Moreover, the World Health Organization's Health and Environment scorecard projects significant increases in negative health outcomes due to the intensification of natural hazards. For instance, heat-related mortality rates in Sri Lanka are projected to rise tenfold by 2050 compared to the period from 1961–1990. Additionally, the current percentage of deaths from ischemic heart disease and stroke is expected to increase by 45% by the year 2050, linked to a 500% increase in air pollution levels caused by climate change [20].

The World Bank's Climate Impact Report further predicts a temperature increase between 1°C to 1.5°C by 2050, which will exacerbate the impacts of climate change already being felt across the nation. Over the past three decades, more than 10 million people have been affected by floods, while significant droughts have compromised the water and food security of over 8 million people in Sri Lanka. Additionally, rising sea levels have contributed to increased land inundation and saline intrusion in coastal areas [17, 21].

Scientific and medical research corroborates these projections, indicating a rise in temperatures and more frequent heavy precipitation events, particularly in the mountainous regions of Sri Lanka. Historical data show an increased incidence of extreme weather events such as floods and droughts [22, 23]. Local medical research has further highlighted that these climatic changes have detrimental effects on public health, increasing physical and psychological stress among workers, and impacting their productivity and well-being. In response, the government has taken proactive measures such as suspending outdoor activities during extreme heat periods to protect vulnerable populations, especially schoolchildren (24, 25).

5 Sri Lankan Apparel Industry

The apparel industry in Sri Lanka is a cornerstone of the national economy, significantly influencing the socio-economic landscape. This sector not only contributes approximately 7% to the GDP but also accounts for 45% of merchandise exports, establishing itself as the largest foreign exchange generator in the country [26]. Furthermore, the industry is a major employment provider, offering jobs to about 1.35 million individuals, with 350,000 employed directly and 1,000,000 indirectly. This substantial workforce underscores the industry's role as a vital employment hub [26]. In addition to its economic contributions, the apparel industry plays a crucial role in social development, particularly in alleviating poverty and advancing gender equality. A significant portion of the workforce is comprised of women, who benefit from employment opportunities that empower them economically and socially [16]. The empowerment of women through employment in this sector not only supports individual and family well-being but also contributes to broader societal advancements.

5.1 Climate Change and the Sri Lankan Apparel Industry

The apparel manufacturing sector in Sri Lanka, a key pillar of the national economy, is confronting numerous challenges magnified by climate change effects [16]. As the global climate becomes warmer and weather patterns more unpredictable, the sector must deal with various climatic threats that may jeopardise its future sustainability and resilience as highlighted in the Fig. 1 and listed below (Sri Lanka Apparel Exporters Association, 2018).

(a) Extreme Weather Events: The susceptibility of Sri Lanka to severe weather phenomena like floods and cyclones significantly endangers the apparel sector. Such events can interrupt supply chains, damage infrastructure, and cease

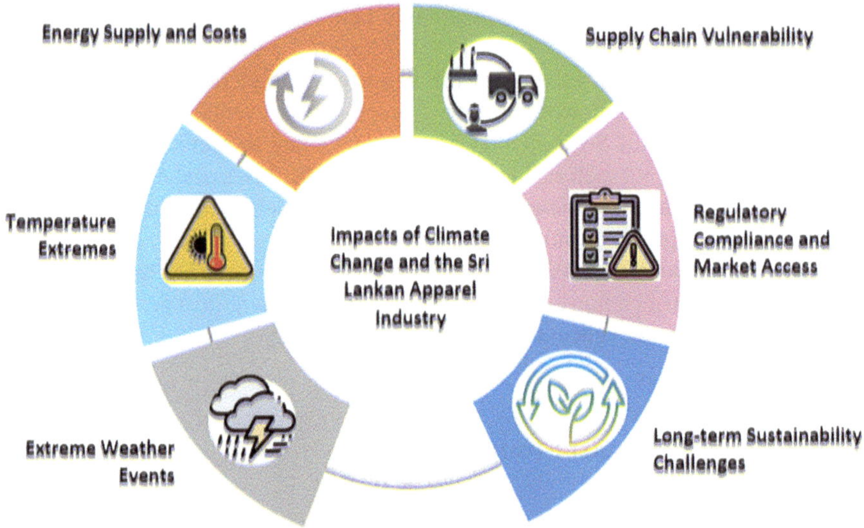

Fig. 1 Impacts of climate change and the Sri Lankan apparel industry based on the information from [16, 28, 29, 31]

production activities. The Asian Development Bank predicts that climate change will heighten the frequency and severity of these extreme weather occurrences in Sri Lanka [16].

(b) Temperature Extremes: The rise in temperatures from climate change presents health risks to apparel factory workers, especially regarding heat stress. Studies show that elevated temperatures may reduce labour productivity and affect the well-being of workers in the garment sector of Sri Lanka [27].

(c) Energy Supply and Costs: Climate changes affect energy infrastructure and lead to higher energy costs, impacting the operational costs of apparel manufacturing. Embracing renewable energy and enhancing energy efficiency are critical steps to mitigate climate-related risks. The International Finance Corporation emphasises the benefits of energy efficiency initiatives in boosting the competitiveness of Sri Lanka's apparel sector [28].

(d) Supply Chain Vulnerability: The apparel industry in Sri Lanka heavily relies on global supply chains for sourcing raw materials and components. Climate-induced disruptions in supplier nations can affect production timelines and profitability. The World Bank highlights the importance of resilient supply chain strategies to combat climate threats in the apparel industry [29].

(e) Regulatory Compliance and Market Access: The growing regulatory demands and consumer preferences for sustainable products pose both challenges and opportunities for the apparel sector in Sri Lanka. Adhering to environmental standards and adopting sustainable practices are vital for maintaining access to markets and competitive advantage. The Sustainable Apparel Coalition offers guidance on sustainable practices within the apparel sector [30].

(f) Long-term Sustainability Challenges: Climate change presents enduring sustainability issues for the apparel industry. Developing strategies for adaptation, such as investing in climate-resilient infrastructure and fostering sustainable production methods, are crucial for the sector's prolonged viability. The United Nations Development Programme provides resources focused on climate adaptation and sustainable growth [31].

Moreover, the National Adaptation Plan for Climate Change Impacts in Sri Lanka for 2016–2025 highlights a significant gap in its coverage, addressing merely nine out of twenty-four potential sectors vulnerable to climate change [32]. These sectors include agriculture, water resources, coastal and marine areas, health, human settlements and infrastructure, ecosystems and biodiversity, tourism and recreation, export agriculture, and energy and transportation. This limited focus raises concerns regarding the plan's overall effectiveness in mitigating the impacts of climate change across a broader economic range.

Additionally, the plan recognises the apparel sector's crucial role in generating foreign income since 1977 and acknowledges its susceptibility to climate impacts. However, it does not specifically include the apparel industry in its adaptation strategies for export sectors. This exclusion is problematic given the significant position of the apparel industry in the economy and its vulnerability to global market shifts, supply chain issues, and the physical risks of climate variability. The lack of targeted adaptation strategies for this vital sector underscores the necessity for a more comprehensive and inclusive national adaptation strategy to boost resilience across all key sectors affected by climate change.

This call for a broader approach is further supported by Sri Lanka's position as the 6th most vulnerable nation according to the Climate Change Vulnerability Index in 2018 [18] and its ranking of 154th in terms of vulnerability exposure as per the ND-GAIN Index 2023 [19]. These rankings demonstrate the inadequacies of the current measures in place to protect against the impacts of climate change, underscoring the need for an integrated strategy that addresses both the immediate vulnerabilities and the long-term resilience of the nation.

6 Climate Change Impacts on Workplace Health and Safety

Climate change poses significant risks to workplace health and safety, manifesting through the increased frequency, intensity, and duration of natural hazards. These hazards, driven by rising global temperatures, have direct and profound impacts on workers' health and safety across various industries [10, 33]. The effects are particularly severe in environments where workers are exposed to excessive heat, extreme weather events, air pollution, ultraviolet radiation, and vector-borne diseases [10]. These climate-induced hazards not only increase the likelihood of occupational injuries, diseases, and deaths but also exacerbate existing vulnerabilities, especially in manufacturing enterprises [10, 34].

6.1 Rising Temperatures and Extreme Weather Affect Worker Health and Safety

Rising temperatures and extreme weather events, both consequences of climate change, have increasingly severe implications for worker health and safety worldwide. The escalation of global temperatures has resulted in more frequent and intense heat waves, which pose significant health risks, particularly to workers in outdoor environments or poorly ventilated indoor spaces. Prolonged exposure to high temperatures can lead to heat stress, dehydration, heat exhaustion, and heatstroke, all of which increase the likelihood of occupational injuries and fatalities. According to the International Labour Organization (ILO), excessive heat currently affects 2.41 billion workers globally, leading to an estimated 22.85 million occupational injuries and 18,970 work-related deaths each year [10].

Extreme weather events such as floods, storms, heavy rainfall, and droughts also present substantial risks to worker safety. These events can disrupt transportation, damage critical infrastructure, create hazardous working conditions, and exacerbate existing vulnerabilities in the workforce. For instance, air pollution, intensified by extreme weather, impacts 1.6 billion outdoor workers, contributing to 860,000 annual deaths (10, 20, 33). Furthermore, ultraviolet (UV) radiation exposure has led to 18,980 deaths annually, primarily due to non-melanoma skin cancer [20]. The interplay between rising temperatures and extreme weather events exacerbates these risks, making it imperative to address workplace health and safety comprehensively in the context of climate change.

6.2 Specific Examples from Sri Lanka and Global Instances

Sri Lanka, as a tropical island nation, is acutely vulnerable to the impacts of rising temperatures and extreme weather events. The country's workforce, especially those in sectors such as agriculture, manufacturing, and construction, faces significant health risks due to these climate-induced hazards.

Historical data indicates a steady rise in temperature levels across all regions of Sri Lanka, leading to increased heat-related illnesses among outdoor workers [35–37]. For example, high atmospheric temperatures have directly affected the physical tasks and psychological well-being of workers, leading to increased instances of heat stress and related health issues [24]. This situation was underscored by the Education Ministry's decision in February 2024 to suspend all strenuous outdoor activities for schoolchildren due to soaring temperatures, highlighting the broader societal impact of rising heat levels [25].

Extreme weather events in Sri Lanka, such as floods and droughts, have also had profound impacts on worker health and safety. Over the past three decades, more than 10 million people in Sri Lanka have been affected by floods, while major droughts from 2008 to 2018 have posed significant challenges to water and food security for over 8 million people [38]. These extreme events disrupt not only the livelihoods of workers but also the broader economic stability of the nation, as seen in the apparel industry, where production delays and infrastructure damage are common during such events [27].

In Pakistan, the devastating floods of 2010 and 2022 underscore the vulnerability of global supply chains to climate-induced extreme weather events. These floods caused extensive damage to the cotton crop, a crucial raw material for the global apparel industry. The disruption in cotton supply led to production delays and significant economic losses for companies like Marks & Spencer, which relies heavily on Pakistani cotton. The situation is a stark reminder of how climate change can destabilise entire industries by affecting the availability and price of essential raw materials (39,40). This impact was not an isolated incident; the 2022 floods in Pakistan, which submerged 33% of the country and affected 33 million people, further disrupted the global cotton industry, causing widespread instability in production and pricing [41].

The apparel industry globally has seen similar vulnerabilities, particularly in regions like Sri Lanka, where climate change poses a direct threat to worker safety and supply chain stability. The country's apparel industry, a significant contributor to its GDP, faces risks from rising temperatures, which exacerbate heat-related illnesses among workers, and extreme weather events like floods, which disrupt production schedules and damage infrastructure [27]. Additionally, the scarcity of raw materials due to climate events in supplier countries can cause production delays and increase costs, as seen in the example of Marks & Spencer's reliance on Pakistani cotton [39].

Moreover, the indirect impacts on supply chains are significant, as climate change affects the availability and quality of raw materials. The cascading effects of such disruptions can be seen in various industries, where delays in one part of the supply chain can lead to broader operational inefficiencies and increased costs. For instance, the body shop has faced challenges in sourcing sesame oil from Nicaragua, where climate change has made agricultural production unpredictable [40]. These examples highlight the interconnectedness of global supply chains and the far-reaching impacts of climate change on worker health and safety, as well as on the economic stability of industries reliant on these supply chains.

The examples global context, demonstrate the severe impact of rising temperatures and extreme weather on worker health and safety. These climate-induced challenges underscore the need for resilient infrastructure, robust supply chain management, and proactive measures to protect workers from the growing threats posed by climate change. The interconnected nature of these impacts highlights the importance of global cooperation and the implementation of comprehensive strategies to mitigate the risks associated with climate change.

6.3 *Limitations of Sri Lankan Strategic Approaches in Enhancing Climate Resilience*

Sri Lanka's efforts to boost climate resilience through various strategies like The National Climate Change Policy, The National Adaptation Plan (2016–2025), The National Disaster Management Plan (2022–2030), and the Climate Prosperity Plan face numerous challenges. These challenges include inadequate policy integration, limited resources for effective implementation, and misalignment with global climate objectives. These constraints impede the effectiveness of the nation's climate strategies and plans (Fig. 2).

Sri Lanka's strategic initiatives like The National Climate Change Policy, The National Adaptation Plan (2016–2025), The National Disaster Management Plan (2022–2030), and the Climate Prosperity Plan face several critical challenges. These include inadequate integration across policies, a lack of detailed financial planning, and misalignment with global climate objectives. Major issues include:

(a) **Undefined Financial Strategies**: While the National Climate Change Policy and National Adaptation Plan acknowledge the need for sustainable financial mechanisms, they fall short in laying out detailed financial strategies for mobilizing, allocating, and monitoring funds. Saja et al. [42] identified similar inefficiencies in community-based disaster risk management projects in Sri Lanka, pointing out their limited effectiveness due to a lack of integration with broader national strategies.

(b) **Policy Integration Deficiencies**: The policies promote integrating climate considerations across sectors but lack actionable guidelines. Siriwardana et al. [43] highlighted significant deficiencies in Sri Lanka's disaster management frameworks, particularly in terms of coordination and integration among various

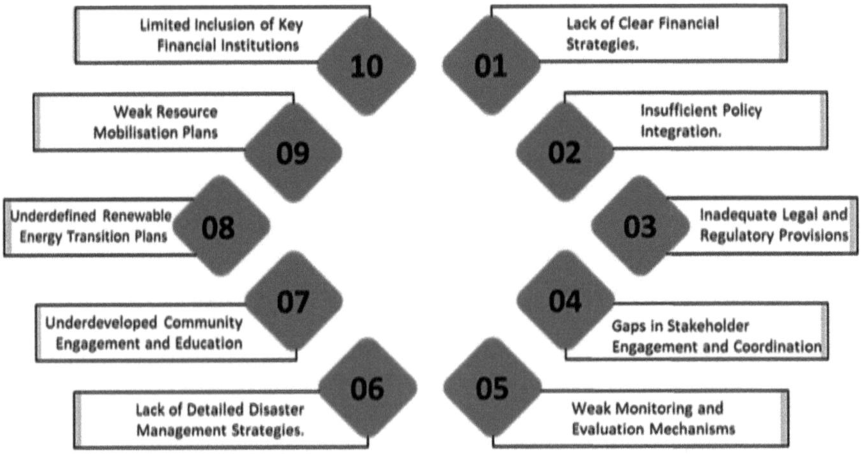

Fig. 2 Limitations in Sri Lanka's national climate change frameworks

stakeholders, which often leads to suboptimal responses to natural disasters such as floods and landslides.

(c) **Inadequate Legal and Regulatory Frameworks**: The existing policies call for enhanced legal and regulatory support but do not specify the necessary reforms or implementation timelines, creating delays in enforcement and effectiveness.

(d) **Limited Stakeholder Engagement**: The strategies emphasise broad stakeholder involvement but lack concrete frameworks for effective collaboration, leading to fragmented efforts in climate resilience measures.

(e) **Weak Monitoring and Evaluation**: General frameworks for monitoring are outlined, but they lack the specific methodologies, indicators, or frequencies needed for effective evaluation, as seen in the general approach to policy monitoring which fails to provide thorough tracking.

(f) **Vague Disaster Management Strategies**: The importance of comprehensive disaster management strategies is acknowledged; however, detailed plans for developing these strategies are missing, making it challenging to effectively manage climate-induced disasters.

(g) **Inadequate Community Involvement**: Policies stress the importance of community education on climate change and disaster preparedness but offer limited execution details. This gap in detailed planning for community engagement can undermine informed decision-making at local levels.

(h) **Non-specific Plans for Renewable Energy Transition**: The Climate Prosperity Plan sets ambitious renewable energy targets without detailed implementation strategies, particularly regarding grid integration and technical challenges like storage and stability, potentially slowing the transition away from fossil fuels.

(i) **Poor Resource Mobilization Strategies**: The need for resource mobilisation is noted, yet there is a lack of robust strategies for securing and managing these resources effectively. This is compounded by the underdeveloped financial planning mechanisms observed in the National Adaptation Plan and Climate Prosperity Plan.

(j) **Limited Financial Institution Involvement**: The minimal involvement of key financial institutions such as the Central Bank of Sri Lanka in the National Adaptation Plan and Climate Prosperity Plan impacts the financial planning necessary for effective policy implementation. This situation reflects a broader issue of preliminary vulnerability management within Sri Lanka's climate policy framework.

Addressing these shortcomings requires a comprehensive planning approach that includes all relevant stakeholders and aligns global sustainability and climate resilience goals.

7 Climate Change Implications for Sustainable Development Goals in Sri Lanka

Sri Lanka's dedication to the Sustainable Development Goals (SDGs) is articulated through its engagement with the 2018–2022 UN Sustainable Development Framework (UNSDF), which is strategically aligned with the nation's priorities and the broader 2030 Agenda for Sustainable Development. The UNSDF emphasises four key strategic focus areas: enhancing data and knowledge management to support evidence-based policymaking, fortifying public institutions to promote enduring peace, bolstering human security and socio-economic resilience, and fortifying resilience against climate change and disasters, alongside strengthening environmental management [44].

Within this framework, climate adaptation and workplace safety emerge as essential facets of sustainable production, highlighting the necessity for industries to navigate the challenges imposed by climate change. Effective climate adaptation entails modifying practices and policies to counteract environmental changes, including rising temperatures and extreme weather events that pose threats to human health and economic stability [45, 46]. The critical need for these adaptation strategies is further emphasized by potential financial losses, projected to surpass $1 trillion for major global corporations unless proactive steps are implemented [45].

Moreover, the synergy between climate adaptation and workplace safety is increasingly recognized as pivotal for achieving sustainable production. As industries grapple with the impacts of climate change, they must incorporate adaptive strategies within their operational frameworks to safeguard workers and sustain productivity [47, 48]. This integration not only addresses climate-related risks, such as heat stress and extreme weather conditions, but also bolsters long-term operational resilience and sustainability across diverse sectors like agriculture and manufacturing [45, 48].

7.1 The Role of Climate Adaptation in Sri Lankan Apparel Manufacturing

In the wake of escalating climate concerns, Sri Lanka's major apparel brands are revamping traditional garment factories into eco-friendly facilities. These revamped structures boast open designs that improve ventilation and diminish both carbon and water footprints, thereby reducing the environmental toll of production activities [49]. Notably, the fashion industry, responsible for around 10% of worldwide carbon emissions, encounters significant emissions during the dyeing and finishing phases [50]. By embracing green architectural methods, the Sri Lankan apparel industry could considerably decrease its environmental impact.

The implementation of sustainable practices in the apparel industry hinges on the availability of funding for energy-efficient solutions, renewable energy adoption, and enhanced waste management systems. Recent dialogues within the apparel sector committee have underscored the critical nature of such financial support [51]. Engaging with international specialists is crucial, providing local firms with bespoke advice to tackle the specific challenges faced by the sector. These international collaborations are key to fostering sustainable practices that contribute to Sri Lanka's climate objectives [52].

With climate-related disasters such as floods and landslides increasingly impacting coastal enterprises, initiatives spearheaded by entities like the United Nations Environment Programme and the University of Denmark Partnership are pivotal [12, 53]. These initiatives aim to bolster the resilience of the Sri Lankan garment industry through the development of disaster risk management and business continuity plans, which are essential in mitigating financial losses and ensuring operational stability during extreme weather conditions [53, 54].

Moreover, there is substantial potential within the Sri Lankan apparel industry to adopt sustainable materials derived from local sources. Utilising traditional resources like plant-based dyes not only honours the nation's cultural legacy but also enhances sustainability [55]. Nevertheless, it is crucial to educate consumers about the environmental repercussions of their fashion choices, as many remain unaware of the link between their purchases and climate change [55, 56].

Finally, technological advancements are pivotal for evolving towards more sustainable manufacturing processes in the apparel sector. Collaborations between textile companies and technology providers are yielding innovations in materials, data management, automation, and supply chain efficiency, which are instrumental in advancing sustainability [56, 57].

These initiatives are directly aligned with Sustainable Development Goal 13 (Climate Action), reflecting the requirement Sri Lankan apparel industry's commitment to integrating climate action into sectoral development. By focusing on areas like financing sustainable practices, technological innovation, and ethical standards, the industry not only mitigates its environmental impact but also sets a precedent for sustainable practices worldwide (Fig. 3).

7.2 Strengthening Resilience: Climate Adaptation and Safety Enhancements in the Sri Lankan Apparel Industry

To reinforce resilience within Sri Lankan apparel industry while advancing decent work and economic growth (SDG 8), several key factors require prioritisation. Firstly, the establishment of comprehensive occupational health and safety systems is crucial.

Historically, the sector has suffered from limited regulatory enforcement and poor awareness among stakeholders, leading to unsafe working conditions. Strengthening health and safety protocols through regular inspections, facility maintenance, and targeted safety training can significantly mitigate workplace hazards, including those intensified by climate change such as heat stress and flood-related risks [10].

Fig. 3 The role of climate adaptation in Sri Lankan apparel manufacturing based on information from the United Nations [44]

Secondly, fostering multi-stakeholder collaborations between apparel brands, suppliers, trade unions, and government institutions is essential to support climate adaptation. These partnerships facilitate the development of binding agreements and investments in infrastructure that enable factory relocation, improved risk management, and climate-smart operations [54, 60, 61].

A third factor lies in the strategic integration of technology and infrastructure upgrades. Innovations such as automation, smart ventilation systems, and energy-efficient machinery not only enhance productivity but also improve safety standards in response to rising temperatures and extreme weather conditions [66].

Additionally, reforms to legal and social protection frameworks are vital. Proposals include implementing robust occupational safety legislation and expanding social security schemes to safeguard the welfare of over 3.5 million private sector employees facing climate-induced risks [63]. Finally, climate literacy and worker training must be prioritised. Empowering workers with knowledge of climate-related hazards and sustainable practices fosters a resilient workforce capable of adapting to evolving threats [10, 48]. These multifaceted strategies are integral to maintaining competitiveness and ensuring ethical labour practices in the face of intensifying climate challenges (Fig. 4).

Fig. 4 Strengthening resilience: climate adaptation and safety enhancements in the Sri Lankan apparel industry manufacturing based on information from United Nations, 2022; ILO 2024

7.3 Enhancing Sustainability in Sri Lankan Apparel Industry: A Pathway to SDG 12

The Sri Lankan apparel industry is at a pivotal juncture, where the integration of sustainable practices is crucial to ensuring long-term competitiveness and resilience. As the sector faces increasing challenges due to climate change, evolving labour standards, and regulatory demands, it is embracing a comprehensive transformation aligned with Sustainable Development Goal 12 (SDG 12), which advocates for responsible consumption and production.

A robust health and safety culture within the apparel sector is essential for ensuring sustainability, operational effectiveness, and ethical production. Historically, the industry has faced challenges in workplace safety due to weak enforcement and limited awareness among employers and workers [58, 59]. However, with increasing global scrutiny of labour conditions, apparel manufacturers are strengthening workplace safety regulations, conducting regular facility inspections, and enhancing employee training programmes to create a more resilient workforce [44].

The integration of safety measures is not just a compliance requirement but a critical component of sustainable production practices, directly influencing worker well-being, retention, and operational efficiency. A secure working environment leads to greater productivity, lower absenteeism, and a positive brand reputation, all of which are vital for achieving SDG 12's objectives of sustainability and ethical manufacturing.

The impact of climate change on the Sri Lankan apparel industry is undeniable, with extreme weather events such as floods, landslides, and heat stress affecting factory operations and supply chains. To enhance resilience, the industry is fostering binding agreements and partnerships among brands, manufacturers, unions, and governments to implement climate adaptation strategies [60]. These strategies include:

- Investments in retrofitting factories or relocating them to lower-risk areas [55, 61].
- Encouraging investors to shift focus from mitigation to adaptation, ensuring long-term operational sustainability [61, 62].
- Integrating disaster risk management plans to prevent production losses during climate-related disruptions [53, 54].

By prioritising climate resilience, the apparel sector is aligning with SDG 12's focus on reducing environmental impacts while securing economic stability.

For sustainability to be fully realised, legal and regulatory frameworks must evolve to accommodate changing safety, environmental, and labour standards. Proposals for institutional reforms, such as the introduction of comprehensive social security programs, aim to protect over 3.5 million private-sector employees [63]. Furthermore, policy-level collaboration between the government, employers, and trade unions ensures that climate resilience is embedded into workplace safety laws [32].

To maintain compliance with international safety and environmental regulations, manufacturers are required to undergo frequent audits and reporting. Strengthening compliance monitoring mechanisms is key to protecting worker rights and ensuring adherence to sustainable manufacturing processes [61].

Achieving sustainable production goals requires a multi-actor approach, where key industry stakeholders—including public and private sector representatives, business associations, and international organisations—collaborate to enhance disaster risk management and sustainable business practices [54]. These partnerships are instrumental in:

- Strengthening policy implementation related to climate adaptation and labour safety [54, 64].
- Encouraging investments in climate-smart technologies that optimise resource efficiency [64].
- Bridging gaps between global climate goals and local industry requirements, ensuring practical, implementable sustainability measures.

By leveraging such partnerships, Sri Lankan apparel industry can scale its sustainability efforts, directly contributing to SDG 12's vision of responsible production.

Innovation and technology play a central role in the industry's transition towards sustainability. The adoption of automation, smart data management, and AI-driven supply chain solutions enhances operational efficiency while reducing waste and energy consumption [65, 66]. Key advancements include:

- Automated production processes that minimise resource waste.

- Smart energy management systems that optimise electricity and water usage.
- Innovative material development, such as bio-based and recycled textiles, reduces reliance on virgin raw materials [66, 67].

These technological advancements directly align with SDG 12's goal of sustainable resource use, ensuring economic growth while minimising environmental impact.

Sri Lanka's apparel industry is making significant strides in embedding sustainability into its operations, integrating strong health and safety standards, climate adaptation strategies, policy reforms, collaborative partnerships, and technological innovation. These elements collectively drive the sector towards a circular economy, ensuring both environmental responsibility and economic resilience.

By fostering multi-stakeholder collaboration, investing in climate resilience, and leveraging technology, the Sri Lankan apparel sector is actively supporting SDG 12, reinforcing its global competitiveness while maintaining ethical and sustainable production practices (Fig. 5)

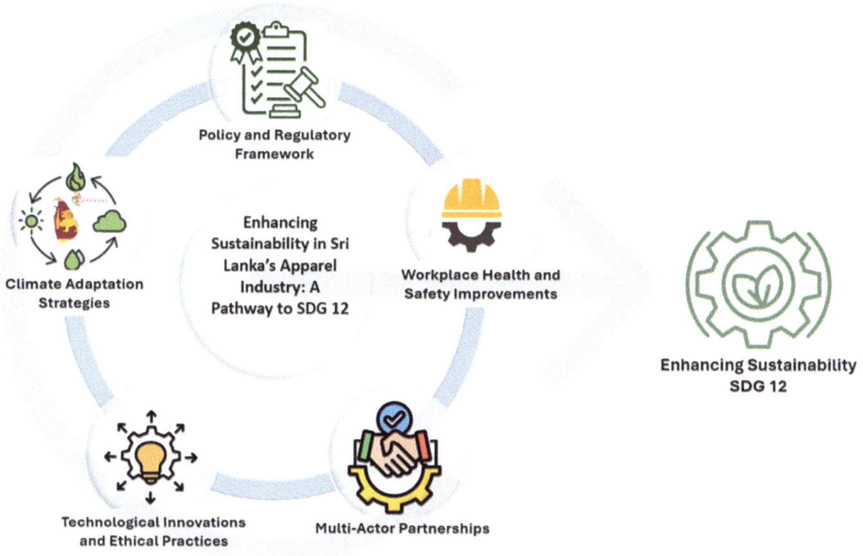

Fig. 5 Enhancing sustainability in Sri Lanka's apparel industry: a pathway to SDG 12 based on information from United Nations, 2022; ILO 2024

8 Conclusion and Recommendations

The Sri Lankan apparel industry's shift towards sustainability and the circular economy plays a critical role in achieving "SDG 12: Responsible Consumption and Production". This transformation is centred around two strategic pillars— (a) Advancing a Circular Economy and (b) Enhancing Sustainability—which are closely interconnected and instrumental in positioning Sri Lanka as a leader in ethical, climate-resilient apparel production. These strategies collectively support broader sustainable development goals by reinforcing economic resilience and environmental responsibility (Fig. 6).

The Sri Lankan apparel sector has made considerable progress in embedding circular economy principles, particularly in areas such as eco-friendly production, renewable energy integration, sustainable material sourcing, waste reduction,

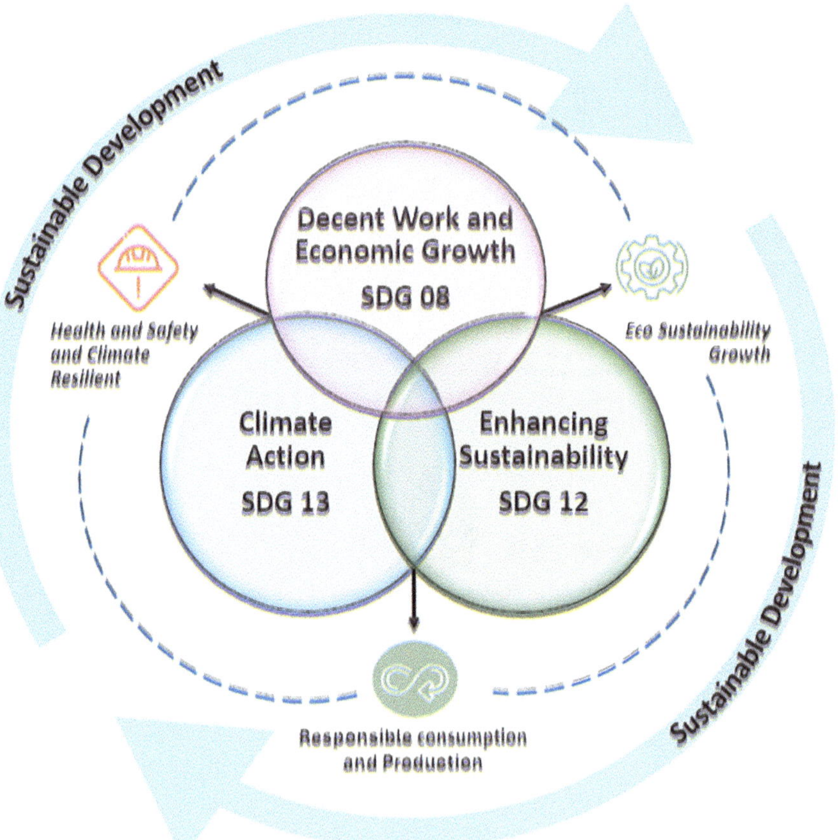

Fig. 6 Strategic integration of SDGs 8, 12, and 13 for sustainable development in Sri Lanka's apparel industry

and consumer engagement. At the same time, efforts to enhance sustainability have focused on improving worker health and safety, adapting to climate-related risks, promoting regulatory reforms, fostering innovation, and encouraging multi-stakeholder collaboration. To align with SDG 12, the apparel industry must continue to focus on:

- Minimising environmental impact through sustainable production models
- Building economic resilience by improving resource efficiency
- Enhancing worker safety and well-being in the face of climate-related challenges
- Strengthening global collaboration to achieve recognised sustainability benchmarks

However, long-term competitiveness and resilience depend on adopting scalable and integrated approaches that combine circular economy practices with broader sustainability frameworks.

Key priorities include expanding green manufacturing, which involves investing in energy-efficient infrastructure, implementing water recycling systems, and developing sustainable materials. Factory design should incorporate passive environmental features such as natural ventilation and low-energy cooling systems to reduce emissions and operational costs. Additionally, promoting closed-loop production and textile recycling initiatives will help limit resource depletion and reduce waste.

Strengthening climate adaptation is equally essential. The industry should conduct climate risk assessments to identify vulnerabilities and develop robust disaster preparedness and risk management strategies. Where necessary, factories may need to retrofit existing infrastructure or consider relocating to less vulnerable locations to ensure continuity of operations.

Ensuring worker well-being is a core component of a sustainable industry. Employers should provide mandatory safety training and awareness campaigns focused on climate-related hazards such as heat stress. Regular workplace audits and stricter enforcement of health and safety standards will contribute to a safer, more productive working environment. Furthermore, climate-responsive working conditions, including adjusted shifts to avoid high-heat periods, can help protect workers and sustain output.

Policy reform plays a vital role in embedding sustainability into industry operations. The development of binding agreements between manufacturers, regulatory bodies, and international stakeholders will be key to ensuring policy compliance and implementation. Enhancing social security frameworks can also offer protection for workers facing climate-related job disruptions. Strengthening environmental and labour regulations will support the industry's ethical standing in the global market.

Technological innovation is a powerful driver of sustainability. The sector should promote partnerships between apparel manufacturers and technology providers to integrate automation, data-driven monitoring systems, and energy-efficient technologies. Innovations in smart energy management and sustainable textile development will significantly improve production efficiency while reducing environmental impact.

Consumer awareness is another critical factor. Brands and policymakers should prioritise consumer education around the impacts of fast fashion and promote more ethical purchasing behaviour. Transparency through eco-labelling and product traceability can enhance consumer trust. Additionally, expanding second-hand clothing markets and upcycling practices can prolong garment lifespans and reduce textile waste.

Finally, the industry's future growth depends on the strength of its global partnerships. Collaborating with international organisations, sustainability platforms, and trade unions can help scale up sustainability initiatives across the sector. Attracting foreign investment and sustainable financing will be crucial for implementing large-scale infrastructure upgrades and maintaining Sri Lanka's competitive edge in an ethical fashion.

By implementing a comprehensive sustainability strategy that integrates circular economy principles, technological advancements, and responsible production practices, the Sri Lankan apparel industry can successfully address climate-related challenges while positioning itself as a global leader in sustainable fashion. This transformation is not only an economic opportunity but also a chance to redefine the nation's industrial identity through ethics, resilience, and innovation.

References

1. National Oceanic and Atmospheric Administration (NOAA), '2023 was the world's warmest year on record, by far', NOAA (2023), https://www.noaa.gov/news/2023-was-worlds-warmest-year-on-record-by-far#:~=Earth's%20average%20land%20and%20ocean,0.15%20of%20a%20degree%20C. Accessed 2 Oct 2024
2. NASA, 'NASA analysis confirms 2023 as warmest year on record', NASA (2024), https://www.nasa.gov/news-release/nasa-analysis-confirms-2023-as-warmest-year-on-record/. Accessed 2 Oct 2024
3. European Parliament, Combating climate change, Fact sheets on the European union (2024), https://www.europarl.europa.eu/factsheets/en. Accessed 2 Oct 2024
4. Intergovernmental Panel on Climate Change (IPCC), Climate change 2021: the physical science basis. contribution of working group i to the sixth assessment report of the intergovernmental panel on climate change, ed. by V. Masson-Delmotte, et al. Cambridge University Press (2021), https://www.ipcc.ch/report/ar6/wg1/downloads/report/IPCC_AR6_WGI_Full_Report.pdf
5. World Meteorological Organization (WMO), 'Climate change indicators reached record levels in 2023: WMO' (2023) https://wmo.int/news/media-centre/climate-change-indicators-reached-record-levels-2023-wmo. Accessed 2 Oct 2024
6. Sri Lanka Apparel Exporter's Association, News (2028), https://srilanka-apparel.com/news. Accessed 10 Aug
7. Intergovernmental Panel on Climate Change (IPCC), Climate change 2023: synthesis report. Contribution of working groups I, II and III to the sixth assessment report of the intergovernmental panel on climate change, ed. by Core Writing Team, H. Lee, J. Romero, IPCC, Geneva, Switzerland, pp. 35–115 (2023)
8. J. MacCarthy, A. Tyukavina, M. Weisse, N. Harris, 'Canada's record-breaking 2023 wildfires released nearly 4 times more carbon than global aviation', World Resources Institute (2024), https://www.wri.org/insights/canada-wildfire-emissions#:~=This%20amount%20of%20tree%20cover,forest%20loss%20combined%20in%202023. Accessed 2 Oct 2024

9. World Health Organization, Climate change and health (2023) https://www.who.int/news-room/fact-sheets/detail/climate-change-and-health. Accessed 21 June 2024
10. International Labour Organization, Ensuring safety and health at work in a changing climate: Global report. Geneva: International Labour Organization (2024), https://www.org/publications/ensuring-safety-and-health-work-changing-climate. Accessed 8 Sept 2024
11. World Trade Organization, World trade report 2023: Re-globalization for a secure, inclusive and sustainable future. WTO, Geneva (2023) https://www.wto.org/english/res_e/statis_e/wts2023_e/wts23_toc_e.htm. Accessed 2 Oct 2024
12. United Nations Industrial Development Organization (UNIDO), Industrial development report 2024: turning challenges into sustainable solutions—the new era of industrial policy, Vienna (2024), https://www.unido.org/. Accessed 2 Oct 2024
13. World Resources Institute, Sustainable and ethical apparel (2022), https://www.wri.org/initiatives/sustainable-and-ethical-apparel. Accessed 17 Jan 2025
14. McKinsey & Company, 'Sustainable style: how fashion can afford and accelerate decarbonization (2023), https://www.mckinsey.com/industries/retail/our-insights/sustainable-style-how-fashion-can-afford-and-accelerate-decarbonization/. Accessed 2 Oct 2024
15. International Labour Organization, Increase in heat stress predicted to bring productivity loss equivalent to 80 million jobs (2019), https://www.ilo.org/resource/news/increase-heat-stress-predicted-bring-productivity-loss-equivalent-80. Accessed 16 Jan 2025
16. Asian Development Bank, ADB annual report 2020. Asian Development Bank (2020), https://www.adb.org/sites/default/files/institutional-document/691766/adb-annual-report-2020.pdf. Accessed 30 Aug
17. World Bank, 'Building Sri Lanka's resilience to climate change', World Bank, 21 September (2018) https://www.worldbank.org/en/news/feature/2018/09/21/building-sri-lankas-resilience-to-climate-change. Accessed 10 Aug
18. D. Eckstein, V. Künzel, L. Schäfer,' Global Climate Risk Index 2021' (2021), https://germanwatch.org/sites/default/files/Global%20Climate%20Risk%20Index%202021_1.pdf. Accessed 12 June 2024
19. Notre Dame Global Adaptation Initiative, ND-GAIN country index. University of Notre Dame (2023), https://gain.nd.edu/our-work/country-index/. Accessed 30 Sept 2024
20. F. Pega, N.C. Momen, K.N. Streicher, M. Leon-Roux, S. Neupane, M.K. Schubauer-Berigan, et al., Global, regional and national burdens of non-melanoma skin cancer attributable to occupational exposure to solar ultraviolet radiation for 183 countries, 2000–2019: a systematic analysis from the WHO/ILO joint estimates of the work-related burden of disease and injury. Environ. Int. (2023). https://doi.org/10.1016/j.envint.2023.108226. Accessed 13 Aug 2024
21. World Bank, Climate knowledge portal—Sri Lanka. World Bank, Climate risk country profile—Sri Lanka (2021), https://climateknowledgeportal.worldbank.org/sites/default/files/2021-05/15507-WB_Sri%20Lanka%20Country%20Profile-WEB.pdf. Accessed 18 July 2024
22. R.V. Cruz, H. Harasawa, M. Lal, S. Wu, Y. Anokhin, B. Punsalmaa, Y. Honda, M. Jafari, C. Li, N. Huu Ninh, Climate Change 2007: impacts, adaptation and vulnerability, ed. by M.L. Parry, et al., *Contribution of Working Group II to the Fourth Assessment Report of the Intergovernmental Panel on Climate Change* (Cambridge University Press, Cambridge, 2007), pp. 469–506
23. U. Ratnayake, S. Herath, Changing rainfall and its impacts on landslides in Sri Lanka. J. Mt. Sci. **2**, 218–224 (2005). https://doi.org/10.1007/BF02973195
24. D. Palihawadana, L.C. Leonidou, P. Christodoulides, L.P. Kyrgidou, Internal drivers and performance consequences of small firm green business strategy: the moderating role of external forces. J. Bus. Ethics **140**(3), 585–606 (2017). https://doi.org/10.1007/s10551-015-2670-9
25. Doloswala,News First, Outdoor activities restricted in Sri Lankan schools due to extreme heat (2024) https://english.newsfirst.lk/2024/02/28/outdoor-activities-restricted-in-sri-lankan-schools-due-to-extreme-heat. Accessed 13 Jan 2025
26. Ministry of Industry, Sector overview: textile and apparel sector (2023), https://www.industry.gov.lk/web/wp-content/uploads/2023/03/Sector-overview-Textile-and-Apparel-Sector.pdf. Accessed 20 Dec 2023

27. Sri Lanka Apparel Exporters Association, News (2023), https://www.srilanka-apparel.com/news. Accessed 16 Jan 2025
28. International Finance Corporation, Strengthening sustainability in the textile industry. Washington, D.C.: International finance corporation (2023), https://www.ifc.org/content/dam/ifc/doc/2023/strengthening-sustainability-in-the-textile-industry-ifc-2023.pdf
29. World Bank, Climate risk country profile—Sri Lanka (2021), https://climateknowledgeportal.worldbank.org/sites/default/files/2021-05/15507-WB_Sri%20Lanka%20Country%20Profile-WEB.pdf. Accessed 18 July 2024
30. GlobeScan and Sustainable Apparel Coalition, Empowering consumers through transparency: report on global consumer research (2019) https://www.apparelcoalition.org
31. United Nations Development Programme (UNDP), *Making our future: new directions for human development in asia and the pacific* (United Nations Development Programme, New York, 2024)
32. Climate Change Secretariat of Sri Lanka, national adaptation plan for climate change impacts in Sri Lanka 2016–2025. Ministry of Mahaweli development and environment (2016), ISBN: 978-955-0033-98-0. www.climatechange.lk
33. L. Siers, The critical link between climate change and occupational health and safety: a new report from the ILO. NYU stern center for business & human rights (2024), https://bhr.stern.nyu.edu/quick-take/the-critical-link-between-climate-change-and-occupational-health-and-safety-a-new-report-from-the-ilo/. Accessed 13 Jan 2025
34. P.A. Schulte, A. Bhattacharya, C.R. Butler, H.K. Chun, B. Jacklitsch, T. Jacobs, M. Kiefer, J. Lincoln, S. Pendergrass, J. Shire, J. Watson, G. Wagner, Advancing the framework for considering the effects of climate change on worker safety and health. J. Occup. Environ. Hyg. **13**(11), 847–865 (2016). https://doi.org/10.1080/15459624.2016.1179388
35. N. Eriyagama, V. Smakhtin, L. Chandrapala, K. Fernando, Impacts of climate change on water resources and agriculture in Sri Lanka: a review and preliminary vulnerability mapping. Colombo: International Water Management Institute. (IWMI Research Report 135) (2010). https://doi.org/10.5337/2010.211
36. S.P. Nissanka, B.V.R. Punyawardena, K.H.M.S. Premalal, R.O. Thattil, Recent trends in annual and growing seasons rainfall of Sri Lanka. Faculty of agriculture, University of Peradeniya, Peradeniya; Natural resource management center, Department of agriculture, Peradeniya; Department of meteorology, Colombo (2011)
37. H.G.A.S. Sathischandra, B. Marambe, R. Punyawardena, Seasonal changes in temperature and rainfall and its relationship with the incidence of weeds and insect pests in rice (Oryza sativa L) cultivation in Sri Lanka. Clim. Chang. Environ. Sustain. **2**(2), 105-115 (2014). https://doi.org/10.5958/2320-642X.2014.00002.7
38. World Bank, Documents & reports (2023), https://documents.worldbank.org/en/publication/documents-reports. Accessed 16 Jan 2025
39. S. Thorpe, M. Fennell, Assessing climate-related risks to business: an introduction to the climate risk assessment tool. J. Bus. Contin. Emer. Plan. **6**(4), 335–347 (2012)
40. Institute of Development Studies, Climate change risks and supply chain responsibility: How should companies respond when extreme weather affects small-scale producers in their supply chain? (2023), https://www.ids.ac.uk/publications/climate-change-risks-and-supply-chain-responsibility-how-should-companies-respond-when-extreme-weather-affects-small-scale-producers-in-their-supply-chain/. Accessed 13 Jan 2025
41. UNICEF, Devastating floods in Pakistan 2022 (2022), https://www.unicef.org/emergencies/devastating-floods-pakistan-2022. Accessed 13 Jan 2025
42. A.M.A. Saja, S.M.L. Sahid, M. Sutharshanan, Capacity-building strategy for creating disaster—and climate-risk-sensitive development plans—a case study of multi-stakeholder engagement in Sri Lanka, ed. by R. Djalante et al., *Integrated Research on Disaster Risks* (Springer Nature Switzerland AG, 2021). Chapter 13, pp. 231–249. https://doi.org/10.1007/978-3-030-55563-4_13
43. C.S.A. Siriwardana, G.P. Jayasiri, S.S.L. Hettiarachchi, Investigation of efficiency and effectiveness of the existing disaster management frameworks in Sri Lanka. Procedia Eng. **212**, 1091–1098 (2018), https://www.sciencedirect.com/science/article/pii/S187770581830167X

44. United Nations Framework Convention on Climate Change, Amendment to the updated nationally determined contributions of Sri Lanka (2022), https://unfccc.int/sites/default/files/NDC/2022-06/Amendmend%20to%20the%20Updated%20Nationally%20Determined%20Contributions%20of%20Sri%20Lanka.pdf. Accessed 23 Jan 2025
45. J. Marsh, 4 climate adaptation strategies from around the world. Earth.Org. (2022) https://earth.org/climate-adaptation/. Accessed 16 Jan 2025
46. E. Cox, A. Varma, E. Cappell, How climate adaptation can protect and grow your business. PwC (2023) https://www.pwc.com/gx/en/issues/esg/how-climate-adaptation-can-both-protect-and-grow-your-business.html. Accessed 16 Jan 2025
47. J. Constible, Can employers keep workers safe from climate change? NRDC (2021) https://www.nrdc.org/bio/juanita-constible/can-employers-keep-workers-safe-climate-change. Accessed 16 Jan 2025
48. European Agency for Safety and Health at Work, Climate change: impact on occupational safety and health (OSH). OSHwiki (2023), https://oshwiki.osha.europa.eu/en/themes/climate-change-impact-occupational-safety-and-health-osh. Accessed 16 Jan 2025
49. Sri Lanka Export Development Board, Environmentally sustainable apparel manufacturing (2022), https://www.srilankabusiness.com/blog/environmentally-sustainable-apparel-manufacturing.html. Accessed 21 Jan 2025
50. M. Bildirici, I. Türkkahraman, Ö.Ö. Ersin, Unravelling the environmental impacts of the fashion industry: a Fourier-based analysis of pollution dynamics and causality across five countries. Sustainability **17**(1), 69 (2025). https://doi.org/10.3390/su17010069
51. Sri Lanka Garments, Accelerating industries' climate response which leads to energy saving (2023), https://www.srilankagarments.com/post/accelerating-industries-climate-response-which-leads-to-energy-saving. Accessed 21 Jan 2025
52. T.S. De Fonseka, Determination of environmental sustainability practices in the apparel sector of Sri Lanka. Eur. J. Sustain. Dev. Res. **7**(4), em0237 (2023). https://doi.org/10.29333/ejosdr/13816
53. United Nations Environment Programme, UNEP establishes Copenhagen climate centre in collaboration with UNOPS (2022) https://www.unep.org/news-and-stories/press-release/unep-establishes-copenhagen-climate-centre-collaboration-unops-0. Accessed 16 Jan 2025
54. SLYCAN Trust, Managing climate risks in Sri Lanka's fashion and apparel sector (2024) https://www.slycantrust.org/knowledge-resources/managing-climate-risks-in-sri-lankas-fashion-and-apparel-sector. Accessed 21 Jan 2025
55. SLYCAN Trust, Climate-friendly fashion in Sri Lanka: perspectives from industry insiders (2023), https://www.slycantrust.org/post/climate-friendly-fashion-in-sri-lanka-perspectives-from-industry-insiders. Accessed 21 Jan 2025
56. Daily FT, Achieving a more competitive Sri Lankan apparel sector through green initiatives (2023), https://www.ft.lk/sustainability__environment/Achieving-a-more-competitive-Sri-Lankan-apparel-sector-through-green-initiatives/10519-720153. Accessed 21 Jan 2025
57. Lanka News Web, Sri Lanka's apparel industry: leading innovation and sustainability in global fashion manufacturing (2024), https://lankanewsweb.net/archives/65185/sri-lankas-apparel-industry-leading-innovation-and-sustainability-in-global-fashion-manufacturing/. Accessed 21 Jan 2025
58. Sri Lanka Export Development Board, Environmentally sustainable apparel manufacturing (2023), https://www.srilankabusiness.com/blog/environmentally-sustainable-apparel-manufacturing.html. Accessed 23 Jan 2025
59. N.R.P. Maduranga, M.D.R. Harshani, The impact of occupational health and safety practices on employees' performance of operational level employees in the apparel industry in Sri Lanka. Department of human resource management, Faculty of commerce and management studies, University of Kelaniya, Sri Lanka (2024). https://doi.org/10.4038/kjhrm.v19i1.134
60. Human Level Communications, The impact of climate change on the apparel sector (2023), https://www.wearehumanlevel.com/content-hub/the-impact-of-climate-change-on-the-apparel-sector. Accessed 23 Jan 2025

61. S. Wijayapala, K.G.R. Samarawickrama, Sustainable development approaches for Sri Lankan textile and apparel industry, ed. by G.M. Ranathunga, *Textiles of Sri Lanka: SDGs and Textiles* (Singapore: Springer, 2024).https://doi.org/10.1007/978-981-97-0569-6_1
62. R. Deyshappriya, A. De Zylva, The future of Sri Lanka's apparel export industry: brexit, the U.S. Election, and other key developments. Lakshman Kadirgamar Institute of international relations and strategic studies (2016), https://www.lki.lk
63. Financial Times Lanka, How five companies are setting the blueprint for sustainable growth in Sri Lanka's north and east (2024), https://www.ft.lk/business/How-five-companies-are-setting-the-blueprint-for-sustainable-growth-in-Sri-Lanka-s-north-and-east/34-751970. Accessed 23 Jan 2025
64. N. Swaris, R.U. Halwatura, D. Amaratunga, Policy coherence for resilience in Sri Lanka: coherence of climate change adaptation (CCA), disaster risk reduction (DRR), and sustainable development goals (SDGs). Int. J. Disaster Resil. Built Environ. **15**(3), 450–473 (2024). https://doi.org/10.1108/IJDRBE-02-2023-0035
65. Sri Lanka Export Development Board, Sri Lankan apparel industry (2016), https://www.srilankabusiness.com/blog/sri-lankan-apparel-industry.html. Accessed 23 Jan 2025
66. Joint Apparel Association Forum Sri Lanka, Innovation (2024), https://www.srilankaapparel.com/innovation/. Accessed 23 Jan 2025
67. A. Withanaarachchi, A.M.H. Silva, Key technologies and critical success factors of industry 4.0 for the Sri Lankan apparel manufacturing sector: a systematic literature review, in: *2023 3rd International Conference on Advanced Research in Computing (ICARC)* (Belihuloya, Sri Lanka, 2023). IEEE, pp. 286–291. https://doi.org/10.1109/ICARC57651.2023.10145696

Chapter 12
SDG 12—Towards Responsible Consumption and Production in the Context of Fashion and Textiles

Saniyat Islam and Jenny Underwood

Abstract This chapter examines the Sustainable Development Goal 12 (SDG 12) in the context of fashion and textiles industry. It critiques how current models of consumption and production are characterised by a dominant linear 'take, make, dispose' model, which is unsustainable and fails to account for planetary boundaries, ecological and social harms. The chapter then argues that for the fashion system to shift towards more sustainable and responsible practices, it is essential to rethink its foundational principles and adopt strategies align to the circular economy that can meaningfully advance SDG12. Drawing on expanded concepts of sustainability, the chapter outlines a multi-stakeholder integrated approach that interlinks economic, ecological, and social well-being to slow down the material flows of fashion. To illustrate how collective action can drive systemic change towards more responsible consumption and production patterns, the case study of Australia's clothing product stewardship scheme, Seamless, is discussed in context to global regulatory changes. As an example of an Extended Producer Responsibility (EPR) scheme, Seamless highlights the potential and challenges of voluntary cooperative industry action and regulatory measures. Holding manufacturers liable for the lifecycle of their products could accelerate the transition to a circular fashion system, thereby reducing waste and fostering sustainability. The chapter highlights that while EPR is important, shifting consumer behaviour towards purchasing less clothing and wearing clothing for longer, will take a fundamental mindset change, will be critical. Ultimately, aligning the fashion and textile industry with SDG 12 is both an urgent necessity and a transformative opportunity to construct a more regenerative and equitable future.

Keywords Sustainable Development Goal 12 · Circular economy · Extended Producer Responsibility (EPR) · Regenerative practices · Production and consumption · Fashion and textiles

---

S. Islam (✉) · J. Underwood
School of Fashion and Textiles, RMIT University, Melbourne, Australia
e-mail: saniyat.islam@rmit.edu.au

A. S. Muhammad Sayem and P. Goswami (eds.), *SDG 12 and Global Fashion Textiles Production*, SDGs and Textiles, https://doi.org/10.1007/978-981-96-9051-0_12

1 Introduction

Sustainable Development Goal 12 (SDG 12) orchestrates a crucial role in addressing the climate and societal challenges of the fashion industry by promoting responsible consumption and production patterns. It highlights the importance of minimising waste, optimising resource management, and implementing sustainable practices across the entire supply chain. By encouraging the use of eco-friendly materials, improving waste management, and ensuring ethical labour practices, SDG 12 aims to mitigate the industry's significant environmental trail and social inequities. Achieving SDG 12 requires a systemic shift towards sustainability, fostering collaboration among stakeholders, and driving innovation to create a more sustainable and equitable fashion industry [1].

The fashion and textile (F&T) industry recognises the critical need of achieving the SDG12 goal and in recent years have been trying to make strides towards sustainability. This has been partly driven by increasing consumer demand and regulatory pressures. However, significant challenges remain, particularly in addressing the environmental and social impacts of the industry. The F&Ts industry is a significant global economic player, yet it is also one of the most resource-intensive and polluting sectors. The industry's environmental footprint is substantial, with high levels of water and energy consumption, chemical pollution, and waste generation. Additionally, the fashion supply chain often involves complex and opaque networks that can contribute to labour rights violations and unethical working conditions. Globally, the fashion industry is responsible for approximately 10% of global carbon emissions and nearly 20% of wastewater [2]. The industry's reliance on non-renewable resources, such as synthetic fibres derived from fossil fuels, exacerbates its environmental footprint. Efforts to address these impacts include using sustainable materials like organic cotton and recycled polyester, as well as implementing circular economy principles to reduce waste and encourage the reuse and recycling of garments [3].

Despite these efforts, the fashion industry still faces significant hurdles in achieving true sustainability. The complexity of global supply chains, the fast-paced nature of fashion cycles, and the economic pressures on manufacturers and brands all influence the difficulty of implementing sustainable practices across the board.

1.1 Background on Sustainable Development Goal 12 and its Context to F&T

SDG 12 is a fundamental component of the United Nations' Agenda 2030, focused on promoting sustainable consumption and production patterns. This goal underscores the urgent need to manage the earth's resources more efficiently and equitably. SDG 12 includes various targets, such as the efficient and sustainable management of natural resources, waste reduction through prevention, recycling, and reuse, and the promotion of sustainable practices across different industries and sectors [1]. The goal also emphasises the importance of sustainable infrastructure, access to basic services, green and decent jobs, and overall improvements in quality of life. At its core, SDG 12 seeks to 'do more and better with less' by decoupling economic growth from environmental degradation and fostering systems that enhance well-being for all. Within the F&T industry, these objectives translate into a pressing need for significant changes in how materials are sourced, how products are manufactured, and how waste is administered throughout the lifecycle of garments and textiles.

SDG 12 is supported by 11 targets that address various aspects of sustainable consumption and production. These targets include the sustainable management and efficient use of natural resources, waste reduction, encouraging companies to adopt sustainable practices, and promoting sustainable public procurement. While some targets are set to be achieved by 2030, others have different timelines or no specific deadlines. For the fashion industry, the most relevant targets under SDG 12 are achieving the sustainable management and efficient use of natural resources (Target 12.2), substantially reducing waste generation through prevention, reduction, recycling, and reuse (Target 12.5), and encouraging companies to adopt sustainable practices and integrate sustainability information into their reporting cycles (Target 12.6).

The concept of sustainability in fashion and textiles (F&T) industry often focuses on balancing economic growth, environmental stewardship, and social responsibility. However, this raises the question of what exactly is being 'sustained' given current industry practices and how responsible patterns of consumption and production can be achieved. While the fashion industry is making some progress in adopting more sustainable materials, reducing water usage, and improving labour conditions, these efforts are often isolated and incremental, failing to address the systemic issues at the heart of the industry [4]. If the F&T industry focus on sustaining the status quo, it risks ignoring and not taking responsibility for the critical issues of overproduction and overconsumption that continue to contribute to environmental degradation, wasteful consumption patterns, and social inequities [5].

For the F&T industry to be truly responsible, a fundamental rethinking of what is sustainable is required. Rather than maintaining the existing system, the focus needs to be on transitioning to a system that prioritises circular 'slow' material flows. This shift is crucial for several reasons: Firstly, the sector's environmental impact necessitates more sustainable approaches to minimise climate impact, protect natural resources and ecosystems. Secondly, ethical imperatives require the industry to address social issues such as fair pay, secure working conditions, and the elimination of exploitative labour practices. Lastly, consumer awareness and behaviour are vital in slowing the cycles of fashion and advocating for brands to adopt better practices.

In this context, the concept of responsibility within fashion production and consumption requires concerted efforts at global, national, and local levels. Responsibility needs to encompass holistic approaches that consider the full impact of fashion on people, the planet, and future generations. This means recognising the interconnectedness of production and consumption patterns and making deliberate choices that prioritise sustainability, ethics, and well-being. Responsible fashion production includes sourcing materials in ways that do not deplete natural resources or harm ecosystems, ensuring fair labour practices, and designing products with longevity and end-of-life considerations in mind. Moreover, it involves transparency and accountability, allowing consumers to make informed choices. For end consumers, responsibility entails mindful purchasing, valuing quality over quantity, and considering the environmental and social footprint of one's wardrobe. Ultimately, responsibility in fashion demands a shift towards extended product lifetimes and design and manufacture textiles and apparels that are appropriate for recycling. However, for fashion industry, the relationship between consumption and production is not binary. There are many players such as brands and retails who act as both producer and consumer but can avoid scrutiny based on their position within the supply chain.

Brands and retailers occupy unique positions within the supply chain, often acting as intermediaries between raw material suppliers and end consumers. As producers, they design, manufacture, and market clothing. As industrial consumers, they purchase raw materials, components, and finished goods from suppliers and manufacturers. This dual role allows them to influence both upstream and downstream activities within the supply chain. For example, a fashion brand might source fabrics from textile manufacturers (acting as an industrial consumer) and then produce garments (acting as a producer). Retailers, on the other hand, purchase finished products from brands or wholesalers (acting as industrial consumers) and sell them to end customers (acting as producers in terms of providing the final product to the market). More broadly, with fashion operated with a global system with different part of the industry operating in different locations. This results in countries and regions being perceived as producers, namely manufacturing and other areas being industrial consumers. Within these locations, there are a raft of different legislations and regulatory frameworks. To address these challenges, transparency and liability across the entire supply chain is crucial. Brands and retailers must take responsibility for the sustainability of their entire supply chain, from material sourcing to disposal of the goods.

In Europe, the European Union's Circular Economy Action Plan includes approaches specifically targeting the textile sector. This plan aims to promote sustainable production and consumption patterns, reduce textile waste, and encourage the use of recycled materials [6]. Additionally, countries like France have implemented extended producer responsibility (EPR) schemes, requiring fashion brands to bear responsibility for the end-of-life strategy of their products [7]. Fashion giants such as H&M and Inditex who have global spread of their supply chain operations, adopted sustainability strategies that include using organic and recycled materials, improving transparency within the supply chains, and promoting circular fashion models.

In the United States, consumer awareness and demand for sustainable fashion have grown significantly. Brands such as Patagonia and Eileen Fisher are leading the way by incorporating eco-friendly materials, ensuring fair labour practices, and promoting product longevity through repair and recycling programs [8]. For example, Patagonia is known for its commitment to circular economy principles, providing repair services for its products and encouraging customers to recycle their garments through its Worn

Wear program [34]. However, the lack of comprehensive federal regulations means that progress is often driven by individual companies and state-level initiatives [8].

In Asia, particularly in countries such as China and Bangladesh, which are major hubs for textile manufacturing, the focus has been on improving labour conditions and reducing environmental pollution. Initiatives such as the Partnership for Cleaner Textile (PaCT) in Bangladesh aim to reduce water and energy consumption in the F&T industry. This initiative has resulted in substantial improvements in resource efficiency and environmental performance (World Economic Forum 2023). However, challenges remain in enforcing labour standards and ensuring compliance with environmental regulations.

These examples illustrate the diverse approaches to implementing sustainable practices in the fashion and textile industry across different regions. By adopting comprehensive strategies and fostering collaboration among stakeholders, the industry can make significant strides towards sustainability.

1.2 Overview of the Current Linear Model of Production and Consumption in the Fashion Industry

The F&T industry has been under increased scrutiny owing to its significant environmental and social impacts. Recent research highlights the industry's contribution to global pollution, resource depletion, and social inequities. A systematic literature review by [9] identifies three main research areas: consumer behaviour towards sustainable clothing, circular economy initiatives, and sustainability challenges throughout the supply chain. The study highlights the necessity for more comprehensive approaches to tackle these issues, including corporate social responsibility (CSR) and circular economy practices. The following section highlights some key concepts that are significant in understanding responsible consumption and production.

The F&T industry plays a pivotal role in the discourse surrounding SDG 12 due to its significant environmental and social impacts. As one of the largest and most resource-intensive industries, fashion is deeply entwined with issues such as water usage, chemical pollution, carbon emissions, and waste generation. The industry's traditional linear model of 'take, make, dispose' (Fig. 1) has led to unsustainable practices that strain natural resources and contribute to environmental degradation. Furthermore, the globalised fragmented nature of fashion production and its long and complex supply chains have resulted in labour exploitation and social inequities, particularly in developing countries where much of the manufacturing takes place.

In this context, the F&T industry is both a major contributor to unsustainable patterns and a critical player for transformative change. The industry's influence on consumption habits, combined with its vast supply chains and production networks, positions it as a key sector where the principles of SDG 12 can be either undermined or upheld. Thus, the fashion industry holds the potential to drive significant

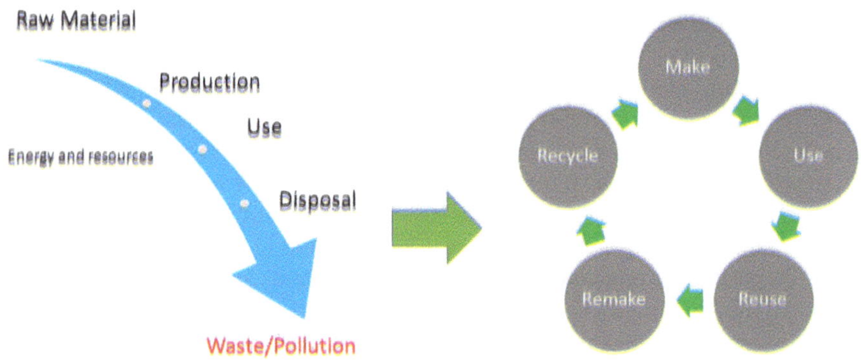

Fig. 1 Linear to circular model in consideration to waste hierarchy

progress towards more sustainable consumption and production practices, but only if it undergoes a fundamental shift in its operations and values.

This chapter highlights that the current practices within the fashion and textile industry are neither sustainable nor responsible, despite increased awareness and efforts to address these issues. Instead, the industry needs to embrace a systemic shift toward sustainability, incorporating concepts such as taming and rewilding, circular economy principles, and regenerative practices. By doing so, the fashion and textile industry can transition from being a part of the problem to becoming a leader in the global effort to achieve SDG 12 and promote a more regenerative future.

1.3 Challenges of the Linear Model of Production and Consumption

The F&T industry operates within a dominant system deeply embedded in the structures of globalised capitalism [10]. Although it plays a major role in the global economy, it is also one of the most resource-intensive and polluting industries. This system is characterised by its complex, fragmented, and lengthy supply chains, as well as the often-opaque network of production and consumption processes. The industry's environmental footprint is substantial, with high levels of water and energy consumption, chemical pollution, and waste generation [4]. Additionally, the fashion supply chain often involves complex and opaque networks that can contribute to labour rights violations and unethical working conditions.

Within the fashion system, the relationship between production and consumption is more than just a simple sequence of producers creating and consumers purchasing; it is a complex, interdependent network where the roles of producers and consumers are intertwined. Consumption in fashion involves not just the purchase of garments, but also their wear, styling, and care, which can be understood as a form of co-production through activities such as altering, repairing, and maintaining garments before disposal. Likewise, manufacturers extract and consume raw fibres to add value by transforming these materials into yarns, fabrics, and eventually garments. These garments pass through many hands in the supply chain and value network long before reaching the consumer. Additionally, once worn and/or disposed of, garments may flow through the system for a second life as hand-me-downs or via second-hand marketplaces. However, most clothing will ultimately be sent to landfill [11].

The prevailing linear fashion system developed during the modern era, fuelled by continuous change and closely tied to the globalised capitalist economy. Payne [21] identifies its consumption and production patterns as consisting of four networks: production, promotion, wearing, and destruction. These interconnected networks highlight how each stage—from the creation and marketing of garments to their wearing and eventual disposal—plays a role in sustaining a system that is fundamentally unsustainable and irresponsible.

The impact of this dominant fashion system is both significant and troubling. As a linear model, it inherently promotes wastefulness and environmental degradation. The focus on continuous production and consumption, driven by trends and the pursuit of profit, leads to the overuse of natural resources, excessive waste, and the perpetuation of exploitative labour practices. Moreover, the speed of fast fashion, characterised by rapid turnover of styles and disposable clothing, exacerbates these issues. The result is a system that is fundamentally misaligned with the principles of sustainability and SDG 12, which focuses on responsible production and consumption.

In context to SDG 12, there is a significant risk that the F&T industry could co-opt the language of sustainability without making substantive changes. This phenomenon, known as greenwashing, occurs when companies present an environmentally responsible image while continuing to engage in unsustainable practices [12]. Greenwashing not only undermines the credibility of sustainability efforts but also diverts attention from the need for systemic change that addresses the limits of growth and the necessity of operating within planetary boundaries. For the industry, these risks being more about sustaining the status quo rather than catalysing genuine change.

If the F&T industry continues to operate within the dominant economic system that prioritises growth over environmental and social considerations, there is a real danger that SDG 12 could be reduced to a marketing tool rather than a driver of genuine transformation. Greenwashing not only undermines the credibility of sustainability efforts but also diverts attention from the need for systemic change that addresses the limits of growth and the necessity of operating within planetary boundaries. For the industry, this approach risks maintaining the status quo rather than driving meaningful change.

2 Concepts to Support Responsible Practices

For the F&T industry there are many interrelated concepts and theoretical framings that are useful to consider when examining responsible practices. These include concepts of circular economy, waste hierarchy, and fashion specific notions of taming and rewilding [21] (Payne 2020) and Earth Logic [13, 19].

2.1 Circular Economy

The Ellen MacArthur Foundation (EMF) describes the circular economy as a system designed to keep materials in use for as long as possible, minimise waste, and regenerate natural systems. This approach contrasts with the traditional linear model of "take, make, dispose" and aims to foster a more sustainable and resilient economy. The circular economy is founded on three principles: eliminating waste and pollution, circulating products and materials, and regenerating nature [14].

2.2 Waste Hierarchy

The waste hierarchy [15] is a framework that prioritises waste management practices to minimise environmental impact. It is structured as a pyramid, with the most preferred options at first and the least preferred at the lowermost as shown in Fig. 2. The hierarchy includes the following levels:

Fig. 2 Waste hierarchy framework

- **Refuse/Prevention**: The most effective way to manage waste is to prevent it from being created in the first place. In the fashion industry, this can be achieved through sustainable design practices, such as using durable materials and creating timeless styles that reduce the need for frequent replacement [16] (Cattermole Consulting, 2020).
- **Reduction**: This involves minimizing the amount of waste generated. Fashion brands can achieve this by optimising production processes to use fewer resources and by encouraging consumers to buy less but better-quality clothing [17].
- **Reuse**: Reusing products extends their life cycle and reduces the need for new resources. This can include second-hand markets, clothing swaps, and rental services that allow consumers to share and reuse garments [18].
- **Recycling**: Recycling involves processing used materials to create new products. In the fashion industry, this can mean turning old textiles into new fabrics or repurposing garments into new designs. However, recycling is less preferred than prevention, reduction, and reuse due to the energy and resources required [20].
- **Recovery**: This includes recovering energy from waste that cannot be reused or recycled. For example, incinerating textile waste to generate energy. While better than disposal, recovery still has environmental impacts and is less preferred [7].
- **Disposal**: The least preferred option in the waste hierarchy, involves sending waste to landfills or incineration without energy recovery. This method should be used only as a last resort when all other options have been exhausted [15].

2.3 Taming and Rewilding

Payne [21] (2020) introduced the concepts of taming and rewilding in her book "Designing Fashion's Future". 'Taming' refers to the process of making fashion production more controlled and sustainable through practices such as reducing waste and using eco-friendly materials. It is considered as short-term solutions of working within the existing system. 'Rewilding', on the other hand, involves creating space for more diverse and inclusive fashion practices that go beyond the commercial and industrial norms. This includes promoting local production, regenerative agriculture, and community-based fashion initiatives [21]. In the context of fashion, rewilding could mean embracing traditional craft techniques, promoting biodiversity in textile production, and allowing for greater variability and flexibility in design and innovative new business models that doesn't require the extraction of materials from nature. By balancing taming with rewilding, the fashion industry could become a more resilient and regenerative system that supports both human and environmental well-being.

2.4 Earth Logic

"Earth Logic: Fashion Action Research Plan" by Kate Fletcher and Mathilda Tham [13] calls for a more radical shift in the fashion industry by prioritising the health of the planet over economic growth. The Earth Logic framework emphasises systemic change, advocating for reduced resource use, local production, and new governance models that support sustainability. The plan outlines six holistic landscapes for transformation: growing out of growth, scaling locally, creating new centres for fashion, developing new knowledge and skills, changing communication, and rethinking governance [13]. This proposition is about challenging the fashion system itself to shift away from business-as-usual practices and towards strategies that are grounded in the realities of the planet's finite resources [19]. It emphasises the need to consider the interconnectedness of all aspects of production and consumption, recognising that local and global actions have ripple effects that influence entire systems.

2.5 Aligning Alterative Concepts for Responsible Practices

These approaches support shifting how industry operated by making incremental changes in the short term to more radical ways of imagining the circular economy by advocating for a mindset shift that prioritises ecological and social health over an economic growth model that is linked extractive and exploitive practices. But how do these concepts align with the SGD12.

Aligning the three circular economy principles of eliminating waste and pollution, circulate materials and products and regenerate nature and the SDG12 targets together begins to highlight that fashion businesses (refer Sect. 3) and the fashion industry (refer Sect. 4) as a whole need to consider a wide range of strategies and tactics to achieve. Circular economy practices such as reducing, redesigning, reusing, repairing, remanufacturing, and recycling are integral to these targets. By keeping products and materials in use for as long as possible and minimising waste, the circular economy helps to decouple economic growth from environmental degradation, thereby supporting the achievement of SDG 12. While most SDG12 targets can support the fashion industry, several, namely Target 12.3 (relating to food waste), Target 12.b (tourism) and Target 12.c. rationalise subsidies (specifically to fossil-fuel subsidies) are only indirectly relevant as shown in Fig. 3.

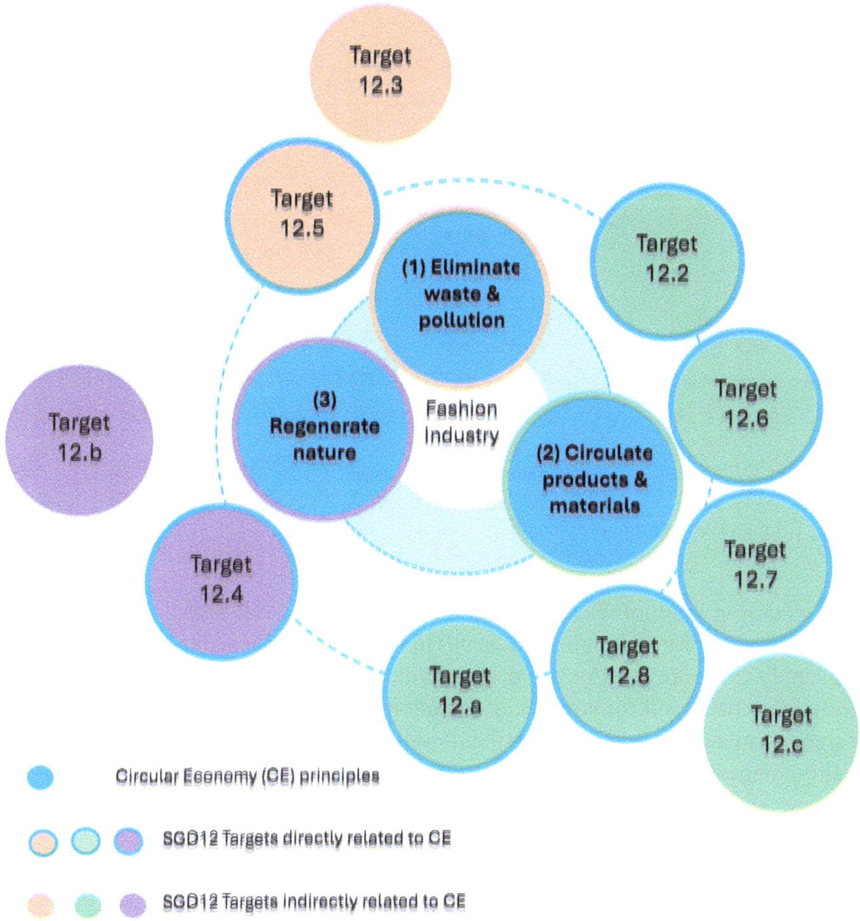

Fig. 3 Synergy diagram of SDG12 and circular economy principles

3 Strategies for Fashion businesses to Conceptualise 'Responsible Consumption and Production' in Fashion

To move the fashion system toward more sustainable and responsible practices, it is essential to rethink its foundational principles and adopt strategies that align with and advance SDG 12 goal of responsible consumption and production. This requires a multi-stakeholder approach to developing new business models, production practices, and changing consumer behaviours to slow down the flow of materials by designing garments for multiple next lives for as long as possible.

What this might mean for a fashion business is the need to rethink all aspects of the supply chain. Critically this starts with design. However, for the design phase to have impact, design teams must understand and design for the products 'end of life'

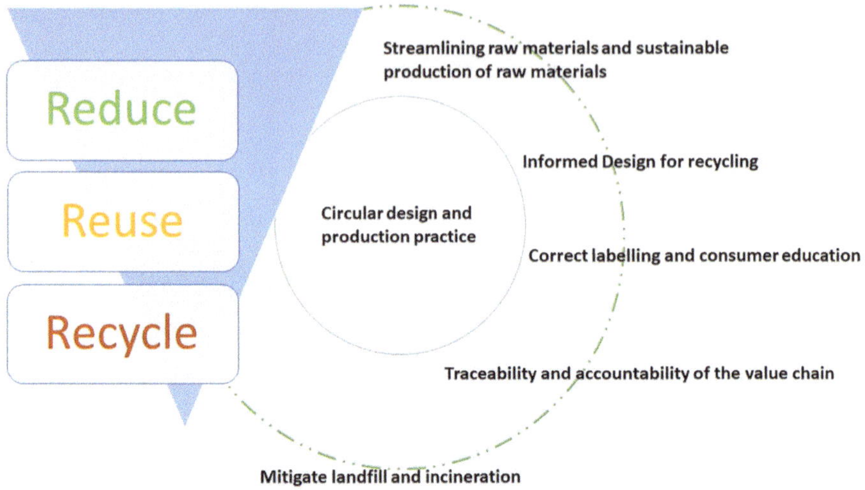

Fig. 4 Circular design principles in context to waste hierarchy

by considering the biological and technical cycles, as well as designing out waste and encouraging for resource efficiency. This requires reorganising how the business and supply chain operates, ensuring communication and feedback loops are built into the supply chain, and for strong collaborations between all stakeholders, including designers, manufacturers, retailers, and consumers, to create a system that values quality, longevity, and regeneration over quantity and disposability.

3.1 Circular Design in Fashion

Within this context, the circular economy offers a transformative approach to how fashion is produced, consumed, and recirculated before eventually being disposed of. In the context of fashion design, this means it is critical to design and create garments designed for durability, repairability, and recyclability, thus extending their life cycle and reducing the need to extract virgin resources as illustrated in Fig. 4.

3.2 Biological and Technical Cycles

A crucial aspect of the circular economy is its focus on both biological and technical cycles [14] (EMF 2024). Biological cycles involve materials that can safely return to the environment, such as natural fibres that biodegrade and nourish the soil. Technical cycles, on the other hand, involve materials that can be continuously cycled through reuse, repair, and recycling without losing their value, such as synthetic fibres and metals. Fashion can adopt a whole-systems approach by understanding these cycles, ensuring that materials are either returned to the earth as nutrients or kept in circulation through thoughtful design and innovative processes.

3.3 Resource Efficiency and Waste Prevention

Resource efficiency and waste prevention are at the core of responsible fashion practices. Strategies to increase resource efficiency include optimising the use of water, energy, and raw materials throughout the production process, as well as adopting technologies that reduce the environmental footprint of manufacturing [14]. Preventing material waste involves not only reducing waste during production but also revaluing post-consumer waste as a new resource. This can be achieved through initiatives such as textile recycling, upcycling, and the creation of closed-loop systems where materials are continually reused in the production of new garments, as well as open-loop systems where materials flow to another industry when no longer appropriate to be garments [14]. By shifting the focus from waste disposal to waste prevention and resource regeneration, the fashion industry can make significant strides toward sustainability.

4 Strategies for Fashion Industry to Support Responsible Practices

To support responsible practices, the F&T industry must adopt a holistic approach that emphasises the transition to a circular economy. This involves designing products for longevity, recyclability, and minimal environmental impact. Multi-stakeholder collaboration is crucial, bringing together brands, suppliers, consumers, and policymakers to create a system where resources are reused, and waste is minimised. By fostering innovation, transparency, and shared responsibility, the industry can move towards sustainable consumption and production patterns, aligning with global sustainability goals. An example of this can be drawn from the EPR scheme currently being implemented in Australia.

4.1 Case Study of 'Seamless' in Australia

Australia's Seamless Clothing Stewardship Scheme, launched by the Australian Fashion Council (AFC) in 2023, represents a pioneering effort to address the environmental impact of the fashion industry. This initiative aims to achieve clothing circularity by 2030, significantly reducing the 200,000 tonnes of clothing that currently end up in landfills each year [22]. The scheme is funded by a levy of 4 cents per garment, contributed by participating brands, which is used to promote durable and recyclable clothing design, expand recycling facilities, and foster new circular business models [22].

Globally, Extended Producer Responsibility (EPR) schemes have been implemented with varying degrees of success. France, for example, has a mandatory EPR scheme for clothing, which has been operational since 2007. This scheme requires producers to finance the collection, sorting, and recycling of used textiles, significantly reducing textile waste (Ellen MacArthur Foundation 2017). In contrast, countries like Sweden and the UK are still in the process of developing similar regulations [23].

The Seamless scheme has seen significant engagement from major Australian brands such as BIG W, Cotton On, and David Jones, which are crucial for its success (Australian Fashion Council 2023). The Australian government has also prioritised clothing textiles on the Federal Minister's Priority List for Product Stewardship, providing a regulatory framework that supports the scheme [23]. However, achieving the target of 60% industry participation to divert 120,000 tonnes of clothing from landfill by 2027 remains a significant challenge [23]. Expanding the infrastructure for clothing collection, sorting, and recycling is essential but requires substantial investment and coordination [23].

In addition to the Seamless scheme, various global initiatives emphasise different approaches to achieving clothing circularity. The Ellen MacArthur Foundation's "Make Fashion Circular" initiative, launched in 2017, unites industry leaders to collaborate on creating a circular economy for fashion. This initiative aims to eliminate waste and pollution, keep products and materials in use, and regenerate natural systems [24]. Similarly, the Global Fashion Agenda's "2020 Circular Fashion System Commitment" encourages brands to set targets for increasing the volume of used garments collected, resold, and recycled [25].

Technological innovations also play a crucial role in advancing clothing circularity. Advances in textile recycling technologies, particularly in chemical recycling, are crucial for achieving circularity but are still in early stages globally [23]. The development of automated sorting technologies and traceability tools can further enhance the efficiency of recycling processes [23].

The Seamless Clothing Stewardship Scheme correspond to a significant stride towards sustainable fashion system in Australia. While it faces challenges in market penetration and infrastructure development, its alignment with global trends in EPR and support from major industry players and the government provide a strong basis for success. The integration of technological innovations and collaboration with global initiatives can further enhance the scheme's effectiveness in achieving clothing circularity.

4.2 Consumer Behaviour and Mindset Change

In the quest for a sustainable future, shifting consumer behaviour towards sustainability is paramount. Consumers hold significant power in driving demand for eco-friendly products and practices. As perception of environmental issues grows, more individuals are seeking to make choices that benefit the planet. This shift is essential for mitigating climate change, conserving natural resources, and ensuring a healthier environment for future generations.

4.2.1 The Power of Consumers in Driving Sustainability

Consumers hold significant power in driving demand for eco-friendly products and practices. As awareness of environmental issues grows, more individuals are seeking to make choices that benefit the planet. This shift is essential for mitigating climate change, conserving natural resources, and ensuring a healthier environment for future generations.

4.2.2 Leveraging Social Influence for Responsible Consumption

One of the most effective strategies to encourage responsible consumption is leveraging social influence. When sustainable behaviours become the norm within communities, more people are likely to adopt them. Campaigns that highlight collective efforts and peer actions can create a powerful ripple effect, encouraging others to follow suit [26].

4.2.3 Incentives and Rewards for Sustainable Choices

Incentives and rewards play a crucial role in motivating consumers to change their habits. Offering discounts for using reusable bags or rewards for recycling can make sustainable choices more attractive and accessible.

4.2.4 Transparency and Informed Decision-Making

Providing transparency about the sustainability of products helps consumers make informed decisions. Labels and certifications indicating eco-friendly practices can guide consumers towards more sustainable purchases [2].

4.2.5 Convenience and Accessibility of Sustainable Products

Convenience and accessibility are key factors in promoting sustainable consumption. Increasing the availability of sustainable products and ensuring they are competitively priced can significantly influence consumer behaviour.

4.2.6 Education and Awareness Campaigns

Education and awareness campaigns are fundamental in informing consumers about the environmental impact of their choices and how they can contribute to sustainability. These campaigns can highlight the benefits of sustainable living and offer practical tips for reducing waste and conserving resources [27].

4.2.7 Achieving Sustainable Development Goal 12 (SDG12)

Education and awareness campaigns are particularly vital for achieving Sustainable Development Goal 12 (SDG12), which focuses on responsible consumption and production. These campaigns can raise awareness about the importance of sustainable consumption and the impact of individual choices on the environment [27].

4.2.8 Promoting Behavioural Change

Promoting behavioural change is another critical aspect of these campaigns. By providing individuals with the knowledge and tools they need, campaigns can encourage the adoption of sustainable practices such as recycling, reducing single-use plastics, and supporting sustainable brands.

4.2.9 Integrating Sustainability into Education

Integrating sustainability into education is crucial. Programs like the World's Largest Lesson and Comics Uniting Nations engage children and youth in learning about the SDGs and how they can contribute to a sustainable future [27].

4.2.10 Supporting Policy and Advocacy Efforts

Education campaigns can support policy and advocacy efforts by building public support for sustainability initiatives. This can lead to stronger regulations and incentives for sustainable practices at both governmental and corporate levels.

Shifting consumer behaviour towards sustainability is essential for achieving environmental goals. Strategies such as leveraging social influence, providing incentives, ensuring product transparency, and making sustainable options convenient can encourage responsible consumption. Education and awareness campaigns play a critical role in promoting these changes and supporting the achievement of SDG12.

4.3 Policy and Regulatory Frameworks

SDG 12 spotlights on ensuring sustainable consumption and production patterns. This goal is supported by a variety of policies and regulations at both national and international levels. For instance, the European Union has applied the Circular Economy Action Plan, which intends to mitigate waste and foster the reuse and recycling of materials [28]. Similarly, Japan's Basic Act on Establishing a Sound Material-Cycle Society provides a comprehensive framework for waste management and recycling [29].

Globally, the United Nations Environment Programme (UNEP) has played a crucial role in advancing sustainable consumption and production (SCP) through initiatives such as the 10-Year Framework of Programmes on Sustainable Consumption and Production Patterns (10YFP). This framework encourages nations to create and implement policies that support SCP, emphasising the reduction of environmental impacts from consumption and production processes [33].

4.3.1 Recommendations for Strengthening Policy Frameworks

To enhance the effectiveness of policies supporting SDG 12, several recommendations have been proposed in this chapter. These recommendations aim to create a cohesive and integrated policy framework that maximises synergies and avoids conflicting objectives.

One of the primary recommendations is to ensure that policies across different sectors are coherent and integrated. This whole-of-government approach requires aligning environmental, economic, and social policies to work towards common goals. By doing so, governments can avoid conflicting objectives and maximise the positive impact of their policies. For instance, integrating environmental considerations into economic policies can help promote sustainable growth while protecting natural resources [30].

Governments can also enhance the effectiveness of policies by providing inducements for fashion businesses and consumers to implement sustainable practices. These incentives can take several forms, such as tax offsets, financial support for green technologies, and penalties for unsustainable practices. By making sustainable choices more financially attractive, governments can encourage a shift towards more responsible consumption and production patterns. For example, subsidies for renewable energy technologies can accelerate the transition to cleaner energy sources [33].

Robust regulatory frameworks are essential for enforcing sustainable practices. This includes setting clear standards for waste management, emissions, and resource use, and ensuring compliance through monitoring and enforcement mechanisms. Strong regulations can drive significant improvements in environmental performance by holding businesses accountable for their impacts. For instance, stringent waste management regulations can reduce landfill use and promote recycling [28].

Inducing public awareness about the significance of sustainable consumption and production is crucial for achieving SDG 12. Education campaigns can inform consumers about the environmental effect of their choices and encourage more sustainable behaviours. By increasing awareness, these campaigns can foster a culture of sustainability and empower individuals to make informed decisions. For example, public service announcements and educational programs can highlight the benefits of reducing single-use plastics and encourage recycling [27].

Investing in research and innovation is another critical recommendation. Developing new technologies and practices that support sustainable consumption, and production can drive significant progress towards SDG 12. Governments can support this by providing funding and creating an enabling environment for innovation. For instance, research grants for developing biodegradable materials can lead to breakthroughs that reduce plastic waste. Encouraging innovation not only addresses current sustainability challenges but also anticipates future needs [30].

4.3.2 Role of International Cooperation and Agreements

International cooperation and agreements are crucial for achieving SDG 12. The Paris Agreement, for instance, is a landmark international treaty aimed at limiting global warming and promoting sustainable development. It encourages countries to adopt sustainable practices and reduce their carbon footprints [31]. Similarly, the Basel Convention on the Control of Transboundary Movements of Hazardous Wastes and Their Disposal is a key international agreement that regulates the movement of hazardous waste between countries and promotes environmentally sound waste management practices [32]. International organisations like the United Nations and the World Bank also play significant roles. These organisations facilitate cooperation between countries, provide technical and financial support, and help build capacity for implementing sustainable practices [33]. Achieving SDG 12 requires a comprehensive approach that includes robust policies and regulations, incentives for sustainable practices, public awareness, and international cooperation. Strengthening these frameworks can help ensure sustainable consumption and production patterns, contributing to a more sustainable future.

5 Conclusion

Achieving SDG12 on responsible consumption and production requires a multi-stakeholder approach and collective action across the level of an individual product, business, industry, and societal level. By understanding and scrutinising all the different interconnect relationships between consumption and production across the fashion supply chain offers a way for every stakeholder to be accountable and to align to more responsible practices.

While individuals can act, it also requires national attention and support. Schemes such as the Seamless Clothing Stewardship Scheme in Australia exemplifies how national initiatives can drive significant progress towards sustainability by engaging industry stakeholders and leveraging government support. However, the success of such schemes hinges on overcoming challenges related to market penetration and infrastructure development.

Shifting consumer behaviour towards sustainability is also equally crucial. Strategies such as leveraging social influence, providing incentives, ensuring product transparency, and making sustainable options convenient can effectively encourage responsible consumption. Education and awareness campaigns play a pivotal role in promoting these changes and supporting the achievement of SDG 12. This also highlights that SDG12 must also be consider in relation to all other SDGs for meaningful change to happen. Ultimately, aligning the F&T industry with SDG 12 provides the transformative opportunity to create a more regenerative and equitable future.

References

1. United Nations, Sustainable Development Goal 12: Ensure sustainable consumption and production patterns (2023), https://sdgs.un.org/goals/goal12
2. Forbes, The state of sustainability in the fashion industry and what it means for brands (2023), https://www.forbes.com/councils/forbescommunicationscouncil/2023/01/02/the-state-of-sustainability-in-the-fashion-industry-and-what-it-means-for-brands/
3. McKinsey, State of fashion archive (2023), https://www.mckinsey.com/industries/retail/our-insights/state-of-fashion-archive
4. S. Islam, Sustainable raw materials: 50 shades of sustainability, in *Sustainable Technologies for Fashion and Textiles*, ed. by R. Nayak (Woodhead Publishing, 2021), pp.343–357
5. S. Islam, Waste management strategies in fashion and textiles industry: Challenges are in governance, materials culture and design-centric, in *Waste Management in the Fashion and Textile Industries*, ed. by R. Nayak, A. Patnaik (Woodhead Publishing, 2023), pp.275–293
6. Global Fashion Agenda, Industry insights: 2022 in review (2022), https://globalfashionagenda.org/news-article/industry-insights-2022-in-review/
7. S. Ray, L. Nayak, Marketing sustainable fashion: trends and future directions. Sustainability **15**(7), 6202 (2023)
8. Global Fashion Agenda, The GFA Monitor 2023 (2023), https://globalfashionagenda.org/resource/the-gfa-monitor-2023/
9. S. Abbate, P. Centobelli, R. Cerchione, S.P. Nadeem, E. Riccio, Sustainability trends and gaps in the textile, apparel and fashion industries. Environ. Dev. Sustain. **26**, 2837–2864 (2023). https://doi.org/10.1007/s10668-022-02887-2

10. A. Briggs, Capitalism's favourite child: the production of fashion. In *Fashion Cultures Revisited: Theories, Explorations and Analysis* (pp. 186-199). Routledge (2013), https://ualresearchonline.arts.ac.uk/id/eprint/6740/
11. A. Payne, Mellick Z (2022), Tackling overproduction? The limits of multistakeholder initiatives in fashion. Int. J. Crime Justice Soc. Democr. **11**(2), 30–46 (2022)
12. A. Badhwar, S. Islam, C.S.L. Tan, T. Panwar, S. Wigley, R. Nayak, Unraveling green marketing and greenwashing: a systematic review in the context of the fashion and textiles industry. Sustainability **16**, 2738 (2024). https://doi.org/10.3390/su16072738
13. K. Fletcher, M. Tham, Earth logic: fashion action research plan (2019), https://katefletcher.com/publications/books/earth-logic-fashion-action-research-plan/
14. Ellen MacArthur Foundation, What is a circular economy? (2023), https://www.ellenmacarthurfoundation.org/topics/circular-economy-introduction/overview
15. F.B. Awino, S.E. Apitz, Solid waste management in the context of the waste hierarchy and circular economy frameworks: an international critical review. Integr. Environ. Assess. Manag. **20**(1), 9–35 (2023). https://doi.org/10.1002/ieam.4774
16. Cattermole Consulting, How the waste hierarchy pyramid is applicable to textiles (2020), https://www.cattermoleconsulting.com/how-the-waste-hierarchy-pyramid-is-applicable-to-textiles/
17. ÀLA.HAUSSE, How to implement the waste hierarchy framework into the fashion industry and our own closet (2021), https://alahausse.ca/how-to-implement-the-waste-hierarchy-in-the-fashion-industry-and-our-own-closet/
18. M.M. Wagner, T. Heinzel, Human perceptions of recycled textiles and circular fashion: a systematic literature review. Sustainability **12**(24), 10599 (2020)
19. K. Fletcher, M. Tham, Earth logic gardening (2023), https://katefletcher.com/publications/books/earth-logic-fashion-action-research-plan/
20. V. Jacometti, Circular economy and waste in the fashion industry. Laws **8**(4), 27 (2019), https://www.mdpi.com/2075-471X/8/4/27
21. A. Payne, Designing fashion's future: present practice and tactics for sustainable change. Bloomsbury visual arts (2020), https://www.bloomsbury.com/uk/designing-fashions-future-9781350092488/
22. Australian Fashion Council, Seamless clothing stewardship scheme (2023), https://ausfashioncouncil.com/program/seamless/
23. Australian Fashion Council, Global scan report (2022), https://ausfashioncouncil.com/wp-content/uploads/2023/05/AFC-NCPSS-Global-Scan-Report.pdf
24. Make Fashion Circular, https://pacecircular.org/make-fashion-circular
25. Global Fashion Agenda, Pulse of the fashion industry. Global Fashion Agenda website (2017)
26. J. Thøgersen, Harnessing social influence to foster sustainable consumption: a review and research agenda. J. Consum. Policy **42**(3), 379–397 (2019). https://doi.org/10.1007/s10603-019-09443-1
27. United Nations Educational, Scientific and Cultural Organization, Environmental sustainability report. UNESCO (2022), https://unesdoc.unesco.org/ark:/48223/pf0000383996
28. European Commission, A new circular economy action plan: for a cleaner and more competitive Europe. Publications Office of the European Union (2020). https://doi.org/10.2779/05068
29. Ministry of the Environment, Japan, Basic act on establishing a sound material-cycle society (2021), https://www.env.go.jp/en/laws/recycle/12.pdf
30. Organisation for Economic Co-operation and development, implementing the OECD recommendation on policy coherence for sustainable development: Guidance note. OECD Publishing (2021). https://doi.org/10.1787/9789264264687-en
31. United Nations Framework Convention on Climate Change, The Paris agreement (2015), https://unfccc.int/process-and-meetings/the-paris-agreement
32. Basel Convention, Basel convention on the control of transboundary movements of hazardous wastes and their disposal (2019), https://www.basel.int/Implementation/Plasticwaste/Amendments/Overview/tabid/8426/Default.aspx

33. United Nations Environment Programme, Annual report 2021. UNEP (2021), https://www.unep.org/resources/annual-report-2021
34. Worn Wear, https://wornwear.patagonia.com/

MIX
Papier aus verantwortungsvollen Quellen
Paper from responsible sources
FSC® C105338
www.fsc.org

If you have any concerns about our products,
you can contact us on
ProductSafety@springernature.com

In case Publisher is established outside the EU,
the EU authorized representative is:
Springer Nature Customer Service Center GmbH
Europaplatz 3, 69115 Heidelberg, Germany

Printed by Libri Plureos GmbH
in Hamburg, Germany